YORKSHIRE COTTON

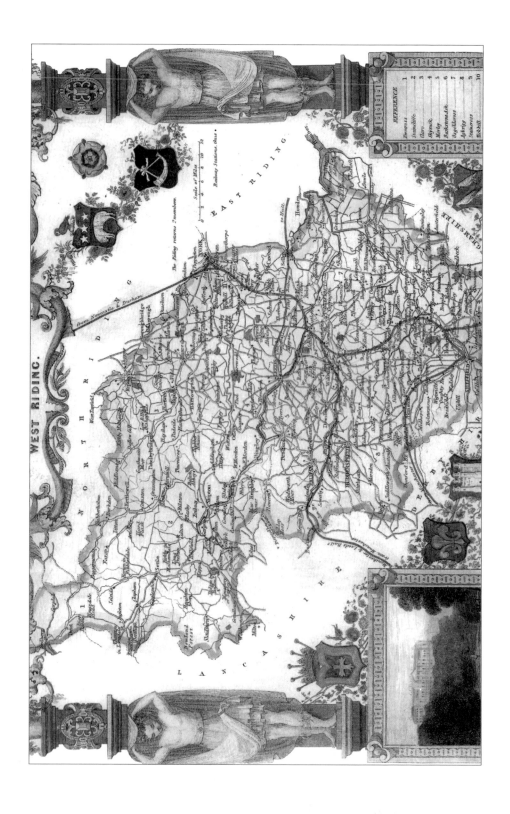

Yorkshire Cotton

The Yorkshire Cotton Industry, 1780–1835

GEORGE INGLE

Carnegie Publishing, 1997

First published in 1997 by Carnegie Publishing Ltd,
18 Maynard Street, Preston PR2 2AL

ISBN 1-85936-028-9

Typeset by Carnegie Publishing, 18 Maynard St, Preston
Printed and bound by Bookcraft (Bath) Ltd

Contents

Acknowledgements   vi

Part 1

1.   Early Development   1
2   The Early Cotton Mills in Yorkshire   12
3.   Sources of Capital   48
4.   The Labour Supply for a New Industry   64
5.   Buying and Selling Cotton and Cloth   77
6.   Cotton Weaving   88
7.   The First Cotton Masters   95

Part 2

8.   Introduction   103
9.   The Leeds Area   105
10.   The Bradford Area   113
11.   The Huddersfield Area   123
12.   The Halifax Area   130
13.   The Keighley Area including Haworth and Morton   161
14.   The Saddleworth Area   180
15.   West Craven   186
16.   East Yorkshire   192
17.   South Yorkshire   195
18.   The Yorkshire Dales   199
19.   Consolidation   243

      Notes   249
      Sources   256
      Abbreviations   257
      Part 2 References   258
      Bibliography   270
      Index   273

Acknowledgements

THE RESEARCH on which this study was originally based was undertaken with help and assistance from the Principal and Governors of Bradford College within the college staff development programme. Preparation of the doctoral thesis was supervised by the late Dr John Iredale at the University of Bradford. I thank them all for their support. Interest in the original work, which was based on the West Riding, was encouraging and it was therefore decided to enlarge the area of study to include the old North and East Ridings of Yorkshire. This has given a more rounded picture of the early cotton industry in Yorkshire.

Most studies in local industrial history rely on previous hard work carried out by a multitude of enthusiasts, both professional and amateur. I would like to thank Dr D. T. Jenkins for his help in tracing cotton mills within the insurance records held at the Guildhall Library. The staff there, and at other libraries and archives have always been most helpful and have shown considerable interest in my area of research. I would also like to thank a number of people who, through their local knowledge, have helped me to trace mills, or records relating to them, which I would inevitably have missed. Shiela Wade compiled an index to mills in the Upper Calder Valley which is now being continued by Malcolm and Freda Heywood. Stanley Lawrence provided information about mills in Lonsdale and Alan Brooke with details of mills in the Huddersfield area. Dr Gillian Cookson's knowledge of early textile machinery makers in Keighley was also useful.

This book is based on the essential written or printed source material but has been enhanced by a certain amount of field work. One of the pleasures of local history is being able to identify sites and places and know something of their history. A surprising feature has been the number of early cotton mill sites where some of the buildings can still be identified. For help with the field work, technological information and general encouragement I would like to thank Chris Aspin. He is currently working on a book on early Arkwright type mills in various parts of the country and has been surprised to see the wealth of remains in Yorkshire.

Without the help of local historians in different parts of Yorkshire much information would have been missed. Despite their assistance mistakes will have been made and errors of fact and interpretation included. I welcome additional information and corrections as the story of this industry can never be complete.

All photographs and illustrations are by the author except: Guildhall Library, London, pp. viii (Insurance Policy), 193; Yorkshire Archaeological Society, pp. viii, 206; Birmingham Central Library, pp. 31, 32, 45, 106; Keighley Library, pp. 167, 171; from original photographs loaned by Miss H. Holmes, pp. 37, 219; from a photograph loaned by Mr W. W. Hannam, p. 213; from a letterhead loaned by Mr C. Aspin, p. 114.

PART I

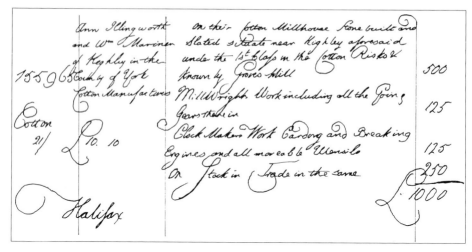

Insurance details with Royal Exchange Fire Office for Illingworth Marriner 1797.

Sale notice for Askrigg Cotton Mill, 1814.

CHAPTER I

Early Development

Introduction

THE EVERY-DAY ASSOCIATION of the wool textile industry with Yorkshire together with the association of cotton with Lancashire, has led to the assumption that one excluded the other. Cotton and Lancashire became synonymous in the late eighteenth and nineteenth centuries to such an extent that the temporary, more widespread nature of the industry, has often been ignored. Efforts have been made to chart the development of the cotton industry in other areas such as the Midlands, Scotland and North Wales but little has been done so far to discuss the significance of the early cotton industry in Yorkshire.[1] The initial and rapid expansion of the industry in the late eighteenth century in Yorkshire, particularly in the West Riding, and then the demise of the more marginal firms was history even by 1835.[2] In that year Edward Baines, the Editor of the Leeds Mercury, which carried many items concerning the early cotton industry, included Yorkshire towns under the heading of Lancashire in his book, *History of the Cotton Manufacture in Great Britain*. In many cases the success of the Yorkshire woollen and worsted industries and the concentration of contemporary and historical attention on them obscured the importance of the local cotton industry particularly where factory production was temporary or where the traditional woollen and worsted industries reasserted their position when their processes became power driven and factory based. A further problem arising from the identification of cotton with Lancashire has been the denigration of the industry outside that county and a lack of appreciation of the impact that mechanised spinning had in such areas as Yorkshire.

In most cases Yorkshire has received little attention in published works on the history of the British cotton industry. The purpose of this survey is to indicate the extent of the industry and to chart its foundation and development, firstly in areas which were and still are remote from the main textile areas, such as the Yorkshire Dales, and secondly in areas where there was transfer of assets, such as mill buildings, from one industry to another. It has therefore been necessary to include references to the woollen, worsted, flax and silk industries. In many ways the textile industries of the region should be considered as a whole and not as separate sections. The mills and power requirements were similar and often interchangeable while capital, labour and entrepreneurship often moved from one to the other. Even the machinery, particularly in the early years, came from the same origins – the cotton industry.

The success of the early cotton industry in Yorkshire had a considerable impact on the traditional wool and worsted textile industries. The result of the intrusion was summed up in Rees' Cyclopaedia in 1808:

> The lightness as well as the cheapness of the calico has rendered it a chief article
> of dress amongst all classes of people, and annihilated the manufacture of many

of the lighter kinds of woollen and worsted stuffs formerly so much in demand. The trade of Halifax and the surrounding country, which consisted almost wholly in such stuffs, has gone entirely to decay, and been replaced by the manufacture of calicoes and other cotton goods, and such are the quantities now manufactured, more especially in the country around Colne and thence to Bradford, that from 16 to 20,000 pieces are brought weekly to the Manchester market.

There are two reasons for the time limit for this survey of fifty-five years from 1780 to 1835.[4] The period starts with the building of the first cotton mill in Yorkshire and finishes with the publication of government statistics on the size of the factory industry. Those statistics gradually become more reliable and more generally available. By 1835 the industry had also settled into a form which was recognisable and remained little changed during the remainder of the century. The second reason is the existence of Dr David Jenkins' work on the West Riding wool textile industries.[3] Dr Jenkins' valuable survey has a different emphasis from this book, but by covering the same geographical area and nearly the same time period, may provide future workers with a basis for comparison.

The geographic area covered in this survey is that of the traditional boundaries of the three Ridings of Yorkshire. The major growth of the industry occurred in the West Riding but there were scattered developments in the North Riding and even in the East Riding. One temptation was to concentrate on the Yorkshire Dales which, if one includes the Craven Dales, experienced their fastest and most extreme industrial-isation with the development of water powered cotton spinning mills. The Dales provide ideal opportunities to see the remains of some of the early mills and will interest the industrial archaeologist and anyone attracted to local history. Indeed, the field work associated with this survey has shown a surprising number of early Arkwright type mills from the eighteenth century to have survived. There are probably more early cotton mills in the Yorkshire Dales than anywhere else in the country. However, the concentration of cotton mills in Keighley and other towns together with the proliferation of cotton mill building between Halifax and the Lancashire border cannot be ignored. In the end the closeness to Lancashire was one of the key determinants as to whether or not a mill continued to be used for cotton spinning in the nineteenth century.

One of the reasons for the possible neglect of the Yorkshire cotton industry by historians has been the dearth of specific records. Those that have survived are in a very incomplete form for a handful of firms and their contribution to an understanding of the industry is of significance only in certain areas. Where these records have been found they are referred to but it is impossible to form a picture of the industry based on the available business records alone.

Many of the late nineteenth-century writers alluded to the existence of an earlier cotton industry in Yorkshire but were so excited by the dominance of the wool and worsted industries that they glossed over the impact of cotton. One exception was John Hodgson, who wrote an account of the textile industries in Keighley up to 1879.[4] Hodgson devoted one chapter to the cotton industry and to twenty-three of the mills which were built for cotton spinning in the Keighley area. As Keighley was an important centre for cotton spinning this work is most useful.

Where topographical texts exist they have been used but with reservations as they are not always accurate.[5] Other sources have been appropriate local government records, estate papers, various archival records and Parliamentary Papers. The two major sources however, have been the registers of fire insurance offices and local newspapers. Throughout the whole of the period covered, the Sun Fire Office was insuring cotton mills in Yorkshire. The number of policies they issued was large but fell off after about 1815 when local offices started to take some of the business. A strong competitor was the Royal Exchange Office, whose records also give details of many policies on Yorkshire cotton mills up to 1801. The place of these records in the task of analysing capital formation for the comparable wool and worsted industries has been discussed elsewhere. In this survey their use has been in establishing the existence of particular mills and providing some indication of their value and relative size. Newspapers have proved to be an invaluable source for confirming the occurrence of mills which only had a few years of life. The marginal nature of many of the firms at times of economic and trade fluctuations prompted many newspaper entries. These were usually notices of auctions where firms were bankrupt or had ceased trading. From these two sources it has been possible to form a picture of the local cotton industry which is more detailed than was first thought likely.

A further source has been contemporary local directories. These were published from the end of the eighteenth century onwards but, at times, have limited use. Information was often out of date or possibly taken from other works. The lack of detail is at times frustrating, particularly when no distinction was made between cotton manufacturers (domestic production) and cotton spinners (factory production). Where there is some doubt as to the accuracy of a source this has been indicated.

Throughout this survey the emphasis has been on the factory based industry. References have been made where possible and where relevant to the extensive but largely unrecorded domestic cotton weaving industry in the county. However, the main purpose has been to chart the progress of the water and steam powered industry which was concerned, for the larger part of the chosen period, with preparing and spinning cotton. Power loom weaving, although added to many Yorkshire cotton mills from the 1820s was, in reality, the start of the next stage of the development of the industry.

The general approach of this book is thematic with chapters based on particular topics covering various aspects of the development of the industry. Most chapters concentrate on the years from 1780 until about 1815 as it is felt that it is this period which has been largely ignored by writers in the past. The widespread introduction of the cotton industry into many towns and villages of Yorkshire in the form of multi-storeyed spinning mills was often the initial stage in the industrialisation of those settlements. Although certain towns were beginning to develop as focal points for industry by 1780, the cotton industry was instrumental in both fostering their further growth and in creating new centres. Thus both Keighley and Todmorden grew with the assistance of the cotton industry while small towns such as Meltham and Burley-in-Wharfedale might have remained tiny hamlets without it. Most of the mill villages and towns of the West Riding in particular had at least one cotton mill which contributed to their growth and development. However, many other towns and villages which today would not be described as 'industrial' also had cotton mills. Included in that number are such places as Ilkley, Malham, Sedbergh, Thorner, Knaresborough and Kettlewell. The main aim of this study has been to locate and describe all the early cotton mills in Yorkshire and these are included in later chapters. In doing so it has been necessary to explore the reasons for their construction, the sources of capital and labour and the problems encountered by the firms which ran them. Thus various themes have been followed in this study rather than a strictly chronological approach.

The reasons why the cotton industry became established in Yorkshire are almost identical to those given for the industry expanding in Lancashire during the last decades of the eighteenth century. Expansion of demand, both at home and abroad, together with advances in technology and a supply of capital and labour resulted in the growth of an industry which was the marvel of the age. The steady growth of the industry in Lancashire up to about 1770 was of a local nature which did not produce the geographical expansion of the next few years. However, the roots of the industry were well established there and by 1750 Manchester was acknowledged nationally as the natural centre of the industry within a circle of other manufacturing towns although production everywhere was by hand.[6]

The next period of growth for the cotton industry brought its expansion into the surrounding areas of North Wales, Cheshire and Yorkshire as well as the Midlands, Scotland and even the West Country. In Yorkshire this was very much a new development with cotton weaving becoming established only in the Upper Calder Valley between Hebden Bridge and Todmorden just before 1780.[7] In the area around Settle it was said that: 'the cotton industry does not seem to have been general in the district until between 1780 and 1790.'

The change to cotton weaving from worsted weaving or the addition of cotton weaving then became very much more widespread in Yorkshire after about 1790 but it was on the power spinning side that the most important steps were initially taken to establish the industry in Yorkshire.

Hargreaves, Arkwright and Crompton

The growth of the cotton industry in Lancashire on a domestic basis over a period of many years led to a certain resistance to change. None of the three major inventors of spinning machines, Hargreaves, Arkwright or Crompton found individual success in Lancashire although their attempts to increase the supply of yarn by mechanical developments were a response to the increased demand for yarn by local weavers. This came about because of the expansion of home and foreign markets and also because of the use of Kay's flying shuttle from the 1750s. This invention had doubled the productivity of individual weavers and further stimulated the demand for yarn.

Hargreaves' hand operated jenny was introduced in Lancashire in 1767 and patented in 1770.[8] The size of the machine was soon increased from the patented model of sixteen spindles to models with eighty spindles by the mid-1780s. In the cotton spinning areas however, the machine developed by Hargreaves was received with mixed feelings. It was felt by many people that it would create unemployment and Hargreaves was forced to leave Blackburn in 1768. He moved to Nottingham where there appeared to be a more favourable attitude to textile inventors and it was from there that he took out his patent in 1770. Nevertheless, before leaving for Nottingham, Hargreaves had sold some jennies in the neighbourhood and the machine became well established in Lancashire.[9] The yarn produced by the jenny was originally only suitable for weft and was of a quality which, unfortunately for Hargreaves at that time, was only suitable for poor quality hosiery so he did not even achieve financial success in Nottingham.[10] In Yorkshire the jenny was used extensively where cotton hand-loom weaving became established as it provided the weft yarn. Cotton manufacturers often supplied their own jenny spun weft, buying the warps from the water frame spinners and a number of firms ran jennies until well after 1811 when Crompton noted that a few firms still used them.

Because of the increasing demand for yarn and the obvious advantages of jennies they were produced in increasing quantities in Lancashire despite the opposition in some quarters. However, in 1779 the opposition to large jennies and also to Arkwright's water frame gathered momentum and large numbers of carding engines, jennies with over twenty spindles and water frames were destroyed by rioters.[11]

This opposition to power driven machinery and large hand powered machines resulted in the destruction of Arkwright's only Lancashire mill at Birkacre near Chorley. This mill was started in 1777 at a site on the river Yarrow. There was already an iron forge and a corn mill on the site when it was leased to Arkwright.[12] He had as his partners Jedediah Strutt, Samuel Need, Thomas Walshman and John Cross. Walshman and Cross went on to have significance for the developments in Yorkshire after the sudden stop to their enterprise in Lancashire. This factory was the first to be built under Arkwright licence outside Derbyshire and Nottinghamshire and thus received a great deal of attention from the rioters in 1779. Ten factories were attacked of which Birkacre was the largest. The rioters achieved their objective as the mill was destroyed and never rebuilt for cotton spinning. The general feeling in the area appears to have been one of support for the rioters, for although some of the assailants were charged, the punishments were lenient for the time.

Arkwright at that time was extremely active in developing his business empire. His role in developing a successful system not only for spinning cotton but for designing a total production unit whereby raw cotton was turned into finished yarn has been well covered elsewhere. It has been said that:

> His genius, coaxed by the voracious appetite of his roller drafting machine, consisted rather in assembling and coordinating a sequence of relatively simple processes to create an integrated system in which power-operated machines superseded manual dexterity at each consecutive stage of the spinning process. In other words, he created order and system from ad hoc developments, and made a commercial success of it.[13]

The setback of Arkwright and his partners in Lancashire was of significance for future developments in Yorkshire. Firstly there was a reluctance to build any more spinning mills in that area of Lancashire.

> This devastating outrage left effects more permanent than have usually resulted from such commotions. Spinners and other capitalists were driven from the neighbourhood of Blackburn to Manchester and other places, and it was many years before cotton spinning was resumed at Blackburn.[14]

The Blackburn area however, remained the centre of the cotton weaving industry so the demand for cotton yarn was still there even though it could not be met locally. Yorkshire, on the other hand, had no inbuilt resistance to new inventions or to power driven cotton mills. These innovations did not offer a threat to any group of workers but promised to provide employment to the wives and children of local woollen and worsted workers. Secondly, two of the partners with Arkright at Birkacre, John Cross and Thomas Walshman, soon after started cotton spinning in Yorkshire in safety from rioting cotton workers. Thirdly, there already existed favourable social and economic values and systems in Yorkshire which motivated individuals to take advantage of the new opportunities.

The third inventor, Samuel Crompton, did not have the business acumen to gain financial success from his invention. He originally wished to spin fine yarn for his own use as a weaver but was eventually persuaded to surrender his invention in return for a very small subscription fund. This fund in 1780 was followed by another in 1800 and a Parliamentary grant in 1812.[15] Prior to the Parliamentary committee meeting in 1812 Crompton had toured the textile districts to find out the extent to which his mule was being used compared with Arkwright's water frame and Hargreaves' jenny. Fortunately he and his supporters collected information from parts of Yorkshire and his survey gives a useful list of mills and machines.

Factors influencing location

Most of the growth of the cotton industry in Yorkshire was a reaction by local entrepreneurs to the overall growth in demand for cotton goods. It was also known that large profits were being made elsewhere by cotton spinners, such as Arkwright and his partners, so attention was turned to the possibility of emulating their success. The motivation of particular individuals enabled them to organise resources to develop an industry in areas where it did not previously exist. Fortunately, in many cases, the necessary resources were available and could be combined together to form production units which became an industry. The problems associated with the establishment of the industry were to do with the management of those resources. Successful entrepreneurs had to adapt to new products, new types of organisation, new labour problems, new techniques of production and new markets. In the first few years they also had to decide whether or not to operate with a licence from Arkwright. To bear the cost of his help or struggle alone.

From the evidence available it would seem that only one Yorkshire mill, Low Mill in Keighley, was built under licence from Arkwright. Although a few mills were built in the following year or two without permission, the overthrowing of his patent in 1783 was the signal for the county wide boom. In addition it was felt that Arkwright's charges were excessive so once they could be avoided with impunity the rush to build mills gathered force.

A number of factors influenced the location of the early cotton mills in Yorkshire. One important trend gleaned from local newspapers during the period from about 1784 to 1800 was the eagerness of landowners to offer millsteads for sale or lease. A small selection of the advertisements shows what the owners thought would attract purchasers. In most cases these sites were specifically designated as being suitable for cotton mills.

1785. Near Nether Mills, Leeds – 'Extensive cotton works may be made, and the labour of children obtained at an easy rate.'[16]

1786. Near Linton Corn Mill near Grassington – 'Linton Mill is the most advantageous situation for a cotton mill, being supplied with water by the great river Wharfe . . . and within half a mile of the several populous towns and villages of Grassington, Threshfield, Linton, etc where from the ring of a bell upwards of three hundred children may be collected in less than half an hour.'[17]

1789. At Ovenden, Halifax – 'a very plentiful spring which runs to the buildings that is sufficient by an overfall to turn a twining or cotton mill.'[18]

1791. At Burnsall – 'where labour is cheap and where the children and upgrown persons are mostly in want of employment.'[19]

1791. At Silkstone near Barnsley – 'The situation is an extremely good one for a cotton mill to be worked with a steam engine . . . fire coal within two or three hundred yards . . . at about two shillings a ton . . . Plenty of hands may be had at reasonable wages.'[20]

1792. At Blubberhouses – '1 mile from the great turnpike road from Knaresborough to Skipton . . . numerous and almost unemployed neighbourhood . . . much stone on the land.'[21]

1791. At Thornton, Bradford – 'A very eligible and commodious situation upon a good constant stream of water with a most capital fall of fifty feet and upwards . . . capable of being worked twice over . . . no work of this kind within some miles, so that plenty of hands may be had very near . . . excellent stone for building is within the premises . . . coals are also plentiful and near at hand . . . the taker – should he wish to establish a factory, may meet with any number of cotton weavers.'[22]

1792. At Bishop Thornton, near Ripley – 'A very commodious situation for erecting an overfall cotton mill . . . plenty of good stone . . . oak timber.'[23]

1792. At Ovenden, Halifax – 'stone quarry in 200 yards.'[24]

1796. In the Washburn Valley – 'Is in a good neighbourhood for getting plenty of hands at low wages.'[25]

1796. At Soyland, Halifax – 'with a fall of water and plenty of good stone.'[26]

The main factor influencing the siting of a mill was the availability of adequate water power. That was an obvious necessity before the widespread use of steam power but there seem to have been some over optimistic entrepreneurs who misjudged the flow of their selected stream. A number of mills were built where there were well known water falls which still attract attention today. Examples are Yore Mill at Aysgarth, Hartlington Mill, Linton Mill, although that was built as a worsted mill, and the mills in the Goit Stock valley near Bingley. Many early mills used the full power of the stream they were sited on and could therefore not be expanded easily as more water could not be provided. At times it was possible, if the site was large enough and the fall sufficient, to build additional mills up or down stream to use the water twice. That was done, for instance, at Greenholme Mill at Burley-in-Wharfedale. In other cases extra dams had to be built to store an extra supply of water, often overnight when the mill wasn't working.

The second factor was the supply of workers and it can be seen from the selection of advertisements above that this was of considerable importance. Where the labour of children was not available, cottages had to be built to accommodate families as at Arncliffe Mill or dormitories had to be provided for pauper children as at Sheffield Cotton Mill.

The third factor was the availability of building materials. The new cotton mills were the largest buildings being constructed at that time in most areas so if stone and timber were available on the site the expense of having to buy and transport them was avoided. Robert Bradley & Co, for instance, were to be allowed to get stone from the Duke of Devonshire's land where they proposed to build Rilston Cotton Mill in 1798.[27]

A fourth factor, which was stressed more when established mills were for sale, was the ease of transporting goods to and from the mill. The nearness to turnpike roads and later canals was often emphasised. For example, the cotton mills in the steep Morton Valley which runs down to the river Aire near Bingley were described as 'adjoining upon the high road from Keighley to Otley . . . within a quarter of a mile of the Leeds and Liverpool Canal,' 'near the Leeds and Liverpool navigation.'[28] The ease with which raw cotton could be brought to the mill and yarn taken away was obviously important when the firm occupying the mill was dealing with Manchester. One reason for the eventual decline in sections of the Yorkshire cotton industry was the cost and difficulty of transport. However, in the early days the establishment of spinning mills in some areas brought about great increases in the volume of road and canal transport. The road across the moors through Stanbury which linked Lancashire and Yorkshire, for example, took much extra traffic from the cotton trade.[29] By 1811 the Craven Navigation Company was advertising that it had a vessel sailing each day on the Leeds and Liverpool canal from Leeds to Blackburn.[30]

Two other factors influencing the siting of mills were related to the immediate market for the yarn that was spun. If there was already a colony of hand-loom weavers in the area they would form a ready market. It is likely that some of the mills in the Craven villages, around Bradford and in the Upper Calder Valley were built with that in mind. A further development for a similar reason occurred when a local cotton manufacturer built his own mill so that he did not have to buy yarn in Manchester or from local spinners.[31] The small scale of many of those early mills and their comparative cheapness meant that they could be built and operated just to serve the local weaving community. When weaving eventually declined or the weavers changed to worsted or linen the established market for the local cotton mill's yarn had gone and at that point they often changed to spin another fibre, reverted to a previous use, perhaps as a corn mill, or became disused. Some of the smaller mills such as Scaw Gill Mill near Grassington became cottages.

Conversely, it was not unusual for a mill to be built by a landowner or investor with no specific use intended. They were built on speculation and they would be used for processing one of the textile fibres. The owner would advertise the lease of the mill when it was nearly completed. It would then be said to be suitable for cotton, woollen, worsted, or in certain areas, flax spinning. If the lease was taken the mill would then be finished according to the requirements of the tenant. He would then supply his own machinery and start production. The first cotton mill to be built in Yorkshire, Low Mill in Keighley, was supposed to have been started by Thomas Ramsden from Halifax and completed by Clayton & Walshman. Another example is Rodmer Clough Mill in Stansfield, four miles from Todmorden. This mill was built in 1792 and advertised to let for a period of years in January 1793. The mill building was then completed and the water wheel and shafting installed but there was no machinery. It was said that the mill could be 'entered into at pleasure' and was suitable 'for working frames for spinning cotton or worsted.'[32]

Changes in agriculture

Of great importance to the development of cotton spinning in the West Riding in particular, was the gradual change during the eighteenth century from arable to pasture farming. This change was partly influenced by the increased possibility of obtaining a living from hand-loom weaving or other textile production in the wool and worsted industries. By the second half of the century much of the common land had been enclosed. High crop prices and increased growth in demand for land and food caused the land enclosure movement to accelerate. In the Bradford area, for instance, the main form of agricultural improvement was the enclosure of uncultivated waste land.[33]

With the eighteenth century revolution in agriculture, which occurred on a national scale, came a great extension of corn land in the Vale of York and other lowland areas. The growth of population in the industrial areas of Lancashire and the West Riding, together with improved communications meant that corn could be brought in to the Pennine valleys from the east. The marginal cultivation in the Dales area gave way to stock breeding and dairying.[34] A fortnightly fair was started in Skipton in 1779 for cattle and sheep which was then convenient for the butchers from the manufacturing towns of Yorkshire and Lancashire.[35] Nine years later a similar fair was started in Settle which by then was said to be in the centre of extensive grazing country.[36]

The building of the turnpike roads, particularly the road from Knaresborough to Skipton, helped the distribution of agricultural products and raw materials and marked a change in the pattern of agriculture in the area. Further south – 'By means of the canal (Calder and Hebble) the corn grown on the rich York Plain became available for Halifax, and Wakefield became a great corn mart for the district.'[37]

The change to grazing in the West Riding brought about a decline in the production of corn which was then brought into the area. One place which developed an extensive corn market was Knaresborough. Dealers from Craven bought there and resold in Skipton.[38] By 1822 it was said that 200 wagons a week went from Knaresborough to Skipton where previously there had been none.[39] The smaller Dales markets were displaced by the larger ones and the larger corn mills took over much of the milling in or near the new industrial centres. One of these, the corn mill at Castle House Hill near Gomersal was steam driven before 1800.[40] The smaller country mills therefore found that they had no local clients wanting corn ground. It was also uneconomic to bring corn from the new markets when flour could be brought for the same effort. Thus there was a surplus of corn mills which became apparent during the latter part of the eighteenth century. These mills would have been left unused if it were not for the demands of the new cotton spinners who were looking for water power sites and for existing buildings to convert. In the Settle area, for instance:

> Many of the old manor corn mills of Giggleswick Parish, which were now practically derelict for lack of custom were at once adapted for cotton spinning. Langcliffe High Mill was taken in 1783 and the old Settle mill at Anley and the Giggleswick Mill at Catteral Hall Gate became cotton mills at about the same time.[41]

As the changes in agriculture resulted in less land being cultivated and less work for the corn mills there was also less work for the farm labourers. As the adult workers were no longer needed in large numbers on the land, and there were few jobs for them in the new mills, many turned to hand-loom weaving while their children went to work as carders or spinners.

Despite the initial surplus of agricultural labour wages were forced upwards by the introduction of cotton mills and the successful development of hand-loom weaving. In the neighbourhood of Settle it was said that day labourers' wages went up from 1s. to 1s. 2d. per day in 1783 to 2s. to 2s. 6d. in 1793 because of the introduction of cotton manufacturing.[42] A contemporary writer who travelled in that area observed that:

> The country beyond Brouton [Broughton near Skipton] consists of large wild pasture grounds, twenty fields laid into one, for the cotton trade, with its high pay, has put an end to all thoughts of husbandry; canals and cotton mills are the only thought.[43]

The rural corn mills were therefore becoming under utilised and often came onto the market at a time when many people were interested in starting cotton spinning. They enabled the new firms to start up with limited capital as the dams and water courses were already in existence as were the water wheels and mill buildings. In many cases additional works and buildings were constructed alongside the existing mill so as to make use of the valuable riparian rights. In some cases, such as at Wreaks Mill in Nidderdale, cotton spinning and corn milling were carried out on the same site.

Early growth

The nearness to Lancashire and the existence of an already flourishing textile industry based on wool also had a part to play. Improvements in transport helped as, in most cases, the cotton industry in Yorkshire was merely a temporary geographic extension of the Lancashire industry. The opening of a section of the Leeds and Liverpool Canal through to Burnley in 1796 eased the transport situation to Manchester. Goods were taken to Burnley by canal and then by road to Manchester and Rochdale.[44] At the same time goods could be taken by canal from Leeds to Bradford in one day.[45] In the more southern dales the building of canals along the Calder and Colne Valleys not only vastly improved the transport of raw materials and finished goods but made it possible to build mills independent of the river as they could be supplied with steam coal by barge.

The contemporary growth of the Yorkshire coal and iron industries was also of great assistance. Although coal was important in some areas to supply steam pumping engines its value increased after 1800 when steam power was developed to supplement or replace water power and even then the take-up varied across the region. Eventually it was nearness to supplies of coal which determined the survival of the industry in individual areas.[46] Deposits of iron ore in the West Riding, in conjunction with coal, provided a valuable asset to the regions textile industries. That was particularly true when iron framed machines were built and steam engines and boilers needed to drive them. The Bradford iron-works of Bowling, started in 1788, and Low Moor, started in 1791, were particularly important to the local textile industry.[47] In addition many of the regions cotton firms bought iron in various forms from the Kirkstall Forge of Beecroft, Butler & Co.

The availability of sites and the possibility of being able to mobilise labour and capital were factors which prompted a sense of opportunity. As with any new industry however, over ambition, lack of knowledge, changing trade and technology all played their part in curbing the growth of particular enterprises. Even by the end of the period under survey many cotton firms that had started with high ambitions had become bankrupt and their assets absorbed into new enterprises. Indeed many firms only lasted a few years and all that remains is a brief entry recording the auction of their machinery and stock. However, in the early years the spirit of enthusiasm was high and in 1788 a Leeds newspaper could report: 'The increase in the above manufacture (cotton) during the last seven years has been most astonishing.'[48]

The rate of growth did not slacken for some time for even in 1802 a rival Leeds newspaper reported that:

> The rapid increase of Cotton Manufactures in the North of England has lately been almost incredible . . . so considerable has its progress been in Yorkshire that the labouring poor in the Western part of this Riding are at present principally employed in it. In the neighbourhood of Halifax, Huddersfield, Bradford and even Dewsbury, several large Manufactures have lately been erected.[49]

Thus in twenty years, up to the turn of the century, the cotton industry had become widely established in Yorkshire and appeared to be ready for further growth. About two hundred and forty mills had been built or adapted for cotton spinning in this time. A small number had already gone out of existence or changed to other uses by 1800 but new mills were added in the following decades, particularly in Calderdale where the cotton industry became well established.

CHAPTER 2

The Early Cotton Mills
in Yorkshire

Listing the mills

FROM THE START of water powered cotton spinning in Yorkshire in 1780 until the 1830s several attempts were made nationally to list the number of mills in existence or the firms running them. These endeavours were carried out for a number of reasons, some to do with legislation relating to the workforce in these new enterprises as the working conditions for the new factory workers roused considerable attention. Other reasons were to do with the commercial and technological growth of the industry as the economic impact of the industry and the new machinery gathered strength. All these attempts were on a national scale but it is possible to identify the Yorkshire mills and concerns from the lists. One feature of all the lists which were compiled is that the Yorkshire mills were under represented, sometimes quite seriously. This lack of information about Yorkshire cotton mills and firms has possibly led to the assumption that the industry was not significantly established in the county.

The major aim in this study has been to trace the development of all power driven mills and factories which were used for the production of cotton yarn or cloth, or occasionally just for a preparatory process such as carding. It is doubtful if there were any premises in Yorkshire which were used for the spinning or weaving of cotton on any scale before the introduction of power driven mills in 1780. After that date there were also jenny shops and weaving shops where spinners and weavers were gathered together to cut transport costs and maximise the use of hand labour. These have been referred to where they have been found but all the main references are to individual mills or factories which were driven by water or steam or a combination of the two. The cotton industry in Yorkshire therefore virtually started from scratch with the multi-storeyed power driven spinning mill as the basic unit of production. The mills, their machinery and power source often involved considerable capital expenditure and the firms occupying them employed the largest workforces in the area. However, it was usually the mills which have survived rather than the firms, although there are exceptions such as the Haggas family where William and John Haggas started cotton spinning at Higher Providence Mill in Keighley in 1801.

Problems with identifying individual mills abound. The close juxtaposition of many mills with consequent absorbtion of one into another as a result of expansion has created many accounting difficulties. Where possible the word 'mill' has been applied to the premises where cotton was processed by power driven machinery. The power was usually supplied by water or steam although horse mills were used in some areas such as Leeds, Otley and Skipton and these could have more spindles than some water powered mills. There has been no trace of wind power being used directly although it was used in the woollen industry.[1] However, there is one example of wind power being used at a cotton mill in Long Preston to pump water back up to the dam so that it could be used over and over again.[2]

Problems of estimating the number of cotton mills at particular periods during the development of the industry have often arisen. Many of the early surveys were inadequate in their coverage which has probably caused the lack of interest in the Yorkshire industry. The various lists which were drawn up between 1788 and 1833 have therefore been examined and additions made to provide as complete a coverage as possible. Several major problems arise with the interpretation of the four lists which were compiled in 1788, 1803, 1811 and 1833–5. The first is that many of the entries only consist of brief details – the name of a person or company or the name of a town. This inclusion in a list of cotton spinners has been taken as accurate unless there is contrary evidence but it can then be impossible to check continuity. For instance, Crompton noted that Samuel & John Sutcliffe were running four mules and six throstles at a mill in Huddersfield in 1811 and it can be confirmed that they still had mules and throstles a few years later. The area was New Street but was that the same mill which was occupied by Joshua Lockwood & Co. and burned down in 1828? A second problem comes with a change of name, either with a change of ownership or rebuilding. At previous times Aireworth Mill in Keighley, was called Screw Mill and also Stubbing House Mill when it was first built for cotton spinning in 1787. The third problem is related to multiple occupancy, either of the same mill or of adjacent mills on the same site. The person compiling a list might have entered one of the tenants at a mill but not the other. When the second name occurs in a different context how can it be located? Colquhoun, for example, noted that Joseph Driver & Co. occupied a mill in Keighley but neglected to say or did not know that the mill was owned by Joseph Smith who was also spinning cotton there from about 1783. Some mills became industrial complexes quite early in this period. For example, High Mill in Addingham was used for worsted spinning, corn grinding and cotton spinning in the 1780s and 1790s. In addition some firms moved premises when they wished to expand or leases were not renewed. Some also ran several mills and it can be difficult to know to which one data should be ascribed.

Colquhoun's list

The first comprehensive attempt to list all the cotton mills in Yorkshire was made by Patrick Colquhoun in 1788. His list was part of a national census compiled to further the case of the cotton manufacturers with the government of the day when trading conditions were difficult. Colquhoun was a Glasgow merchant who had traded with the American colonies but became more interested in promoting trade with the Continent, especially in cotton goods. Eventually he became involved in trade politics and wrote his famous pamphlet 'An Important Crisis in the Calico and Muslin Manufactory Explained'. A detailed examination of the complete list from that pamphlet has already been made and the deficiencies and problems highlighted.[3] One of the areas where omissions occurred was Yorkshire, particularly with the Yorkshire Dales being well away from the main manufacturing areas. Some entries were allocated to other counties and these can be dealt with easily. The problem comes with the other mills which may or may not have existed in 1788 but were left out of Colquhoun's list. The life of many of the early firms was often quite short and some had gone out of existence by 1788. A reference shortly after that date to a mill which was obviously well established is a temptation to suggest that it was probably missing from Colquhoun's list. What has been done therefore, is to quote Colquhoun's list with later additions provided by Chapman, together with some amendments, and some further additions. Details of all these mills have been included in Part II and the location of the mill, where known, has been added below to help with identification. The aim has been to supplement Colquhoun's list with the names of all the other cotton mills which were in existence in Yorkshire in 1788 together with others which had operated in the previous eight years. In that way the scope of Colquhoun's list can be established and the scale of the industry's growth can be measured. However, there is no doubt that further research will bring more additions to the list in the future.

Table 2.1

COLQUHOUN'S ORIGINAL LIST
YORKSHIRE

Askwith & Thompson	Steeton near Keighley
Claytons & Walshman	Langcliffe near Settle
Claytons & Walshman	Low Mill, Keighley
Craven & Co	Walk Mill, Keighley
Crossley & Co★	Todmorden
Davidson & Co	Settle Bridge Mill, Settle
Joseph Driver & Co	Castle Mill (part) Keighley
Greenwood & Co	Northbrook Mill, Keighley
Garforth & Sidgwick	High Mill, Skipton
Watsons, Blakey & Co	Greengate Mill, Keighley
Wells, Middleton & Co	Sheffield
Whitehead (Ralph) & Co★	? Todmorden

WRONGLY PLACED IN OTHER COUNTIES

Buckley, Ogden & Co	Gatehead Mill, Saddleworth
Priestley & Co	Bridgehouse Mill, Haworth
Weatherall & Co	Morley
Wigglesworth, Armitstead	Clapham
Petty & Co	

ADDITIONS BY S. D. CHAPMAN

Carr & Paley	Gomersal Hall Mill
Joseph Smith	Castle Mill (part) Keighley
Thomas Binns	Stubbing Mill, Keighley
Brayshaw, Hartleys & Co	Malham
Hawksworth & Curtis	Ilkley
Mounsey & Marshall	Eller Gill, Otley
Walker & Co	Silver Mill, Otley
Richard Hargreaves	High Mill (part) Addingham
Robert Pearson?	Draughton Mill

T Mitchell	Higgin Chamber, Sowerby
D Bottom	Jumble Mill, Hebden Bridge
C Rawden	Spa Mill, Hebden Bridge
Elkanah Hoyle	Hollings Mill, Ripponden
Winstanley, Harrison & Co	Yore Mill, Aysgarth
James Brenand	Runley Bridge Mill, Settle
Ard Walker	Waterloo Mill, Leeds
M & J Richardson	Slithero Mill, Halifax
Edmund Lodge	Willow Hall Mill, Halifax
FURTHER ADDITIONS	
Drivers & Dinsdale	Askrigg Mill
O & T Routh	Gayle Mill, Hawes
William Myers	Hartlington Mill, Burnsall
Cockshott & Lister	High Mill (part)Addingham
Garforth & Sidgwick	Otley Mills
King, Turner, Paley & Varley	Mytholme Mill, Hebden Bridge
George Widdop & Co	Hudson Mill, Hebden Bridge
Thomas Eastwood	Eastwood Mill, Hebden Bridge
John Varley	Shaw Carr, Huddersfield
John Heaton	Spring Head Mill, Keighley
G A & H Salvan	Hull Mill, Saddleworth
J & J Buckley	Shore Mill, Saddleworth
Seville & Lees	Lowbrook Mill, Saddleworth
Livesey, Hargreave & Co	Low Moor Mill, Clitheroe
John Brown	Airton Mill

★ These two mills were on the Lancashire side of Todmorden

It is easy to see from the table above that Colquhoun identified fewer than half the cotton mills in Yorkshire and also that the eight years from 1780 had seen a spectacular growth of cotton mill building in the county. Over forty mills had been built or converted to cotton spinning, most of them in the few years just before 1788. The national figures are as illusive as those for Yorkshire but it is felt that only the West Riding, Cumberland and Westmorland had the most omissions. As Cumberland and Westmorland had only a few mills the forty or so mills in Yorkshire, out of a total of just over 200 constituted a significant proportion.

With Arkwright's dominance over the early industry in most parts it is intriguing to know how far the individual Yorkshire mills were built on the Arkwright principle and were running his original standard of 1,000 spindles on water frames. The influence of Arkwright, from the need to use his licence, or at least to follow his design is interesting to trace. It appears to have led to mills of fairly standard dimensions, about 70ft × 30ft and three to four storeys high. Fieldwork, insurance valuations and other sources suggest that many of the early mills in Yorkshire were quite small but many of them did fit the Arkwright pattern and can still be seen, particularly in the Yorkshire Dales where the need to convert them to other uses did not occur.

From the list of mills above three locations stand out, the Keighley area with eight mills mainly on the river Worth and North Beck, Calderdale with eight mills and the dispersed area of the Yorkshire Dales with seventeen mills. Of the seventeen, Colquhoun knew of four but it is no wonder that he missed the rest in places like Hawes, Ingleton and Ilkley. Some of these were small. Malham Mill was only twenty-seven feet by twenty-one feet but four storeys high while Runley Bridge Mill is about the same size but only three storeys and today could easily be mistaken for a farm building which is its current use. However, several of the other mills have the dimensions of an Arkwright type mill with examples being Silver Mill in Otley and Ingleton Cotton Mill.

In the Upper Calder Valley the early use of jennies and mules instead of water frames meant that early mills there had different dimensions. Many were small, often only the size of two or three cottages, and built in the narrow cloughs above the main valley. Examples are Castle Clough Mill and Clough Hole Mill.

Cotton mill returns, 1803–1806

The second attempt to list some of the larger cotton mills in Yorkshire came in 1803 but for a very different reason. The first Factory Act of 1802 applied to cotton mills with three or more apprentices or twenty or more employees and laid down certain regulations.[4] Mills had to be whitewashed inside at least twice a year, openings for fresh air had to be provided and the apprentices had to have two suits of clothing provided during their period of apprenticeship. Employers had to provide education facilities and the opportunity to attend Sunday School and church. One of the main enactments was that the apprentices should not work for more than twelve hours a day. To ensure that this new legislation was put into action the justices of the peace for every county had to appoint two persons 'not interested in, or in any way connected with, any such mills or factories, to be visitors . . . one of whom shall be a justice of peace . . . and the other shall be a clergyman of the Established Church.'

Mill owners were notified of their responsibility to register their mill and details of their apprentices through advertisements in the newspapers and had to pay 1s. to the Justices.[5] Only the larger factories with over 20 workpeople and over 3 apprentices had to be registered thus excluding the smaller mills. It would seem from the records for the West Riding, and the letters sent by the mill owners, that many ignored the Act or were not aware of their responsibility. Of those that did reply many promised to pay the Justice of the Peace, whom they obviously knew, when they saw him next.

The firms which did register their mills are listed below. (The first of any duplicate entries has been taken).

Table 2.2
Cotton Mills Registered with West Riding Magistrates 1803–1806

Firm	Mill or Town	Other Details
Anthony Fentiman	Addingham	
James Brown & Co	Hartlington, Burnsall	
Sidgwick & Garforth	Sedbergh	
Walker & Co	West End, High Mill	516 spindles
Walker & Co	West End, Low Mill	1,540 spindles
Thos Driffield	Knaresborough	750 spindles

Mr Willett	Raikes Mill	1,400 spindles
Messrs Clayton	Langcliffe Mill, Settle	
J J & T Thornber	Settle	
Edmund Armitstead	Settle Bridge	
Samuel Westerman	Howgill Cotton Mill	
Thomas Danson & Co	Bentham	
Sidgwick & Garforth	Skipton	
Helliwell & Garforth	Bell Busk	
John Broughton	Booth Bridge, Thornton in Craven	
Clayton & Walshman	Keighley	200 employees
Watson, Blakey, Marriner & Ellis	Keighley	130 employees
John Greenwood	Keighley	93 employees
John Greenwood	Keighley	110 employees
Craven, Brigg & Shackleton	Keighley	30 employees
Newsolm, Sugden & Wright	Keighley	45 employees
John Heaton	Keighley	40 employees
Robert Heaton	Keighley	25 employees
Watson & Binns	Keighley	30 employees
Illingworth & Marriner	Keighley	
William Ellis	Haworth	25 employees
Hollings & Ross	Stanbury	
John Webster	Hartshead Cotton Mill	
Henry Hirst	Heckmondwyke Cotton Mill	
J & H Barker	Morton Mill	60 employees
Robson, Edmondson & Co	Idle, Bradford	27 employees
Smith, Tetley & Co	Wilsden, Bradford	53 employees
Garforth	Castlefields, Bingley	
Edmund Eastwood & Co	Slaithwaite	
Thomas Varley	Slaithwaite	(two mills)
Thomas Gill	Lingards, Almondbury	
John Haigh	Crow Hill, Marsden	
John Haigh	Franks Mill, Marsden	
John Haigh	Old Corn Mill, Marsden	
John Haigh	Upper End Mill, Marsden	
Joseph Steel	Tickhill	
Heathfield, Middleton & Martin	Sheffield	3–4,000 spindles
James Holdforth	Leeds	400 employees
Turner, Bent & Co	Stansfield, Halifax	
Henry Lodge	Willow Hall, Halifax (two mills)	

Total: 47 mills

Source: Returns of cotton mills 1803–6 WYAS Wakefield

The number of mills here is obviously far less than the total at the period as it excludes the smaller mills and those which did not employ apprentices but it does indicate the spread of the industry across the county. The mills in the Yorkshire and Craven Dales are well represented as are those near Bradford and Huddersfield. However, it is the Keighley area which dominates with about a quarter of the mills making a return being from that town or the immediate neighbourhood.

As the list above in Table 2.2 is far from complete an attempt has been made to estimate the total number of mills which had been built by 1800. Such a list gives a useful indication of the growth of the industry in the first twenty years and can be taken from the details in the references to Part 2. The great majority of the mills mentioned above can be subsumed in the longer list as they were running in 1800. A further reason for choosing the year 1800 is that it was the year when Watt's patents expired and many people took the opportunity to build steam powered cotton mills which then started to change the nature of the industry. The two years 1800 and 1801 were years when a number of new steam powered mills were built but it is interesting to note that none of the firms operating these mills either knew about or saw fit to register with the magistrates. Perhaps they did not employ apprentices but many would have had more than twenty employees. One example in the centre of the industry was Hope Mill in Keighley where, in 1800, Thomas Corlas was preparing a building 'intended for a cotton mill' and was at the same time constructing his machinery and a steam engine. Another, just a few miles away, was Providence Mill which was built in 1801 and was the first steam powered mill in Bingley Parish.

At a reasonable estimate about 240 cotton mills had been built in Yorkshire by 1800 but only about a fifth of that number were registered with the magistrates if the West Riding example is true for all three Ridings. These 240 mills were spread widely across the region from Sedbergh to Sheffield and Driffield to Todmorden. The previous concentrations in the Halifax and Keighley areas continued with widespread use of water power throughout the Craven Dales, Wharfedale and north to Wensleydale. The growth of the industry was such that it was difficult to find a town or village of any size in West Yorkshire which did not have a cotton mill by 1800.

Crompton's survey, 1811

The next survey was undertaken by Samuel Crompton in 1811.[6] He toured the textile areas of England and Scotland to collect evidence of the use of the spinning mule which he invented and, for comparison, the use of water frames, throstles and jennies.

Hargreaves' jenny was the first successful spinning machine with multiple spindles and, in his 1770 patent, had sixteen spindles. This machine was always hand powered and had certain technical limitations which meant that it could only produce softly twisted yarn suitable for weft.[7] However, as the first of the machines on which yarn could be mass produced, it became very popular and larger models with many more spindles were widely used despite the need for skilled operatives. Even in 1811 some were still in use.

Arkwright's water frame was always power driven and so necessitated a larger investment than Hargreaves' machine. When automatic raising and lowering of the bobbins had been developed this machine gave continuous spinning and winding on.

Nevertheless it had to be replenished with rovings and the full bobbins taken off. Also any broken ends had to be pieced together but these were all unskilled jobs and so the machine could be looked after by children with very little training. The yarn it produced by the nature of the machine had a hard twist and was therefore stronger and so suitable for warps. It thus complemented Hargreaves' jenny and many early cotton spinners, who were also manufacturers, had both machines. For example a cotton mill at Conisborough had frames, jennies and mules in 1795 and there were also some looms on the premises.

Crompton's mule combined features of the two other spinning machines but until the self acting mule was developed in the 1830s this machine required a good deal of skill from its operator. However, it became very popular as it was a versatile machine which could spin both hard and soft yarns and, more important, would spin the finest counts if required. Unlike the water frame or its successor the throstle, mule spun yarn came straight off the machine as a cop which could be used immediately for weaving and not on bobbins which had to be rewound. The mule therefore had certain advantages over the throstle and it may be significant that many Yorkshire firms which survived when others failed operated mules and so could adjust their output as markets changed. On the other hand the throstle remained in competition with the mule for certain types of yarn throughout the entire period of this survey.

Crompton was anxious to show the widespread use of his mule to increase his chance of financial reward and his notebooks contain long lists of mills, firms and places where his mules were being used, as well as the number of spindles. Unfortunately his notes are often meagre with just the name of the town or of the firm so other sources have to be used to identify the exact location. For example in his list headed from 'Todmorden to Hepton Bridge' are the names:

	Mules	*Throstles*
Malt Kiln	1,296^6/$_{18}$	600^6/$_{100}$
Hudson Mill	720^3/$_{20}$	

In his list for the Ripponden-Sowerby-Halifax area he was at times more explicit:

Watts Wrigley, Hightown	1,008^4/$_{252}$

For some reason there was a change to giving the name of the firm when the mills in Keighley were listed. This may have been because a different person collected the data on Crompton's behalf:

Marriner & Co.	1,872^{36}/$_{52}$

His fractions for mules indicated the number of mules as the numerator and usually the number of spindles in dozens on each machine as the denominator. For throstles or the older water frames the numerator is the number of frames and the denominator the number of spindles on each frame. It is thus possible, most of the time, to determine the size of each machine as well as the total number of spindles running in each mill.

Crompton's survey is useful in that it was an attempt to list all the mills in the main cotton spinning areas and also illustrates the distinction between those areas where mule spinning was common and those where water frames and throstles dominated. Many significant cotton spinning towns in Yorkshire were omitted such as Skipton and Settle, but the Halifax, Bradford and Keighley parishes were well covered. With the help of other sources it has been possible to locate nearly all of the mills noted in Crompton's survey.

As the purpose of his investigations was to find out how far the spinning mule was being used his classification of spindle capacity is valuable. There is some reason to doubt the accuracy of his survey although the care with which details of each machine were recorded would indicate a high level of precision. Crompton received considerable help and assistance from prominent people in the trade in collecting the statistics.[8] Nevertheless, in light of the fact that this survey was undertaken to justify an appeal to Parliament that the mule was widely used, and in the absence of other evidence, there may be some limitations to his report. Few details of the machinery in any of the mills Crompton named have been found for the same period but his survey does appear to be generally satisfactory. The problem has been locating all the mills and firms mentioned by him and there is some doubt, not as to the use of the mule, but as to its actual use in 1811. It is possible that where a mill had been in use for cotton spinning but had recently changed over to worsted or woollen spinning it was still included as a cotton mill because it helped Crompton's case. One example is Gomersal Hall Mill in Birstall. Crompton noted that 'Carr' at 'Gomersall' had 4,032 mule spindles and 1,920 throstle spindles. However, Thomas Carr was bankrupt in 1803.[9] Gomersal Hall Mill was for sale in 1804 and again in 1808 when the next owners, Swaine Brothers of Halifax, were also bankrupt. It has been suggested that this mill was used as a woollen mill from 1803 so it may be that Crompton's survey is accurate as far as detail is concerned but that some of it may be for years prior to 1811. In the case of Gomersal Hall, Crompton may have been given information about the cotton machinery which had been in use there previously. Another example where he may have been provided with information rather than checking for himself was also in Birstall. Crompton listed the machinery for 'Burkenshaw Factory' in one place in his notebooks but in another listed the machinery of Rangeley & Tetley who also occupied part of Birkenshaw Mill. According to Crompton, Rangeley & Tetley had 16 throstles in 1811 with 1,142 spindles. However, when their machinery was for sale later that year their spinning machinery included eight mules.[10] These mules were probably from the twelve he had listed under the heading of 'Burkenshaw Factory'. Thus there can be serious problems in interpreting Crompton's data in too detailed a manner but it is felt that it is accurate enough to enable some analysis to be attempted.

The mills surveyed by Crompton, or people working on his behalf, have been put into groups based on geographical areas. The details collected by Crompton have been listed in Table 2.3. It should be noted that where the heading 'Throstles and Water Frames' has been used, the small number of spindles on some of the machines suggests that they are the older water frames. Thus the entry for Ponden Mill, near Haworth, gives $^{16}/_{48}$, which would be 16 water frames, each with 48 spindles giving a total of 768 spindles. On the other hand Temple Mill in Rishworth, part of Halifax Parish, is recorded as having 16 mules with 300 spindles each which gives a total of 4,800 spindles.

Table 2.3

Samuel Crompton's Spindle Enquiry 1811 – Yorkshire Cotton Mills

Mill or Firm	Mules	Throstles or Water Frames
TODMORDEN TO HEBDEN BRIDGE		
Waterside Mill	$2280^{10}/_{19}$	$864^{6}/_{144}$
Pudsey Mill	$1080^{5}/_{18}$	$144^{1}/_{144}$
Fieldhurst Mill	$1440^{6}/_{20}$	$480^{4}/_{120}$
Kitson Wood Mill	$1440^{6}/_{20}$	
Line Holme	$4560^{19}/_{20}$	
Cross Lee Mill	$648^{3}/_{18}$	
Holme Mill	$432^{2}/_{18}$	
Malt Kiln	$1296^{6}/_{18}$	$600^{6}/_{100}$
Hough Stone Mill	$646^{3}/_{18}$	
Hole Bottom Mill	$1440^{6}/_{20}$	
York Field	$1296^{6}/_{18}$	
Lumbutts – Higher Factory	$2400^{10}/_{20}$	
Ditto	$2400^{10}/_{20}$	
Ditto	$2400^{10}/_{20}$	
Folly Mill	$864^{4}/_{18}$	
Eastwood Mill	$960^{4}/_{20}$	
Burnt Acres Mill	$720^{3}/_{20}$	
Staups Mill	$432^{2}/_{18}$	
Clough Factory	$2400^{10}/_{20}$	
Land Mill	$960^{4}/_{20}$	
Hudson Mill	$720^{3}/_{20}$	
Lumb Mill	$3588^{13}/_{23}$	$4680^{39}/_{120}$
Mytholm Mill	$2880^{12}/_{20}$	$1440^{12}/_{120}$
Upper Bankfoot Mill	$2964^{13}/_{19}$	
Lower Bankfoot Mill	$1728^{8}/_{18}$	$200^{2}/_{100}$
Hebden Bridge Lanes	$648^{3}/_{18}$	
Greenwood Lee		$720^{6}/_{120}$
New Bridge Mill		$576^{12}/_{48}$
Midge Hole Mill	$864^{4}/_{18}$	
Nut Clough Mill	$2640^{10}/_{22}$	
RIPPONDEN, SOWERBY, HALIFAX		
Temple Mill	$4800^{16}/_{300}$	
Booth Factory	$2800^{13}/_{240}$	
Lambert Factory	$2856^{14}/_{204}$	$960^{8}/_{120}$
Slithero Mill	$2448^{12}/_{204}$	$480^{4}/_{120}$
Kebroyd Middle Mill – Hadwen & Wilson	$4880^{20}/_{240}$	$960^{8}/_{120}$
Kebroyd Low Mill – Holroyd & Denton	$2880^{17}/_{240}$	$480^{4}/_{120}$
Soyland Mill	$1224^{6}/_{204}$	$240^{2}/_{120}$
Severall Mill	$3024^{12}/_{252}$	
Clough Factory	$1512^{6}/_{252}$	

Mill or Firm	Mules	Throstles or Water Frames
Luddenden Factory	6000^{20}_{300}	1680^{14}_{120}
Wharfe Mill	7200^{24}_{300}	
Willow Hall Mill (Stone)	4320^{18}_{240}	
Willow Hall Mill (Brick)	5760^{48}_{120}	
Stern Mill	$2048^{8}_{240}.^{1}_{120}.^{4}_{126}$ 630 Jennies	
Copley Mill	$4752^{12}_{228}.^{8}_{252}$	
North Bridge Mill	$2016^{6}_{228}.^{2}_{324}$	
Bowling Dyke Mill	2400^{8}_{300}	
Mathewsons	$2232^{6}_{300}.^{2}_{216}$	
Old Lane Mill	1800^{6}_{300}	1080^{2}_{216}
Greenwoods	2880^{12}_{240}	
Shibden Mill	1200^{5}_{240}	
Holmefield Mill	2400^{10}_{240}	
New House Mill	816^{4}_{204}	360^{3}_{120}
Wainstalls Mill	2016^{8}_{252}	480^{4}_{120}
Lumb Mill	2016^{8}_{252}	600^{5}_{120}
Spring Mill	2520^{10}_{252}	600^{5}_{120}
Hortons	1824^{8}_{228}	1800^{15}_{120}
Dean Mill	1920^{8}_{240}	720^{6}_{120}
Teasdale	960^{4}_{240}	480^{4}_{120}
Higgin Chamber Mill	1440^{8}_{180}	
Jowler Mill		630^{5}_{126} Jennies
Ludding Dale	960^{4}_{240}	
Watts Wrigley, Hightown	1008^{4}_{252}	
Samuel Broadbent, Brow Bridge	2880^{12}_{240}	
Thorn Hill Bridge	2400^{20}_{120}	
Thrum Hall	2880^{12}_{240}	
William & Thomas Rushworth	2400^{10}_{240}	
Lumb Factory	2880^{12}_{240}	

KEIGHLEY AREA

Mill or Firm	Mules	Throstles or Water Frames
Aireworth Mill		384^{8}_{48}
Dalton Mill		576^{12}_{48}
Low Mill		1872^{36}_{52}
Damside Mill		1536^{16}_{96}
Cabbage Mill		1248^{24}_{52}
Greengate Mill		1872^{36}_{52}
Grove Mill		624^{12}_{52}
Ingrow Mill	864^{4}_{216}	312^{6}_{52}
Damens Mill	864^{4}_{216}	312^{6}_{52}
Vale Mill		1664^{32}_{52}
Higher Providence Mill	1584^{6}_{264}	
Mytholm Mill		832^{16}_{52}
Walk Mill		624^{12}_{52}

Mill or Firm	Mules	Throstles or Water Frames
Springhead Mill		$832^{16}/_{52}$
Griffe Mill		$520^{10}/_{52}$
Ponden Mill		$768^{16}/_{48}$
Castle Mill		$624^{12}/_{52}$
Low Bridge Mill	$1368^{6}/_{228}$	
BINGLEY AND BRADFORD AREA		
Castlefields Mill, Bingley		$1892^{36}/_{72}$
Providence Mill, Bingley		$1680^{20}/_{84}$
George Tweedy, Wilsden	$1428^{7}/_{204}$	
Smith, Tetley & Co., Wilsden	$4080^{12}/_{252}$	$312^{6}/_{52}$
Goitstock Mill, Harden		$1680^{10}/_{96}.^{10}/_{72}$
Hallas Bridge Mill, Harden		$624^{12}/_{52}$
Bent Mill, Harden		$1440^{20}/_{72}$
Morton Bridge Mill, Morton	$1224^{6}/_{204}$	
Knight & Co., Great Horton, Bradford	$4320^{20}/_{216}$	
Robson & Co. Idle, Bradford		$1512^{18}/_{84}$
Birkenshaw Factory, Birstall	$2952^{12}/_{204}.^{2}/_{252}$	
Rangeley & Tetley, Birkenshaw Mill, Birstall		$1142^{16}/_{76}$
Gomersal Hall Mill	$4032^{16}/_{252}$	$1920^{20}/_{96}$
WHARFEDALE AND NIDDERDALE		
George Foster (Hey), Otley		$2520^{20}/_{126}$
Jonathan Cawood, Otley		$1260^{10}/_{126}$
Townend Mill, Addingham		$1260^{10}/_{126}$
Greenholme Mill, Burley-in-Wharfedale		18000
Wreaks Mill, Hampsthwaite		12000
Little Mill, West End, Washburn Valley		$1440^{20}/_{72}$
SHEFFIELD		
Taylor		1600
HUDDERSFIELD		
Samuel & John Sutcliffe	$1200^{4}/_{300}$	$720^{6}/_{120}$
Thomas Varley near Marsden	$1080^{5}/_{216}$	$240^{2}/_{120}$
Thomas Varley	$5184^{21}/_{216}$	
Waterside Mill	$3888^{18}/_{216}$	
LEEDS		
Wilkinson & Co.		$1440^{20}/_{72}$
Ormroyds		$1440^{20}/_{72}$
Holbeck Factory?		$2160^{30}/_{72}$
SADDLEWORTH		
Wrigley	$1080^{5}/_{216}$	
Joseph Wrigley	$1512^{7}/_{216}$	
Abraham Seville	$2160^{10}/_{216}$	
Edward Ratcliffe	$864^{4}/_{216}$	
George Buckley	$1008^{4}/_{252}$	
William Buckley	$432^{2}/_{216}$	

Mill or Firm	Mules	Throstles or Water Frames
Thomas Wrigley	2400¹⁰/240	

Crompton listed one hundred and nineteen mills in Yorkshire but left out many large concerns in the Skipton and Craven areas. However, a list such as the one above is significant in establishing the size of the industry at the time and possibly the type and number of spindles run by particular firms. It is also significant in that divisions in the industry can be seen from the types of machinery which were used in the various major production areas. There was extensive use of Crompton's mule in the Halifax area and also around Saddleworth. In both districts cotton spinning survived and had further developed by 1835. Elsewhere in Yorkshire cotton spinning declined as the wool and worsted industries regained supremacy in the years after 1811. The early division of the industry based on the two spinning machines indicates a degree of specialisation which had an impact on the survival of the industry in those areas near the Lancashire border. However, there is no clear evidence to show that firms in the Yorkshire hinterland engaged in mule spinning were more likely to survive than those with throstles. By 1811 the mills that had been established in many of the Pennine valleys, because of the availability of cheap labour and water power were specialising in the production of cheap yarns. Their profitability was low because of the limitations of the throstle with which most of them were equipped. The mule, which was more economical in terms of power, was being developed rapidly and was the spinning machine used by the successful Lancashire firms. It was the mule that was to carry forward the Yorkshire cotton industry in the period after Crompton's survey.

Factories Enquiry Commission Reports

The final survey of the cotton industry in Yorkshire before 1835 was carried out by the Factories Enquiry Commissioners in 1833/4. This was then followed by the appointment of inspectors who collected further information and the size and distribution of the industry was then well documented on a national scale into modern times. However, if we draw on both these official sources the extent of the industry in Yorkshire can be ascertained fifty-five years after the building of Low Mill in Keighley. Remnants of the early industry still existed. William Clayton & Son were still running Low Mill and spinning cotton twist with power supplied by a 17 hp water wheel. They also still owned Langcliffe Mill near Settle where they employed more people than at Keighley and were engaged in cotton spinning and power loom weaving. A number of other firms at old established mills also responded to the Commissioners questions. Amongst those were the Greenwoods who occupied cotton mills at Burley-in-Wharfedale, Birstwith, Keighley and Haworth. The new distribution of the industry was indicated though, with the numerous returns from the Saddleworth area and the upper Calder Valley from Sowerby Bridge to Todmorden.

The Commissioners received replies from about sixty-two cotton mill tenants or owners who were asked to give the information relating to the situation on the 1st May 1833. Some of the mills were also used for other purposes besides the preparing and spinning of cotton or cotton weaving. These other uses included wool scribbling, silk waste spinning and corn grinding. There was the same spread of machinery noted by Crompton twenty-two years previously. A few mills still had jennies but where it was stated, the more modern mills tended to use mules rather than throstles. Where the mills could be located the answers relating to the mill premises have been included with the description of the mills in Part 2.

The final table, which summarises the scale of the Yorkshire industry in 1835, is taken from the returns related to the Factory Act of 1833. The returns were aggregated into townships thus making detailed analysis of the Halifax or Saddleworth areas difficult. This table is regarded as being fairly accurate with only ten to fifteen mills being omitted.

Table 2.4
Cotton Mills in Yorkshire, 1835

Leeds	3
Bradford	9
Huddersfield	11
Halifax	57
Keighley	5
Saddleworth	21
Craven	7
East Yorkshire	0
South Yorkshire	0
Yorkshire Dales	31
Total	**144**

Water power

The use of water power in the Yorkshire textile industry was widespread by the 1770s.[11] It was extensively used to drive fulling and scribbling mills to process woollen cloth and from 1780 and 1787 respectively to drive cotton and worsted spinning mills. The use of water power was also well established for corn milling, seed crushing, dye making, forging and grinding. One of the main reasons why the cotton textile industry expanded from its base in Lancashire was the availability of water power sites in Yorkshire on the eastern and western slopes of the Pennines. The sites were either supporting redundant corn mills or were suitable in other ways for the building of cotton mills. Up to 1800, when about two hundred and forty cotton mills had been built in Yorkshire, only twelve were steam powered.

From details of some of the mills it is possible to see the development of larger water wheels as the demands for power increased. Certainly the mill owners were interested in any means of increasing the power of their wheels and a number of books were published and advertised in the local newspapers claiming to give information of a technical nature. Two of these, 'A Treatise on Mills', London 1795, and 'On the Power of Machines', Kendal 1803, were by John Banks. He was aware of the growing interest in the efficient use of water power and gave public lectures in Yorkshire towns. For 5s. it was possible to hear Banks speak in Halifax in 1793 on water wheels when it was said that he:

Intends to deliver a lecture on the properties of circular motion, on the quantity of motion produced by a given power in a given time; and also passing over a given space; on the application of water; on the velocity of wheels; on the size compared with the fall; on the greatest effect that a given fall can produce.[12]

A few months later in Bradford, John Banks was going to show experiments with water wheels which would take four to five hours to prepare. The starting time of his lecture and demonstration at the Nag's Head was therefore set for 10.30 a.m.

John Banks' books based on his lectures were scorned by some practical millwrights. John Sutcliffe, the Halifax millwright who designed and built many mills in Yorkshire, wrote:

By these paltry models and representations made of them the public have been greatly mislead: and large sums of money have been expended to no purpose in making water wheels after such imperfect models.[13]

Some details of the type, size and horse power of the water wheels used in Yorkshire cotton mills have been found. This information was often quoted in advertisements when cotton and other mills were for sale. The usual information given was the size of the wheel, which appeared to be enough to give any prospective purchaser an indication of the potential power of the wheel. From the examples which have been found two types of wheel were extensively used. Firstly, the large diameter but narrow wheel, which was probably backshot, was found in mills in steep valleys where there was a good fall of water. This was the type of wheel found in some of the early and smaller mills. Secondly, the breast wheel where the width was about the same as the

Water wheel at Settle Bridge Mill of the type used in a number of early cotton mills.

diameter, was suitable for lower falls of water as the water was fed onto the wheel at a point just above axle level. These tended to be used in the later, usually larger mills, built on the main rivers and entailed more extensive dams and works.

Two examples of the first type of wheel were used to power Crossland Mills in Almondbury in 1811. The two, four storey mills had water wheels which were thirty-four feet by four feet and fifteen feet by three feet.[14] Another mill in Almondbury, Franks Mill, had a thirty foot by two foot three inch water wheel in 1808.[15] In the same year Eller Carr Mill near Bingley had a twenty-seven foot by two foot six inch wheel which was used to drive six cotton spinning frames and two worsted frames.[16] Another mill in the steep Morton Valley across on the other side of the Aire, Dimples Mill, had a twenty-seven foot by five foot wheel in 1818 while Austwick Cotton Mill in the village of Wharfe was built with a thirty foot by four foot wheel in 1793.[17] One of the narrowest wheels found was at Tom Hole Mill in Soyland, near Halifax, which in 1803 measured twenty-seven foot by two foot.[18]

The disadvantage of the larger diameter wheels was that their speed had to be low or centrifugal force could start to throw the water out of the buckets. A slow wheel also involved complicated gearing to bring the axle speed up to the speed required to drive the machinery. The number of mill sites where large falls of water could be obtained was also limited unless long goits were built well upstream from the mill. When more power and speed were required wheels were constructed of smaller diameter and greater width, sometimes with these measurements being on a 1 : 1 ratio. Greater power and width also imposed additional stress on the axles so it became normal to take the drive from a circle of gear teeth attached to the circumference of the wheel. This primary drive also gave a much higher shaft speed for transmission through the mill. The teeth were made up of iron segments bolted to the edge of the water wheel.

Castlefields Mill near Bingley was one of the larger cotton mills built before 1800. In 1804 the water wheel measured eighteen feet by eighteen feet and drove 2,232 water frame spindles as well as all the preparing machinery. When Markland, Cookson & Fawcett enlarged their cotton and worsted mill in Leeds in 1796 they offered their old water wheel for sale. It measured fourteen feet in diameter but was seventeen feet long.[19] A more usual size for this type of wheel was eighteen feet by twelve feet which was the size of the wheel at a cotton mill in Otley in 1809.[20]

Some mills, because of their locations, were built to take advantage of the steep slopes of their particular Pennine valley. Upper Mill, in the narrow valley above the village of Morton near Bingley, was one of those. According to replies given to the Factory Enquiry Commissioners in 1833 it was built for cotton spinning in 1779, but there is no corroborating evidence for that. However, as the first mill to be built on the stream advantage was taken of a prime site. A short goit was taken from a small weir across the stream. This led to a dam on the valley side which was very near to the mill. The gradient of the valley was such that the outlet from the dam gave a fall of sixty feet to the mill which enabled two thirty foot wheels to be driven, one mounted above the other.[21] Further down the same valley Morton Mill, which was built for cotton spinning about 1792, had a fall of fifty feet which was used to drive two wheels of thirty feet and fifteen feet diameters.[22] Lothersdale Cotton Mill, which was built about

High breast
wheel at
Skyreholme
New Mill, built
in 1832 for
cotton spinning.

1792, had a large forty-five foot diameter water wheel which was one of the largest
internal wheels to be made and can still be seen by special arrangement.

About 1830 Fielden Brothers built new dams for their mills in the Lumbutts Valley
in the Todmorden area. In the top mill they then installed three, thirty foot water
wheels, one on top of another in a vertical arrangement. The water from Pearson and
Healey Dams was put on the top wheel, from Lee Dam on the middle wheel and
from the Old Dam on the bottom wheel. The combined amounts of water were
estimated to develop a maximum of 53.75 hp although the normal working power
was about 28 hp.[23]

There are records of three dwelling houses which were partially built or converted
for cotton spinning or preparing by water power. One was Ryburn House in Soyland
which had a twenty-two foot by two foot wheel in 1792.[24] The second was a house
called Greenwood Lee in Heptonstall where a twenty-four foot by one foot six inch
wheel was attached to the house to drive a cotton picker, a billy and a thirty inch
double carding engine. The wheel and water supply to it were added about 1802 but

Tower built
about 1830 to
house three
water wheels at
Lumbutts Mill
in Langfield.

the wheel was offered for sale in 1805 when the cotton machinery was removed from the house.[25]

Sandywood House in Keighley was built partially as a house and partially as a mill. Water was fed onto the wheel, which was in the basement of the house, from a culvert which ran under Skipton Road. The middle storeys of the building were used for residential purposes so the drive from the wheel was taken up through them to the top floor where the spinning frames were operated.[26]

Within the first twenty years of the establishment of cotton mills in Yorkshire few inroads were made into the dominance of the water wheel as the prime mover. The high running expenses of the pumping engines on the one hand and the cost of Watt's premium payments on the other, stopped firms changing to steam. In addition the costs that had already been incurred in installing water wheels with all their necessary capital works had probably been recovered so annual running costs for water power were low.

Steam power

Low Mill in Keighley was the first mill to be built in Yorkshire in 1780 and was
equipped with an auxiliary pumping engine in 1785 which was used to pump water
back into the dam to increase the flow to the water wheel.[27] This method of utilising
the early type of steam power was adopted by a number of textile mills until they
were able to buy Boulton & Watt engines or other manufacturers rotary engines after
1800. The early 'fire engines' were pumping engines of Savery or Newcomen design
and were used, not as the prime mover, but to augment the mill stream when water
was in short supply. One of these engines was installed at Old Lane Mills in Halifax
in 1792. The engine was made by Booth & Co. and it was estimated that it would
pump 500 gallons of water at each stroke.[28] Its power was such that it worked one
mill until 1793 but had the potential to work a further mill which was in the course
of construction on the same site. Another pumping engine was installed in nearby
Jumples Mill at Ovenden. That engine had been made by the firm of Emmetts at
Birkenshaw Ironworks in 1785.

The greatest concentration of pumping engines and early Boulton & Watt engines
used by the local cotton industry was in Leeds. The unsatisfactory nature of the early
pumping engines prompted most cotton spinning firms to replace them with rotary
engines. When Beverley, Cross & Billiam built their cotton mill in Leeds in 1792 they
were advised to install a pumping engine which they did. However, they found that
the engine was not satisfactory and within a few months decided to buy a Boulton &
Watt engine. The pumping engine was inadequate and unreliable to the extent that
they found:

> our situation really pitiable. We have stopped most of this week to cobble up the
> old engine and cannot get to work at all when we have done.[29]

Beverley, Cross & Billiam also had problems with their boiler and appeared to be
determined to have a Boulton & Watt engine.

> . . . for the last month or six weeks we have not wrought 6 whole days and on
> Saturday instant the boiler's bottom, which was a very bad one, coming out we
> thought it most advisable not to be at any more expense about it accordingly
> discharged most of our hands that we could, some however and those of the
> highest wages we are obliged to retain.[30]

> We have a high opinion of the power of mechanism of it and as it will be the
> first that will be erected in this country for the spinning of cotton, we would
> wish to have it completed in such a manner as will do you credit and answer
> our purpose.[31]

Unfortunately the cotton trade was in temporary decline after the new engine was
installed and Beverley, Cross & Billiam found difficulty in paying for it and also paying
the premium Boulton & Watt expected. On the one hand it was explained that the
engine was not up to their expectations and on the other that 'We are sorry to pay
you so slowly but the times have been so awkward we could not well avoid it.'[32]
Boulton & Watt charged the premium from when the engine started work but that

Beam engine by Boulton and Watt for Beverley Cross of Billiam's Cotton Mill in
Leeds, 1792.

was thought to be unfair as: 'The engine never wrought to our satisfaction till near
Christmas since which time it has gone very well except that the cold water pump is
rather troublesome.'[33]

Another Leeds cotton spinner who had difficulties with the firm of Boulton & Watt
was Richard Paley. He was a partner in several enterprises including the iron foundries
at Hall Ings, Wakefield and Bowling near Bradford. Because of his business connections
he installed a Bowling Iron Works engine in his cotton mill in Leeds about 1790.
When the lease of the mill was advertised in 1796 the steam engine was then said to
be a 'fire-engine'.[34] However, that was possibly a subterfuge, for Boulton & Watt felt
that Paley was infringing their patent, in which case the engine would have been of
the rotary type. James Watt Junior visited the Leeds mill and then threatened to take
Paley to court. Paley decided to install a Boulton & Watt engine, partly because that
would remove their threats and partly because the Bowling engine was expensive on

coal. It did not develop the claimed horse power and as a result he was only able to run 1,400 spindles instead of 2,000 spindles as planned.[35]

After the erection of the Boulton & Watt engine at Beverly, Cross & Billiam's mill in Leeds, a number of orders for other engines came from the same town. Apparently Boulton & Watt were not enthusiastic about taking orders from textile concerns but they did sell several engines to the Leeds cotton spinners.[36] The Boulton & Watt engines sold to the Yorkshire cotton spinners are listed in Table 2.6. From this table it can be seen how orders were sparse after the Leeds firms had bought their engines.

> Gentlemen
>
> We was favored with Yours Yesterday and am sorry to find you have not neived the St Stretch of the Premises on which the 6 Horse Engine is to be fixt and which we sent you the 4th of last Month and as this delay has happend begs you will as soon as posable send drawing necessary for preparing the Building which can be finished here in a very little time and we beg you will send of the Engine as speedaly as Posable you may direct the Engine to Mr Ambrose Dean Addingham and send it by way of Leeds to be left at the Cannal Warehouse Shsden the drawings please to send to us at Baildon and as soon as posable You may either place it to the account of the person it is orderd for or to us it will be paid for punctually   we are your most Obedi Servt
>
>   Thomas Halliday & Co
>
>   Baildon Feby 14 1803
>
>   I hope you will be able to understand this Sketch

Order for Boulton and Watt engine for Townhead Mill, Addingham.

Table 2.5
Cotton Mills in Yorkshire with Pumping Engines

Firm	Place	Date	Reference
Claytons & Walshman	Keighley	1785	B & W MSS
Ard Walker	Leeds	1789	Sun OS365/562205/1789
Coupland & Wilkinson	Leeds	1791	RE20/120909/1791
Beverley, Cross & Billiam	Leeds	1792	B & W MSS
Markland, Cookson & Fawcett	Leeds	1792	LI 19.9.1796
Samuel Blagborough	Leeds	1793	SunOS392/610103/1793
William Mitchell	Ovenden	1793	LI 28.1.1793
Wells, Heathfield & Co.	Sheffield	1793	Sun CR1/622019/1793
Tarboton & Carr	Thorner	1794	SunCR3/629212/1794
Wilkinson, Holdforth & Paley	Leeds	1796	LI 4.7.1796
George Woodhead	Marsden	1796	RE32/153590/1796
Firm not known	Conisborough	1802	LI 19.4.1802

From Tables 2.5 and 2.6 it can be seen that apart from the Leeds area steam power had made little impact on the local cotton industry up to 1800. Only fifteen cotton mills have been traced which had steam power, compared with about eighty-one in the woollen and worsted industry.[37] After 1800 however, a number of local firms started building steam engines as Boulton & Watt's patent had expired and new cotton mills were likely to be equipped with these engines. That was particularly true where most of the possible water power sites had been taken. For example, three new cotton mills built in Keighley about 1800, Damside, Hope and Low Bridge Mills were all steam powered.

Table 2.6
Boulton & Watt Engines Supplied to Yorkshire Cotton Mills 1792–1835

SUN AND PLANET ENGINES:

Beverley, Cross & Billiam	Leeds	Ordered 1792
		22 hp cylinder 25.75in × 5ft
Markland, Cookson & Fawcett (Half power used for wool)	Leeds	Ordered 1792
		30 hp cylinder 28in × 6ft
Gowland & Clark	Leeds	Ordered 1795
		20 hp cylinder 24in × 5ft
M & J Bateson (Soon converted to wool)	Leeds	Ordered 1796
		20 hp cylinder 23.75in × 5ft
T & H Lodge	Halifax	Ordered 1802
		30 hp cylinder 28.12in × 6ft
T & H Lodge (Bought second-hand)	Halifax	Installed 1802
		10 hp cylinder 17in × 4ft

BEAM (CRANK TYPE) ENGINES:

Blagborough & Holroyd	Leeds	Ordered 1796
		20 hp cylinder 23.75in × 5ft

Holdforth, Wilkinson & Paley	Leeds	Ordered 1796 36 hp cylinder 23.75in × 5ft
Heathfield & Co.	Sheffield	Ordered 1812 36 hp cylinder 30.75in × 6ft
Heathfield & Co.	Sheffield	Ordered 1813 20 hp cylinder 23.75in × 5ft

SIDE LEVER TYPE ENGINES:

Ambrose Dean	Addingham	Ordered 1803 6 hp

Source: Boulton & Watt MSS. Birmingham Reference Library.

The relative position of water and power is shown in two further tables – 2.7 and 2.8 About the same amount of power was generated by water and steam in 1835 but there were nearly twice as many water wheels as steam engines.

One event which cast doubts on the future of water power was the severe draught in the summer of 1826. This caused grave problems for firms which depended solely on water power. Just to the west of Halifax, the Luddenden Valley mill owners had already built dams at Cold Edge to ensure their water supplies. In 1825 the owners of the mills driven by the neighbouring Mixenden and Wheatley streams were also still satisfied with water power and similarly decided to make a reservoir at Ogden to ensure a more constant flow of water. However, in 1826 the long draught made the mill owners abandon their reservoir scheme and several equipped their mills with steam engines.[38]

Table 2.7
Power Generated by Water Wheels in the Yorkshire Cotton Industry 1835*

Power of Wheel	Total Power	Number of Wheels
50 hp and over	—	—
40–49 hp	125	3
30–39 hp	150	5
20–29 hp	229	11
10–19 hp	508	40
Under 10 hp	368	75
Total	1,380	134

The average power of the wheels was just over 10 hp.
Source: Factory Inspectors Returns for 1835 and 1839* (where omitted)

Table 2.8
Power Generated by Steam Engines in the Yorkshire Cotton Industry 1835*

Power of Engine	Total Power	Number of Engines
50 hp and over	150	3
40–49 hp	166	4
30–39 hp	186	6
20–29 hp	328	15
10–19 hp	393	31
Under 10 hp	111	18
Total	1,334	77

The average power of the engines was over 17 hp
Source: Factory Inspectors Returns for 1835 and 1839* (where omitted)

Horse mills

As more and more water and steam powered cotton mills were built in Yorkshire a third type of mill was built. This was the horse mill where the motive power for some of the machines was provided by one or two horses turning a large wheel from which the drive was taken. These mills have only been traced in Skipton, Otley, Horsforth and Leeds but were in the areas where much of the yarn was produced on water frames or throstles. As these horse mills normally housed jennies or mules they were probably used to produce weft while the larger power driven mills in the area supplied the warp yarn needed by the local weavers.

Of the three types of spinning machine which were available, two, the jenny and the mule, could be worked by hand. However, it was still necessary to have power driven cards to supply cardings in quantity but this could be achieved with a horse mill. It was therefore possible to set up small spinning workshops which required a minimum of power but could produce quantities of yarn with little capital expense. It has even been suggested that the jenny workshops provided considerable competition for the early Arkwright type mills using power driven water frames.[39]

Within Yorkshire a number of these small workshops have been found. For a small cotton manufacturer they provided a controlled source of yarn where work and standards could be supervised. They were of particular advantage to a small manufacturer for the ownership of such a workshop gave him control of all or part of his yarn supply in terms of cost, quantity, quality and availability.[40] Their small size and limited capital requirements however, have made their numbers impossible to quantify. In addition they probably only existed for a few years when localised cotton hand-loom weaving was important and before wider markets developed in a wide range of yarns of different qualities.

Two market towns, Otley and Skipton, provide useful examples of these workshops which existed within a short distance of large water powered mills producing twist for warps. There were at least three of these workshops in Otley, two of them owned by a Mr Bucktrout. The first appears to have been purpose built about 1795. In 1798 it was offered for sale:

A substantial new built dwelling house situate in Bondgate in Otley, in Mr Bucktrout's own occupation, with a stable and other convenient outbuildings, and a garden behind the same, also a newly erected factory for the spinning of cotton attached to the house and partly over it, containing upwards of three hundred and ten square yards.

The advertisement then went on to give the dimensions of the various rooms in the 'factory'.

1	=	42ft × 20ft
1	=	40ft × 12ft
2	=	22ft × 12ft

1	=	18ft × 13ft
1	=	13ft × 10ft

Also for sale was the machinery which consisted of:

2 mules	168 spindles
2 mules	156 spindles
3 mules	144 spindles
1 mule	132 spindles
Total	1,212 spindles

All the machinery was nearly new, well constructed and in good condition.[41] Bucktrout appeared to have the preparing carried out at two other horse mills which were situated near his house in Bondgate. The machinery in these consisted of :

Horse Mill	–	2 pair 18in cards for mules
		1 drawing frame
Horse Mill	–	2 pair 18in fine cards
		1 drawing frames
		1 roving frame 60 spindles

The interesting point about Bucktrout's factory was that he was operating a total of 1,200 mule spindles which was a larger number than in many water powered mills.

In the Skipton area similar factories for hand and horse powered production existed. One of them was to let in 1793:

'A cotton manufactury, situated in the New Market at Skipton, in Craven, Yorkshire, containing one tumming and two carding engines, with drawing and roving frames, and every other article necessary for preparing cotton ready to spin.

Also six machines commonly called mules with four reels, a scale and every other article necessary for spinning, reeling, weighing and making for sale muslin twist.

Also a large horse wheel, with two good horses, for turning the carding engines, drawing and roving frames.[42]

In 1801 William Chamberlain, who was an ironmonger and cotton spinner, with a medium sized cotton mill at Eastby, also had three mule spinning shops in Skipton.[43] Chamberlain also used these premises for weaving cotton.

These small factories were not only found in the smaller towns. Thompson & Naylor were cotton manufacturers in Leeds who also operated a horse mill. This had a thirteen foot diameter horse wheel which would have been used to drive the carding and breaking engines. The main items of machinery in this horse mill consisted of:

1 18in breaking engine	
1 40in finishing engine	
1 double 18in carder	
2 mules	204 spindles each
3 mules	216 spindles each
1 stretching frame	78 spindles
6 heads of drawing	
19 pairs of cotton looms	

2 dressing frames
1 twining jenny 106 spindles
3 spinning jennies[44]

Benjamin Thompson and Thomas Naylor insured their property for the following sums in 1803:

	£
Millwright work, horse spinning rooms and weaving shops	25
Machinery	275
Stock	200
	500

The property was owned by a gentleman called Edward Armytage who insured it for £700.

Thompson & Naylor were also fustian manufacturers as well as cotton spinners. Thompson died in 1805 and the property of the partners was then assigned to George Thompson of Thorner who was also a cotton manufacturer. The spinning machinery then included:

2 mules × 204 spindles
3 mules × 216 spindles
3 jennies

They also had nineteen pairs of cotton looms.

High Mill, Addingham *c.* 1950.

In terms of power and number of spindles there was little difference between the horse mills and the smaller water mills. Marshall & Lister ran a small jenny mill at Burley Woodhead. When the mill was for sale in 1800 the machinery consisted of:

    2 carding engines
    2 billies
    12 jennies[45]

The small dam at the back of the mill shows that water power was used to drive the cards. Hetton Mill, which had limited water power, had similar machinery:

    1 cotton beater driven by water
    1 carding engine
    1 roving billy 42 spindles
    6 jennies × 100 spindles[46]

Where power was limited the spinning machinery was usually jennies and mules with only the heavy machines being power driven. The small factories working hand spinning machines did not pre-date the larger mills but were built and run at the same time. The jenny mills, however powered, produced weft while the early Arkwright type mills produced twist for warps. In this way they complemented each other and provided both types of yarn the local weavers needed.

Machinery

During the period 1780–1835 the types of cotton spinning machinery in use in Yorkshire were at first the jenny and water frame and later the mule and throstle. There was a fair degree of overlap in time however, for some mills, such as Old Lane Mill at Ovenden near Halifax had been equipped with mules by 1799,[47] whereas Thomas Haigh at Colne Bridge near Huddersfield was still using jennies in 1833.[48] Some mills, such as Lower Mill at Long Preston ran mules and throstles,[49] but in most cases where that was done there were more mule spindles than throstle spindles, eg Bent Mill near Wilsden had 4,248 mule spindles and 2,312 throstle spindles in 1826.[50]

   The relatively simple nature of much of the early textile machinery, together with its development from a few basic designs, meant that it was eventually possible to process other fibres on the same machine with certain modifications. In some cases it was therefore not necessary to re-equip a mill completely when it changed from say, wool scribbling and carding to cotton carding and spinning. When Firth House Mill near Scammonden was to let in 1802 it was suggested, optimistically, that the four wool cards in the mill could easily be used for carding cotton.[51] A more usual change was in weaving where worsted looms could easily be used for weaving cotton. It was even suggested that some carpet looms 'may be altered at a very small expense for the cotton business'.[52] At High Mill in Addingham, Robert Hargreaves experimented using cotton machinery to spin worsted yarn in 1787 and was eventually very successful.

Water frames and throstles

The earliest Yorkshire cotton mills used Arkwright's water frame or local adaptations. These had separate gearing to each set of spindles. As the number of spindles on the early frames rarely exceeded forty-eight or fifty-two they gradually went out of use after 1800 although some firms retained them for many years. Thomas Parker of Damside Mill in Keighley had twelve water twist frames with sixty-four spindles each up to 1807.[53] Those frames had more spindles than some throstles but the frames were said to be new. When Strong Close Mill in Keighley was sold in 1805 it contained twelve water twist frames with four hundred and ninety-two spindles between them.[54] Six years later those frames had been replaced by twelve throstles with forty-eight spindles each.[55] Joseph Driver of Castle Mill in Keighley was bankrupt in 1807 and his spinning machinery was for sale. Seven of the machines were described as frames and had forty-eight spindles each. The two larger machines were also described as frames and had ninety-six spindles each although they were probably throstles. William Sidgwick of High Mill in Skipton retained water frames until 1814 when he changed over to throstles.[56]

The new throstle which replaced the water frame appears to have come into general use in Yorkshire after 1800. This type of frame had a simpler drive system with a tin roller for driving the spindles running the length of the frame instead of a drive to each set of four spindles. The throstle also ran at a higher spindle speed than the water frame. One of the first throstles made in Keighley in 1798 by William Carr, was supplied to Blakeys & Marriner at Greengate Mill for cotton spinning.[57] There were a number of textile machine making firms in Keighley by that time and many of the Keighley cotton spinners eventually changed over to the throstles which were made by them. However, there appeared to be no rush to install the larger machines and in 1811 only two of the mills had them as can be seen in Table 2.9. It should be noted however that both Damside Mill and Cabbage Mill were run by the successful Greenwood family who had the capital resources to buy up-to-date machinery.

Table 2.9
Examples of the Size of Frames and Throstles in Keighley Mills, 1811

Firm	Mill	Throstles
Dinah Heaton & Sons	Spring Head Mill	16 × 52 spindles
Craven, Brigg & Shackleton	Walk Mill	12 × 52 spindles
Wrights & Sugden	Mytholm Mill	16 × 52 spindles
Clayton & Walshman	Low Mill	36 × 52 spindles
Illingworth & Marriner	Grove Mill	12 × 52 spindles
Blakeys & Marriner	Greengate Mill	36 × 52 spindles
Greenwood & Ellis	Damside Mill	16 × 96 spindles
John Roper	Damens Mill	6 × 52 spindles
Thomas Binns	Stubbing House Mill	8 × 48 spindles
John Greenwood	Cabbage Mill	16 × 96 spindles

Source: Samuel Crompton's Spindle Survey 1811. Bolton Civic Museum

Mules

Although mule spun yarn first became available in 1780, the year the first cotton mill was built in Yorkshire, mules did not come to be installed in any large numbers in Yorkshire mills until much later as the early firms concentrated on spinning the coarser yarns. Even by 1806 it was said that they had only recently been introduced. The advantage of mules was that they could spin yarn suitable for the manufacture of fine cloths such as muslins which had previously been imported from India but the disadvantage was that they needed wider buildings. With the improvements to the basic mule and the use of Sea Island cotton in the late 1790s it then became possible to attack the cheaper end of the muslin market. When mules were adopted in Yorkshire their use was far from uniform. The main concentration was in Halifax Parish, but there were several other more isolated concentrations. Although it is not possible to calculate the fraction of mule spun yarn out of the total yarn production in Yorkshire it is likely that by 1812 it amounted to the half or three quarters which was claimed nationally by Crompton.[58]

Crompton's figures, based on the spindle capacity of the majority of the Yorkshire mills at that time, illustrate the importance of mule spinning in the Halifax area and the importance of water frame and throstle spinning in the Keighley area. Around Halifax, sixty-four mills contained 131,522 mule spindles, but only 21,312 frame and throstle spindles. By contrast the figures for Keighley were that nineteen mills contained 4680 mule spindles and 15,184 frame and throstle spindles in 1811. Out of the nineteen mills only four actually had mules. Those differences between the two areas were related to the pattern of development of the local industry. The Keighley mills were older and, as many had been built and equipped in the 1780s, they were only suitable for water frame spinning. One exception was Low Bridge Mill which was steam powered and relatively modern as it was built about 1800. In 1811 Low Bridge Mill was used for mule spinning with six mules with 228 spindles each.[59] Two further exceptions only two or three miles from the centre of Keighley were Upper Mill and Morton Bridge Mill, both in Morton, where the two mills were running 5,544 mule spindles between them. Of the eighteen mills listed in Keighley most had turned over to worsted spinning within the next ten years.

After about 1796 mule spinning developed rapidly elsewhere when power was applied to some of the motions.[60] This spinning machine then started to gain the ascendency over water frames although the development of the throstle curtailed some of its lead. In the Halifax area, where there was a relatively slow development of cotton mills until about 1800, there were still sites available for mule spinning mills to be built. In many areas of Lancashire the best water power sites had gone so the new mule spinning mills were built in towns and were driven by steam, which ensured their future success.[61] However, the Calder Valley and its tributaries still provided many sites, often occupied by a corn or fulling mill, so new mills which were built in the area were fitted out with mules. An example of an early mill was Longbottom Mill in Warley which was converted for cotton spinning about 1792. The spinning machinery consisted of frames which were taken out and sold to Ard Walker of Leeds in 1800. On the other hand most of the mills built after about 1800 were built specifically for mule spinning. A good example was Temple Mill at Rishworth which was built in 1799 and illustrates the change from the smaller water frame mills. Temple Mill was seven storeys high with 'each storey consisting of one light spacious room capable of containing eight mules of three hundred spindles each'.[62] Thus with a capacity of over 16,000 spindles, this mill, and others which were to follow, was to set a pattern for constructing mule spinning mills which was to establish the cotton industry firmly in the Calder Valley.

It should be noted however, that even by 1811, Temple Mill only held 4,800 mule spindles.[63]

In accounting for the eventual success of the cotton industry in the Calder Valley based on mule spinning it should be recognised that there were other areas of Yorkshire that had concentrations of cotton mills equipped with mules. One of those areas was Embsay, a small village outside Skipton where five cotton mills were built on a minor stream. Millholm Mill was built in 1793 and an additional building was being used for mule spinning by 1795.[64] Sandbeds Mill across the road was built about 1794 and had been used for mule spinning prior to 1805 when the mill and seventeen mules with three thousand spindles were for sale.[65] Whitfield Syke Mill, further upstream, was built about 1795 and had been used for mule spinning when it was for sale in 1809. The spinning machinery then consisted of six mules with 1,144 spindles.[66] Hammond & Tattersall, who had occupied the mill, also had four mules with 854 spindles in a building in New Market in Skipton. Good Intent Mill in Embsay and Eastby Cotton Mill a mile or so away were also used for mule spinning. In this particular area there was therefore an early adoption of mule spinning which may have been of significance in leading to the firm establishment of the cotton industry until recent times.

The expansion of steam power to 1835

One Yorkshire millwright who had much work from the local woollen and worsted mills was John Jubb.[67] He was also a partner in an early cotton mill at Morley in 1785[68] and appears to have made cotton machinery for in 1789 he was advertising for workmen 'who understand the making and fitting up of cotton and worsted machinery'.[69] Despite his partnership in the cotton mill and later in a woollen mill he concentrated on his millwright' business. He would supply other items than machinery for in 1792 he had two oak axle trees for sale, one 22ft × 2ft 8in and the other 18ft × 2ft 4in. A year later he was offering a complete 14ft × 11ft water wheel.

Despite the size and weight of second hand water wheels, which must have caused considerable transport problems, several were offered for sale as firms replaced them. Marshall, Fenton & Co, a flax spinning firm in Leeds were selling one in 1790.[70] Blagborough & Holroyd, cotton spinners in Leeds, also had one in 1796,[71] while a third Leeds firm, Fenton, Murray & Wood, the engineers, were offering several wheels and a steam engine in 1800.[72] Even by 1814, wood suitable for 'mill axles' was being offered from an estate near Brimham.[73]

In addition to second-hand water wheels it was also possible to buy second-hand steam engines and horse wheels. A firm in Heckmondwyke was offering a 15in cylinder crank engine in 1796[74] and in the following year T Lockwood & Co. of Skipton were offering a large horse wheel, 'nearly new and little used'.[75]

When Boulton & Watt's patent expired in 1800 the way was open for local engineering firms to supply the power needs of local textile concerns. Thomas Glover & Sons advertised promptly the following year that they wanted orders for steam engines as did Nichols & Bennett, who were brass and iron founders in Keighley.[76] The principal local manufacturers of steam engines were Fenton, Murray & Wood of Leeds, the Bowling Iron Company and Low Moor Ironworks near Bradford. However, the early development of the Yorkshire cotton mills with their strong reliance on water power curbed the need to take advantage of the now easily available steam power. Where steam engines were bought in the decade or so after 1800 it appeared to be in cases where all the river-side sites had been taken or where water power needed to be supplemented in conditions of draught or where extra productive capacity was being added to a mill. Providence Mill in Bingley had a steam engine when it was built in 1801 as most of the sites on the Aire and its tributaries had already been taken by cotton mills. Townend Mill, which was on a very small stream in Addingham, had the power from its water wheel supplemented with a 6 hp Boulton & Watt engine in 1803 while a steam engine was added to power the new machinery in the extension to Birkenshaw Mill in 1806.[77]

Table 2.10 shows the Yorkshire cotton mills which have been identified as having steam power between 1780 and 1835. By 1835 water power was still more important than steam power in more than two thirds of the West Riding parishes. However, many of the parishes where steam power was dominant were to be the places where the cotton industry would continue to flourish. They included Halifax, Saddleworth, Huddersfield, Almondbury and Skipton. Elsewhere, small water powered mills were still running but were destined to go out of business or be converted to other uses within a few years.

By 1824 there were twelve steam engines installed in mills in Todmorden with 150 hp between them. That figure was shortley to be increased by a 60 hp engine for Fieldens which was intended to drive eight hundred new power looms.

Table 2.10
Yorkshire Cotton Mills With Steam Power, 1780–1835

Mill/firm	Location	Engine Details	Date
Low Mill	Keighley	Pumping	1785
Ard Walker	Leeds	Pumping	1789
Coupland & Wilkinson	Leeds	Pumping	1791
(Replaced with rotary engine 1803)			
Beverley, Cross & Billiam	Leeds	Pumping	1792
(Replaced with Boulton & Watt 22 hp)			
Markland, Cookson & Fawcett	Leeds	Pumping	1792
(Replaced with Boulton & Watt 30 hp)			
Samuel Blagborough	Leeds	Pumping	1793
(Replaced with Boulton & Watt 20 hp)			
Old Lane Mills	Halifax	Pumping	1793
Sheffield Cotton Mill	Sheffield	Pumping	1793
(Replaced with Boulton & Watt 36 hp 1812, 20 hp 1813)			
Thorner Cotton Mill	Thorner	Pumping	1794
Gowland & Clark	Leeds	Boulton & Watt 20 hp	1795
Wilkinson, Holdforth & Paley	Leeds	Pumping	1796
George Woodhead	Marsden	Pumping	1796
J & J Sutcliffe	Huddersfield		1800
Hope Mill	Keighley		1800
Damside Mill	Keighley		1800
Low Bridge Mill	Keighley		1800

Holme Mill	Bradford	(Part use)	1800
Providence Mill	Bingley		1801
Gomersal Hall Mill	Birkenshaw	16 hp	1801
		plus 24 hp	1803
Bowling Dyke Mill	Northowram		1801
Willow Hall Mill	Skircoat	Boulton & Watt 30 hp	1802
	(Second-hand Boulton & Watt 10 hp)		1802
?	Conisborough	Pumping	1802
Rand's Mill	Bradford	10 hp	1803
	(Fenton, Murray & Wood)		
Ambrose Dean	Addingham	Boulton & Watt 6 hp	1803
Mellor & Robinson	Ecclesfield		1803
R Lobley	Leeds	(Part use)	1804
Hebble Mill	Ovenden		1804
Shepards Factory	Dewsbury	(Part use)	1804
?	Sandal	10 hp	1804
Knights Mill	Bradford		1806
Birkenshaw Mill	Birstall		1806
Upper End Mill	Marsden	24 hp	1806
Jumble Mill	Almondbury	16½ in cylinder	1807
The Factory	Almondbury?	24hp 27½ in cylinder	1808
Bankfoot Mill	Heptonstall		1808
Woodhead Mill	Keighley		1808
Idle Cotton Mill	Bradford	25 hp (Part use)	1809
George Moss	Ovenden		1811
At Littletown	Birstall	(Part use)	1811
Lower Mill	Long Preston	5 hp	1812
Grove Mill	Ovenden		1816
Bell Busk Mill	Gargrave		1816
Jumples Mill	Ovenden		1816
Wilsden Cotton Mill	Wilsden		1818
Bent Mill	Wilsden	Fenton, Murray	
		& Wood 4 hp	1818
G Stansfield			1818
Strines Mill	Saddleworth		1821
Higgin Chamber Mill	Halifax		1822
J Howard	Leeds	24 hp	1823
R Ingham & Sons	Stansfield	20 hp	1824
William Emmett	Birstall	10 hp (Part use)	1824
Fieldens	Todmorden	60 hp	1824
High Mill	Skipton	30 hp	1825
Stoodley Bridge Mill	Langfield	18 hp	1825
Hallas Bridge Mill	Wilsden	4 hp	1826
J Cartledge	Hipperholme	30 hp	1826
Old Cotton Mill	Barnoldswick		1827
J Cummins	Huddersfield	35 hp	1829

J Dewhirst	Skipton		1832
Haigh & Bros	Halifax		1832
Stones Mill	Soyland	12 hp	1833
Langcliffe Mill	Giggleswick		1833
Smallees Mill	Soyland	20 hp	1833
Shaw Carr Wood Mill	Slaithwaite	12 hp	1833
Dyson Lane Mill	Soyland	13 hp	1833
William Kenworthy & Sons	Saddleworth	20 hp	1833
J Buckley	Saddleworth	12 hp	1833
J Robinson & Son	Saddleworth	20 hp	1833
J Waring & Sons	Saddleworth	12 hp	1833
W Beaumont & Co.	Saddleworth	16 hp	1833
J Neild & Sons	Saddleworth	8 hp	1833
G & M Andrew	Saddleworth	20 hp	1833
W Hegginbottom	Saddleworth	6 hp	1833
S Wrigley & Bros	Saddleworth	24 hp	1833
J Hegginbottom	Ovenden	8 hp	1833
T & L Threlfall	Morton		1833
R Ingham	Stansfield	10 hp	1833
J Greenwood & Sons	Keighley	30 hp	1833
J Brook & Bros	Meltham	50 hp, 26 h 8 hp	1833
P Bold	Ovenden		1833
Causey Wood Mill	Langfield	8 hp	1833
Shaw Lane Mill	Ovenden	14 hp	1833
G Bramall	Saddleworth	10 hp	1833
Varley & Dyson	Huddersfield	12 hp	1833
J Crossley & Sons	Hebden Bridge	20 hp, 30 h (part)	1833
Hanging Lee Mill	Soyland		1833
Thrum Hall Mill	Soyland	10 hp	1833
T Ramsden	Stansfield	18 hp	1834
Goit Stock Mill	Wilsden		1834
Swift Place Mill	Soyland	12 hp	1834
Kebroyd Mill	Soyland	30 hp	1834

It has been estimated that by 1800 the cotton industry in Great Britain and Ireland accounted for 84 out of the total of 321 steam engines which were in use. In the Yorkshire cotton industry there were seventeen of these engines thus giving Yorkshire about 20% of the total number of engines.

Mill construction

In most cases where records have been found, the responsibility for organising the building and equipping of a mill was taken by the owner or tenant who was to run the mill. This responsibility covered a wide variety of tasks. Firstly materials had to be bought from a variety of sources and transported to the mill site although local stone and wood were sometimes used. Secondly, the work involved dealing with a new technology where there was often very little information available and expertise was closely guarded. During this time money had to be laid out but there was no return until the first machines started running.

An example of these activities in Leeds concerned a man called Ard Walker who built a cotton mill in Hunslet.[78] Construction of the mill started in 1800 but it was not completed until 1804. Walker had inherited an existing oil and cotton mill and so had some experience of the work before he started. Arrangement were made to buy a Fenton, Murray & Wood, 30 hp steam engine in August 1800 for which the sum of £100 was advanced. Payments were made for removing earth for the foundations of the boiler house, bringing 1,000 fire bricks from a pottery, buying timber and for a boiler which cost £64 10s. od. In December 1800 ten spinning frames and other

Drawing for gas apparatus at Greenholme Mill, Burley-in-Wharfedale.

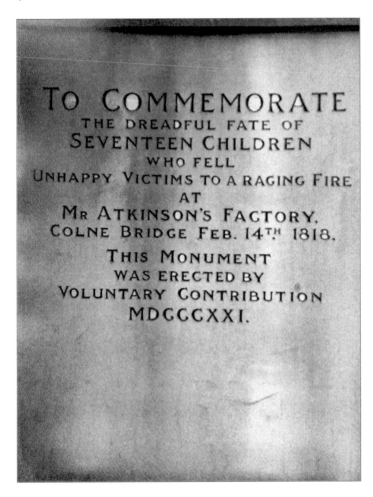

The memorial
to the children
killed at Colne
Bridge Mill fire
in 1818.

machinery were bought second-hand from Samuel Milne of Longbotton Mill at Sowerby Bridge in Warley, near Halifax at a cost of £168. During 1801 a further £100 was paid to Fenton, Murray & Wood and several thousand bricks were bought and laid to build the new mill. The dam was cleaned out, models of machines were bought and the joiners started to build the full size versions. Construction of the mill and machines continued during 1803 but the sums of money paid for making the machines exceeded those paid for finishing the mill. For example, from April 1803 to April 1804, £140 was paid for building work but £773 for machinery. This is reflected in the insurance valuation placed on the mill and contents in 1804 when it was completed:

	£
Mill	1,000
Mill work	400
Machinery	2,500
Stock	200

Engine and boilerhouses	300
Steam Engine	600
	5,000 [79]

Fire was the great danger to many of the early cotton mills and a number of newspaper entries recorded their destruction. The reasons for the excessive fire hazard were related to the materials used in the construction of the early mills, the nature of the machinery and the productive processes. Large amounts of wood were used in the construction of the first mills with the floor joists, floors and roof supports all made of timber which then became saturated with oil, which was flung or dripped from the machinery. All the early machines were made with wooden frames and there are cases where the machines themselves caught fire, possibly through frictional heat. The processing of cotton, particularly the early cleaning and opening operations, resulted in the production of a great deal of dust and 'fly' which clung to the machines and walls. Steps were taken to extract the dust but the machinery used was not very efficient. Poor ventilation, which was necessary for processing the cotton, did not help. A final hazard came from lighting and heating. Open stoves and lights were forbidden by the insurance companies in areas of high fire risk. However, oil lamps and candles were necessary and were widely used even after the introduction of gas lighting. This was only used in the larger mills and then only after about 1810.

When fires did break out, the fire fighting equipment and drill were often inadequate. One sad example in 1818 occurred when a fire at Colne Bridge Mills in Kirkheaton destroyed the mill and killed seventeen girls ages between nine and eighteen who were working there on the night shift.[80]

'Fire proof' cotton mills were eventually built with iron frames supporting stone floors and minimum wood was used in their construction. Wider roof spans became possible and the size of mills increased to enable more and larger machines to be accommodated. One of the first of these mills was the 'New Mill' of Greenwood & Whitaker at Burley-in-Wharfedale which was built in 1811 and said to be 'fireproof' and was lit with gas.[81]

CHAPTER 3

Sources of Capital

THE RAPID GROWTH of the cotton industry in Yorkshire from 1780 to about 1810 indicates that capital could be mobilised to satisfy an urgent demand. Although the capital raised was for a new industry, much of it was transferred from existing textile, merchanting and other businesses while there was widespread transfer of assets from other forms of production. The overall growth of the industry in such a short time and its widespread geographic location hides a number of difficulties which faced individual firms.

Within England at that time, there was a certain flow of wealth from increased overseas trade and expanding agricultural interests, but the channels to direct capital to industrial needs were inadequate. Some cotton firms found that they had insufficient resources when they were almost ready to start trading. In many ways the problems of finding working capital were greater than those of being able to finance the construction of fixed assets where some security could be provided to investors. According to one recent writer on the northern cotton industry:

> Investment in spinning mills on Arkwright's technique was characterised by a preponderance of merchant capital, by the migration of capital from other branches of the textile industry, and by the expansion of the industry into new peripheral areas in search of water-power sites and colonies of hand-loom weavers.[1]

This was certainly true for much of the early Yorkshire cotton industry which was based on the Arkwright type of mill either by licence or more likely as an illicit copy. However, it does not give the complete picture. There was a large investment by merchants in cotton spinning in some areas, particularly Leeds, but elsewhere capital came from many sources which often had no textile connection. Most of the expansion of the industry across Yorkshire was financed outside the Manchester heart land of the industry and for many years there was a Yorkshire cotton industry which was additional to the existing wool textile industry.

Capital requirements

The capital requirements of the Yorkshire cotton firms varied tremendously. There was a need for capital to cover the cost of both fixed assets and current assets, but this was alleviated to some extent by the common situation of capital transfer when firms running failing corn or fulling mills sold them to be converted into cotton mills. Apart from towns where there was a heavy concentration of mill construction, such as Keighley, the majority of the country mills were built on the sites of earlier corn mills. In many cases it was possible to start cotton spinning with minimal capital. In the first place buildings could be converted. Throughout the period from 1780 to 1815 the owners of corn mills often wished to sell or lease their mills because of the declining milling trade. Often, when they advertised their mills, they specified that they could be used for textile purposes and often mentioned cotton spinning. Examples of where it was suggested that corn mills could be converted for spinning cotton have been found for Holmefirth, Linton, Threshfield, Coley near Hipperholme, Kirkby Malzeard, Osmotherley, Easingwold, Blubberhouses, Thornton Arch near Leeds and Aldwick-on-the-Street near Doncaster, to name but a few. Many other corn mills were sold for the value of the site in terms of its relatively easy conversion to cotton spinning. Thus the cost of making dams, weirs and goits was already covered although it is likely that those were allowed for when the price of the mill was decided. A further point is that access roads and such out buildings as stables were already in existence. However, the power requirements for cotton spinning were usually greater than those for corn milling so it was often necessary to raise the level of weirs, increase the capacity of dams and put in new water wheels. Within a few years the old corn mill in so many places was just being used for storage while alongside was the new purpose built cotton mill.

Another type of mill to be converted for cotton spinning was the fulling mill.[2] These mills had existed in the county from the twelfth century and with the steady development of the woollen industry during the eighteenth century more mills were built on the banks of the Wharfe, Aire, Calder and Colne.[3] In addition to housing fulling stocks, these mills provided the premises for the new wool scribbling and carding machines. From the installation of carding machines for wool it was then only another step to introduce cotton carding and spinning. There were many instances where wool scribbling, carding and fulling mills were converted to cotton spinning, particularly in the Ryburn valley and Saddleworth areas.[4] These water powered mills with power and transmission systems suitable for textile production could easily be turned over to cotton spinning. Several, such as Slitheroe Mills, were used for cotton spinning and fulling at the same time.[5]

The owners of other buildings which could in some way be converted to cotton spinning also offered them for sale for this purpose. These buildings had previous textile uses such as the Cloth Hall in Gomersal and the White Cloth Hall in Leeds.[6] Others had been used for a variety of purposes,eg an iron foundry at Seacroft in Leeds,[7] a malt kiln at Farnhill near Kildwick,[8] a saddler's shop in Bradford[9] and a warehouse in Bingley.[10]

There were also several cases of corn millers and wool scribblers putting in a few cotton machines and financing this new development from their existing business. One of these was at Brearley Mill in Midgley near Halifax, where cotton machinery had been installed by 1802 but was probably taken out a few years later.[11] Similarly, Anthony Fentiman who built a new mill on a stream in the centre of Addingham had previously run some cotton machinery for a short time at Beamsley corn mill on the other side of the river Wharfe.

Secondly, an entry to the cotton spinning business could be made in a small way. There are a number of examples of firms which used horse wheels or very small water powered mills to drive only the carding engines while spinning was done by hand on jennies or small mules. Henry Marshall and William Lister had a small mill at Burley Woodhead in 1795 where the 'cotton factory' was insured for £250 and the utensils for £90.[12] The mill still stands today and is divided into two houses with the small dam behind them. Hannah Brookes occupied a room over an iron turners shop in Marsh Lane, Leeds in 1808 and had machinery valued at £90. John Holroyds cotton mill at Ripponden Wood was only valued at £50 in 1797.[13]

At times it was possible for a person or firm embarking on the business of cotton spinning to pay only a proportion of the cost of the premises. The rest could be paid in instalments on the security of the buildings. Examples of that system occurred with Old Lane Mills in Ovenden, Halifax, where in 1799 'The sellers have no objection to accommodate the purchaser of the new mill and machinery with part of the purchase money, on security of the premises'.[14] Or Hebden Mill in 1800 where 'part of the purchase money may remain in the hand of the purchaser',[15] and Burton-in-Lonsdale Cotton Mill where 'Payment of the purchase money (if required) may be made by instalments and at periods to suit the purchaser'.[16]

The normal form of loan capital was the mortgage. The money was provided by merchants, land-owners, manufacturers and others. In 1811 for instance, James, Jonas, William and Joseph Brook of Meltham Mills mortgaged a warehouse, hand spinning room and tenements to Thomas Leigh of Honley.[17] Heathfield & Co. were prepared to sell their large cotton mill in Sheffield in 1818 and receive only half of the money with the rest covered by a mortgage on the premises.[18]

Although it is not possible to ascertain their importance, the Leeds papers carried several advertisements for money to lend during the early 1790s.[19] This money was available on security while the interest on the mortgage was from 4% to 5%. These advertisements were unusual for it is likely that the majority of mortgages were arranged privately through business or family connections or through solicitors. For example Joseph Mason bought Airebank Mill in Gargrave about 1818 with the help of a mortgage.[20] One of the mortgagees was John Carr who was a lawyer in Skipton but the mortgage was with the Skipton bankers Chippendale, Netherwood & Carr.

Another way for firms to start cotton spinning with little capital was to rent mills which had been built for them by landlords. In prosperous times people even speculated by building mills and then looked for tenants. Thus at the height of the cotton spinning boom in the early 1790s, a mill being built in Halifax could be 'adapted to the convenience of the taker or purchaser'.[21] Other mills were advertised as being suitable for cotton, worsted, woollen or flax spinning for the owners provided the building and power but not the machinery. When Old Lane Mill near Halifax was built in 1793, it was offered to let and 'the foundation whereof is laid . . . may yet be adapted to the convenience of the taker or purchaser'.[22] It was also possible to buy the remaining part of a lease in a mill and often the machinery the previous tenants had been using. Thus sixteen years of the lease of Castle Mill, Knaresborough could be bought in 1793, together with twenty-six spinning frames as well as the stock of the previous tenants, Lomas, Thornton, Lomas & Co.[23]

In many cases the assignees of bankrupt firms wanted tenants to take over empty mills where the machinery might have been sold. In that case it was not unusual for the assignees to advance some capital to start a new venture. Certainly there were many cases of assignees selling mills or leases on mills and in the case of Richard Paley's cotton mill in Leeds, of running the mill to the annoyance of the creditors who wanted their money paid to them by the sale of the mill.

The amount of rent paid by the tenants depended on a number of factors. One of these was who financed the building of the mill and therefore expected a greater return. If the mill was constructed by the landlord the rent was higher. The agreement between Robert Bradley & Co. and the Duke of Devonshire for the building of Rilston Mill in the Parish of Burnsall is an illustration:

> The said partners propose to erect and furnish in a workmanlike manner according to a plan to be mutually fixed upon a building to be used as a manufactury for carding and spinning cotton wool upon the ground where the old mill at Rilston now stands or as near thereto as may be thought convenient and to pay every expense attending the same and an annual rent of £20 on having a lease granted thereof for the term of 21 years and His Grace allowing timber for the roof of such mill.
>
> They further propose that in case His Grace should be desirous of building such a mill at his own expense which they estimate at about £600 that they will take a lease thereof for the aforesaid term at the rent of £60.[24]

Partnerships

A further way to enter the business of cotton spinning for those who had some money was to join an existing firm as a partner. The partnership was the normal form of business organisation but the membership was often fluid with a number of partners leaving or joining depending on a range of circumstances. Within families it was normal for sons, brothers, brothers-in-law and even mothers and sisters to join and leave the partnership which ran the family mill. Thus Peter Garforth and John Sidgwick who occupied High Mill in Skipton were brothers-in-law; Springhead Mill in Haworth was run by Dinah Heaton & Sons after the death of John Heaton and Lord Holme Mill near Heptonstall was occupied by the Gibson family for so many years that it became known as Gibson Mill.

It was often necessary for a firm to take more partners. Death, or the wish of one or more partners to leave left the remaining partners with the problem of paying out to the deceased's family or to those who wished to withdraw their share of the capital. For a successful firm which had traded for some years and extended its property and machinery that meant considerable sums of money had to be found. New partners were therefore welcome and were also sought when a firm wished to expand but did not have sufficient resources.

In the early years of cotton spinning other attributes than the possession of capital were sometimes required:

COTTON SPINNERS

Wanted, an active partner in a cotton mill on a constant stream of water that will work twenty frames, etc.

A person who has been accustomed to the business and has a competent skill in mechanics, and possessed of a few hundreds will be treated with on very liberal terms.[25]

A few weeks later partners were wanted to help convert the corn mill at Thornton

Arch near Clifton outside Leeds. As inducement it was said that the 16ft fall would provide power to drive about 600 spindles and that there were plenty of hands to be had at low wages.[26] However, there is no record that this corn mill was ever turned into a cotton mill.

At times larger amounts of money were needed. For example, in 1803 a partner in a cotton factory was wanted with £5,000.[27] A partner was wanted to help run Sandal Cotton Mill near Wakefield in 1804, who would have to bring £1,000 to £1,200 into the business.[28] In 1809 a partner was wanted in an 'extensive cotton factory in the neighbourhood of Huddersfield. A person wishing to take a share in the above concern will be required to advance £5,000 to £10,000'.[29]

Despite the opportunities to join existing partnerships they were not always taken up. John Ritchie had been a cotton spinner and calico manufacturer at Silver Mill in Otley up to his death in 1799. The manager of his spinning mill wanted a partner so that he could continue to run the mill. He was prepared to put up half the capital required himself and wanted a partner to 'produce a few hundred pounds'. However, a partner did not materialise and Silver Mill was sold to Thomas Butler.[30] Another example of being able to join a partnership in cotton spinning, this time without any capital, came in 1797. So long as the person understood the 'art of carding and spinning . . . no money will be required to be advanced'.[31]

Machinery

After finding a suitable mill the machinery and other equipment could always be bought. This included such large items as water wheels, steam engines and boilers as well as picking, carding, roving, drawing and spinning machinery. Even such incidental items such as cans, scales and weights were available. The water wheels came on to the market when they were being replaced by larger versions or steam power. Several of the Leeds cotton spinning firms such as Blagborough & Holroyd and Markland, Cookson & Fawcett tried to sell their water wheels when they bought engines from Boulton & Watt. Markland, Cookson & Fawcett also tried to sell their old steam engine complete with flywheel and boiler.

Second-hand cotton preparing and spinning machines were frequently offered for sale after about 1790. This occurred when they were replaced as the average size of throstles and mules increased over the period and so the earlier, smaller machines came on to the market. A more usual source was the demise of one of the existing cotton firms, sometimes with the death of the owner but more frequently through insolvency or bankruptcy.

An early example comes from Bradford in 1792:

TO BE SOLD

The engines for the working of cotton which were late the property of Mr Thomas White of Brick Lane near Bradford deceased, consisting of one scribbling machine, one carding machine, a drawing and roving frame, with about ninety tin cans, a ratching frame and a reel, all nearly new.[32]

A few weeks later the cotton machinery at Balby Mill near Doncaster was for sale and

> Betty Hudson
> Rebuilt this Mill
> in the year 1802
>
> And that ye study
> to be quiet and to do
> your own business
> and to work with
> your own hands

Inscription in stone commemorating the rebuilding of Damside Mill.

included a cotton picker, cards, roving and drawing frames, a throstle, mules and jennies.[33] Few weeks went by without some second-hand machinery being offered for sale in the Leeds and Halifax papers. It was therefore possible for aspiring cotton spinners to buy the machinery they wanted at one of these frequent public sales.

A further way of entering cotton spinning with minimum capital was to specialise. Firms that did so concentrated their resources on particular types and qualities of yarn for which they hoped there was a regular market. Many of the Yorkshire firms came into the category of 'country spinners' who spun low counts and were not concerned with changes in fashion. Others started cotton spinning as an addition to cloth manufacturing so that they could secure their own yarn supplies. As they were organising the weaving of their own yarn they would know exactly what to produce and could cut out some costs by vertical integration. Many of the local domestic cotton manufacturers built or rented spinning mills for that purpose.

The heavy capital cost of constructing a mill with its power supply, shafting and millwork could be avoided by renting. This was a way by which many firms started as they only had to supply the machinery and cover their initial stock and operating costs. Existing textile mills and other buildings were always coming on to the market to rent. For instance, Gomersal Cloth Hall was advertised as being of interest to 'woollen and cotton manufacturers' in 1786, and was leased for cotton spinning from about 1788. Bridgehouse Mill in Haworth was to let in 1789 and could be used for linen, wool or cotton. It was well supplied with coal and water while 'any number of hands may be had at reasonable prices'.[34] From then on the Leeds and Halifax papers carried numerous advertisements for cotton mills to rent. One interesting example was Midgehole Mill near Hebden Bridge which had been used for flax spinning until 1811 but was then offered to let with the rents being:

£18 15s. 0d. pa if woollen
£23 15s. 0d. pa if worsted

£37 10s. 0d.　　pa if cotton[35]

Presumably profits from cotton spinning justified the higher rent. It was also always possible for a small firm with limited capital to rent part of a mill and start cotton spinning. This occurred in most parts of Yorkshire but was particularly prevalent in the Saddleworth area when the cotton industry became established there. The opportunity to let off 'room and power' also allowed firms which owned or leased a complete mill to reduce their capacity during times of bad trade. The earliest example found was at Castle Mill in Keighley where Joseph Driver & Co. and Joseph Smith were cotton twist spinners in 1788.

Raising capital

With the amounts of capital required to start cotton spinning varying from a few hundreds to several thousand pounds it was inevitable that a number of sources of capital were tapped. The evidence available does not give an overall picture or balance of these sources but the examples that have been found give some indication of the people, firms and institutions who were prepared to back cotton spinning enterprises. Briefly these were landowners, family, firms in another trade, banks, partners and private investors. The transfer of capital was usually achieved by mortgage. An additional source of capital available to successful firms was the plough back of profits. In many cases a combination of sources was used.

Ownership of land provided a variety of ways into cotton spinning. With the development of the water powered mill came a rush to find sites and many people found that their land had suddenly become more valuable. One area where it is easy to find examples is Keighley where over twenty cotton mills were built in as many years. Land owners were as eager as others to embark on the business of cotton spinning and they became aware that they possessed useful assets with possible advantages. The first was that land which already included an existing corn or fulling mill with the dams and goits had building costs substantially reduced. Walk Mill in Keighley had, as its name implies, been used as a fulling mill but was bought by John Craven in 1776. In 1783 he went into partnership with Thomas Brigg and Abraham Shackleton to start cotton spinning at the mill and he was to receive £12 a year rent for the mill together with the water and watercourse.[36] Even if there was no mill, which might have to be replaced anyway, there could be other advantages which were becoming apparent. These were a fall of water, a site which could easily be prepared for a mill, stone and timber readily available, easy access to a main road or canal, nearness to coal pits, within a short distance of a labour supply or in the centre of a cotton weaving area. These were all points which were stressed when land was offered for sale. Those who already owned land with these benefits could take advantage of its enhanced value and usefulness. Further examples from Keighley are Holme House Mill which was built about 1793 on land owned by a gentleman farmer called John Horsfall. Another gentleman farmer called John Shackleton built Wood Mill about a quarter of a mile from his own house on a stream with a good water fall. John Greenwood bought a small farm near Haworth about 1790 and as he gave a very liberal price for it, people were at their wits end to know why he had given so much money for so little a farm, but he had no sooner got the conveyance signed and the purchase money paid than people had their eyes opened, as he proceeded to take in a very excellent and valuable waterfall which he had bought along with the farm. He conducted the water down to the mill, at the same time putting in another water wheel, and thus using the water from the Haworth Beck twice over.[37]

Landowners were usually people with disposable income so besides the possibility that they might own suitable sites for cotton mills they also had money which they could use to finance the capital requirements of a new enterprise. This they could do in several ways. Firstly they could build mills to rent, secondly they could supply money to new firms, perhaps as a mortgage, thirdly they could go into partnership with others, taking advantage of their own land and capital and fourthly they could go into production for themselves. In many cases some combination of these methods was used.

The Parish of Keighley provides several more examples of land owners becoming involved in cotton spinning. The two main streams, the river Worth and North Beck developed enhanced value after the success of Clayton & Walshman at Low Mill from 1780 onwards. Those who owned land alongside these streams were not slow to take advantage of the possibilities.

Abraham Smith, a gentleman from Kildwick, bought the West Greengate Estate in Keighley in 1775. This bordered the river Worth and about the same time he bought land further up the valley. In 1784 Smith was joined in partnership By Rowland Watson, a solicitor, Joseph Blakey, a wool stapler, John Blakey, a stuff maker and James Greenwood who appears to have been a millwright and a person with some experience in cotton spinning.[38] Abraham Smith was possibly the driving force behind this partnership for it was said that it was he who built the mill. Smith was paid £43 0s. 8d. by the other partners (except Greenwood) for the 3,443 square yards of land on which the mill was to be built. He was also paid £210 for the firm's right to take water from the river and through his land further up the valley. There are no further details of the cost of building the mill, but there was apparently a need for extra capital. Smith mortgaged land in Keighley in 1786 to Jonas Whitaker a cotton spinner who was later a partner in a mill at Burley-in-Wharfedale. Again in the following year Smith mortgaged land to Mary Wright of Skipton. Further evidence of the slender resources of the partners and their reliance on Abraham Smith is shown by the fact that the money owing to him by the other partners was not paid until 6 May 1788.

Writing in 1879, John Hodgson said of Keighley:

the rage was so great to embark in the business of cotton spinning, that not only the leading gentlemen and land owners of the neighbourhood, but ladies also embarked in the enterprise, doubtless being lured by the prospect of acquiring wealth.

Hodgson was referring to Mrs Betty Hudson who built Damside Mill, Miss Rachael Leach who built Dalton Mill and Mrs Ann Illingworth who built Grove Mill at Ingrow near Keighley for her two sons David and William. Other mills built by landowners included Damens Mill built in 1789 by the Roper family, Ponden Mill and Royds House Mill built by the Heaton family and Newsholm Higher and Lower mills built by Robert Hall. These last two mills still exist but are now houses.

Large landowners were also involved in financing the new cotton mills. Lord Dartmouth owned large estates, particularly in the Slaithwaite area and some extracts from his Terriers illustrate his involvement:

> But the spirit of erecting cotton mills being high at the time, Varley, Eastwood & Co. proposed to take the bare stone walls as they were at the old rent of £60 per annum provided Lord Dartmouth advanced £600 at 8% and gave £200 worth of timber; which being done they erected a large cotton mill and are to have a lease for 31 years from May 1803 at the rent of £108 0s. 0d.[39]

Lord Dartmouth also lent a further £100 when the mill dam was enlarged.

Other cotton mills were built on Lord Dartmouth's land sometimes at the expense of the firms involved. These were New Mill (Messrs Shaw & Haigh), Shaw Carr Wood Mill (John Varley), for which Lord Dartmouth lent £250, Black Moor Holme Mill (Townend & Varley), Waterside Mill (Eastwood & Co), Lingards Mill (James Garside), Steps Mill (William Beaumont) and Meltham Mill (William Brook).

The Duke of Devonshire's Yorkshire estates also provided sites for cotton mills such as Low Mill, Keighley, Castle Mill, Keighley and Rilston Mill in Burnsall Parish.

Although it has been found that West Riding woollen merchants were not extensively involved in the development of the early woollen and worsted mills there was some mercantile involvement in the cotton industry, particularly in the Leeds area.[40] Dr Chapman, for instance lists seven examples of Leeds merchants who provided some or all of the capital to build cotton mills in the West Riding. These were:

Firm	Location of Mill
Markland, Cookson & Fawcett	Leeds
Blagborough & Holroyds	Leeds
Ard Walker	Leeds
Blesard & Arthington	Hampsthwaite (not Leeds)
Peter Garforth	Otley, Skipton, Sedbergh, Carlton, Kirkby Malham
William Burrows	Leeds
Atherton & Rawstornes	Colne Bridge

A further three mills were built by investors with mercantile connections:

Firm	Location of Mill
Gowland & Clark	Leeds
Wilkinson, Holdforth & Paley	Leeds
Beverley, Cross & Billiam	Leeds

The following list gives examples of other firms with strong merchant connections who built cotton mills elsewhere in Yorkshire:

Firm	Location of Mill
Atkinson, Wood, Sidgwick & Willett	Bingley
George Woodhead	Marsden

King, Turner, Paley & Varley	Stansfield, Halifax
Haigh Brothers	Marsden
Edmund Lodge	Skircoat, Halifax
Rangeley & Tetley	Birkenshaw
Greenwood & Ellis	Keighley, Burley etc
Wilkinson, Bucks, Jay & Co	Settle

In some cases it is difficult to discriminate between the manufacturing and merchanting activities of a firm. On one hand the three Haigh brothers, John, Thomas and Samuel appear to have been primarily merchants before starting cotton spinning in Marsden about 1792. John was based in London while his two brothers managed their warehouse and sales in Peel Street, Manchester, where in 1795 they had £10,000 worth of stock insured.[41] They were also cotton manufacturers with a large weaving shop near Ashton in Lancashire in 1797.[42] Their expansion into cotton spinning in Yorkshire with at least three mills was possibly seen as a way of controlling their own supply of yarn for their weaving interests.

Another large firm of merchants was that run by John Greenwood and Lister Ellis. The Greenwoods started cotton spinning in Keighley about 1782 and went on to build, buy or lease the following mills, sometimes with partners:

Northbrook Mill, Keighley	John Greenwood	1782–1800?
Vale Mill, Haworth	John Greenwood	1792–1835+
Cabbage Mill, Keighley	John Greenwood	1800–1835+
Damside Mill, Keighley	Greenwood & Ellis	1808–1813?
Castlefield Mill, Bingley	Greenwood & Ellis	1808–1835+
Airton Mill, Kirkby Malham	Greenwood & Ellis	1808–1822
Wreaks Mill, Hampsthwaite	Greenwood & Ellis	1804–1835+
Greenholme Mill, Burley	Greenwood, Ellis & Whitaker	1811–1835+
Eller Carr Mill, Bingley	Greenwood & Craven	1816–1835+

Although Greenwood & Ellis were often described as merchants in contemporary records it is likely that merchanting developed alongside their manufacturing interests. They were therefore probably an example of a successful firm ploughing back their profits and adding merchanting to their spinning and weaving interests.

Another example of a cotton manufacturer and merchant who ventured into cotton spinning was John Holroyd of Soyland, Halifax. A man called John Broadhead, who was a corn miller, had started to build or convert Severhills Mills for cotton spinning about 1799. Broadhead quickly used up his limited capital and had to borrow further sums on mortgage. In 1802, to bring in more capital, he formed a partnership with John Holroyd who was already established as a cotton manufacturer and merchant. The partners drew up an agreement about the finance that was necessary to finish the mill and pay for the machinery. Broadhead had to supply two water wheels with gearing and also some coal houses. Holroyd had to supply the mill work, machinery and stock. As manager of the spinning side only, Broadhead was to have sixty guineas a year. Holroyd, who of course already paid regular visits to Manchester to sell cloth was to receive expenses in proportion to the business done by the new concern. The mill, machinery and stock were to be insured and the profits were to be used to repay the money which had been borrowed.[43] Unfortunately Broadhead died in 1804 and so Holroyd had to run the mill himself.[44]

It is not clear how far banks provided finance for fixed and circulating capital. Certain local banks and bankers were closely involved in the cotton industry. Bradford's first bank was formed in 1777 when three local men formed a partnership as bankers. These three men were William Pollard, William Hardcastle and Thomas Leach. Leach lived at West Riddlesden Hall near Keighley and owned estates at Morton and Halifax. He was later a partner in a large coal mine in Riddlesden which had two steam engines for working the five foot nine inch seam there.[45] However, prior to that he had been a partner with Robert Smithson and William Leach as cotton manufacturers. Besides manufacturing they built Dimples Mill in Morton for cotton spinning sometime between 1791 and 1793 and ran the mill until 1818.[46] How far Leach had been involved in cotton manufacturing before 1791 is difficult to say but he may have turned to it after the Bradford bank run by Leach & Co. was bankrupt in 1781.[47] Further family involvement in cotton spinning came when Thomas Leach's sister, Rachael, built Strongclose Mill in Keighley in or prior to 1793.[48]

In the Skipton and Craven area two banking firms had interests in cotton spinning. William Birkbeck, who was a partner in the banking firm of Birkbeck & Co. of Settle, had a share in a number of cotton mills in England and Scotland. The first was probably Yore Mill at Aysgarth where building commenced in 1784.[49] By 1800 he also had a financial interest in mills in Bentham, Settle and Montrose as well as Linton Mill which was used for worsted spinning. The partnership of Sidgwick, Chippendale, Netherwood & Carr was formed to provide additional banking facilities in Skipton in 1802. William Sidgwick withdrew from the partnership in 1804, possibly to concentrate his attention on the running of High Mill in Skipton where his family had been spinning cotton since 1782. Another partner, Christopher Netherwood, built Scalegill Mill near Kirkby Malham in 1795. Netherwood was involved with cotton spinning there for many years although he leased the mill to tenants whenever possible. In 1819 Chippendale, Netherwood & Carr provided the mortgage for Joseph Mason to buy a cotton mill at Gargrave.[50]

In Halifax, Swaine Brothers & Co. opened Halifax Commercial Bank in January 1801.[51] The Swaines were merchants whose venture into banking did not last long. However, in the six years their bank was in operation they supported cotton spinning ventures in two places. The first of these was at Gomersal Hall Mill. The building had originally been a cloth hall for the local manufacturers, but was not successful. The property was owned by Sir James Ibbotson who, after trying to sell the building in 1786, leased it to his agent Thomas Carr. A partnership between Carr and Richard Paley, a merchant from Leeds, started cotton spinning at the Hall about 1788 and after Paley left the partnership in 1796 Carr continued until his bankruptcy in 1803.[52] The mill was then assigned to James and Samuel Bayley who were owed money by Carr but they in turn sold the mill in 1803 to Henry Ramsbotham for £2,360.[53] Ramsbotham was the Swaine Brothers agent so it was probably their money which was used to buy the mill.

The banking partnership consisted of Robert and John Swaine of Halifax who were merchants, John's sons – Edward of London, merchant, Joseph of Halifax and also

Hannah Swaine, together with Henry Ramsbotham of Bradford. Ramsbotham had built Holme Mill in Bradford for worsted spinning about 1800. After a fire in 1803 he was joined by the Swaine brothers and the mill was used for cotton and worsted spinning until 1807.[54] On the collapse of Swaines & Ramsbotham's banking and commercial interests in 1807 the various mills were offered for sale. Swaine Brothers were prepared to take their own notes and bills in payment for the Gomersal Hall Mill from those unfortunate enough to have them. By some means the two younger Swaine half-brothers, Joseph and Edward managed to salvage Gomersal Hall Mill from the financial crisis but they converted it to woollen manufacturing.

From these few examples it can be seen that local banks and bankers did supply capital for the new cotton spinning industry. However, as the individuals concerned were active in other commercial concerns such as merchanting as well as banking they would possibly have ventured into cotton spinning anyway. How far these early banks provided funds for other early cotton firms cannot be estimated. No doubt the firms had accounts with the local banks. They would then be able to borrow money and have their bills discounted. In that way the banks may have been more important for providing short term credit than in providing the long term capital for mill construction which they did not see as their role.

The other need for capital was to finance the purchase of raw materials which would be transformed into cash after processing and secondly to pay for other necessary materials and wages. It has been suggested that circulating capital amounted to a greater sum than fixed capital and was in many cases harder to raise.[55] Robert Heaton the younger, for instance spent £285 of the £493 insurance money rebuilding Ponden Mill in 1795 but spent £650 on 6,200 lbs of cotton later that year. It is difficult to calculate the circulating capital requirements of the early Yorkshire cotton mills. Certain items, such as the stocks of raw materials and finished goods, can be estimated from insurance records. These were nearly always included in the mill policies but would have varied over time as raw cotton, for example, appears to have been bought in large amounts a few times a year but the stock valuations would have been constant. Stock was also held in warehouses apart from mill premises and not insured as it was considered that the fire risk was less.

Although a certain amount of credit was given by cotton merchants other raw materials and wages had to be paid for as soon as the mill started operating. From the examples found it seems that several firms did not have adequate capital to finance the first few months of production. Two mills came onto the market in 1800 which had been run by Soloman Lumb for only a short time. Lumb bought Cockroft Mill in Rishworth near Halifax for £1,200. The mill had been in production for four years and had cost £1,050 to build. By 1798 Lumb owed £1,602 to James & Joseph Kershaw who were cotton merchants in Manchester. He also owed mortgage repayments to Mr W Brisco of Marylebone, Middlesex, who had lent him the money to buy the mill.[56] Despite these debts Soloman Lumb built Temple Mill, also in Rishworth, and this mill was finished in 1799. There is no evidence of where he raised the capital, which was possibly about £1,800, to build this mill but he was in such financial difficulties that Kershaws forced him to transfer his assets to them and eventually brought bankruptcy proceedings against him. What may have happened was that Lumb bought cotton on long credit terms, bought Cockroft Mill on a mortgage and tried to complete

Temple Mill with his own capital and any profits he was making from running Cockroft Mill. Unfortunately for him his creditors would not allow him to use their money any longer and so his two mills were sold.

Another example of a firm with inadequate finance was the partnership between Elkanah Hoyle of Swift Place and Joshua Bates of Soyland Mill. Hoyle was a cotton manufacturer who bought Smallees Fulling Mill in 1796, presumably with the intention of converting it for cotton spinning. Bates, who was a millwright, joined Hoyle in a partnership in 1801 when they raised a mortgage with Swaine Brothers, the Halifax merchants and bankers. Building of the four storey mill at Smallees commenced and part of it was fitted out with mules by 1803. At that time £1,500 had been borrowed from Swaines who would not lend any more money. The partially finished mill was therefore offered for sale and eventually Elkanah Hoyle lost control of the mill.

Fixed capital formation

Although it was possible to reduce the cost of starting cotton spinning by the methods discussed above, considerable sums were required by certain firms. In the first instance finance for the fixed assets had to be provided. For cotton spinning concerns these are fairly easy to identify. They included the mill buildings used for the various production processes as well as the buildings used as warehouses, counting houses and workshops. Other buildings such as barns and stables may have been needed if the firm had to carry raw materials or its own yarn or cloth. Within the mill was a power source and transmission system, usually of horizontal and vertical shafts with belt drive to the various machines. The machinery itself included that which was under construction or repair together with all the other utensils necessary for producing the yarn for market. The mill buildings also needed heating and lighting systems which had to be included. Outside, access roads and other works, particularly the weirs, dams and goits of water powered mills had to be taken into account. Many firms also had to provide cottages and houses to accommodate their workers and overlookers, particularly if the mill was in a remote area. These were so necessary that it was unusual to find a mill for sale which did not have several cottages attached or in the vicinity. Some firms also had to provide apprentice houses for the children they employed. These however, have not been included in calculations of fixed capital costs.

Dr Jenkins has carefully considered the valuation of fixed capital stock in the contemporary woollen and worsted industry in the West Riding where the great majority of the textile mills were to be found. His methods have, where possible, been replicated so that some comparisons can be drawn for the same area. Much of the data used in this survey comes from the same sources and refers to industrial premises which were more or less identical to the wool textile mills. As has been explained in Chapter One, there was constant change and movement between the textile industries in Yorkshire with mills frequently being altered to process different fibres depending on market forces.

Although there are a few accounts of the costs of building cotton mills and some records of the sums involved when mills were sold, the main sources used by Jenkins and for this survey have been the records of the fire insurance offices. The meaning and reliability of these valuations has been explained at length by Jenkins and is summarised here with examples for the cotton industry. Within the fire insurance business there was competition between the dominant London Offices with their networks of local agents and the provincial offices. The three main London Offices, the Sun, Phoenix and Royal Exchange were well established and provided cover for Yorkshire cotton mills from the 1780s. However, soon after the turn of the century the Halifax Journal was also carrying advertisements for the Imperial, Norwich Union, Globe, Albion, Atlas, Eagle, Hope, County and Union fire offices. Unfortunately records for most of the minor offices are not available but those of the Sun office had current policies for twenty-one West Riding cotton mills and the Royal Exchange for seventy-two.

The cover provided by a single policy was not always a true indication of the value of a firm's assets. Policies for mills did not necessarily cover all a firm's property. The risk could also be shared between two or more companies, particularly if the sums provided for were large. For instance, Whitaker & Merryweather insured their cotton mill at Burley-in-Wharfedale in 1805 for £10,000 with this sum divided between the Sun and the Imperial offices.[57]

A number of factors affected the value of the premises and property which could be damaged by fire. As machinery was used and started to wear or become obsolete its value would decline, but on the other hand a period of good trade would be expected to increase the disposable value of mills and also possibly of machinery. The replacement cost of property might also be considerably more than its current worth. Mill owners also did not need to insure all their property as only part of it could be destroyed by fire. The mill and machinery might have to be replaced, but dams and goits would not.

The large insurance companies were careful to check all fire hazards within the insured property and would adjust the policy when a mill was taken out of use. There was no point in mill owners over insuring their property as the companies would only allow cover up to the amount that it would cost to rebuild or reinstate after damage. The original valuation of the property was not usually done by the insurance company but they wanted to know all details of processes and in particular what measures were taken to separate buildings so that fire could not spread.

There were cases of dishonest owners or tenants who burnt down their own property and hoped to cheat the insurance company. Samuel Collier rented part of a cotton mill in Saddleworth from James Lees in 1804. Collier used his section for carding and roving cotton and insured his machinery and stock for £500 and £50 with the Royal Exchange company. On the 14th September he set fire to the whole mill but was caught and sentenced to death, although he was later reprieved.[58]

The cost of claims against the insurance companies rose as fires on mills were not infrequent so in turn premiums had to be adjusted upwards. Some mill owners may therefore have underinsured to save money, but if they had a serious fire they only received a stated sum to cover part of the value of the property. Unintentional under insurance may have occurred when prices were rising or when more machinery was being added to an existing mill. When Gibraltar Mill, which was a woollen mill, burned down in 1812, it was suggested that 'it is feared the property is much undercovered, as is too often the case in large concerns'.[59] The use of insurance valuations as a measure of fixed capital may therefore show more error for larger firms than smaller ones.

Overall, Jenkins suggests that:

> The meaning and reliability of insurance values will undoubtedly be the subject of much further inquiry and debate. Instances of under-insurance are evident and more will come to light. There are clearly reservations about the use of the insurance figures for estimating fixed capital values but it would seem that they do provide by far the best and most detailed information that is available. Any inaccuracy that does result from their use will be an under-estimate of the value of the capital stock.

He suggests that where possible insurance value should be compared with other values to check their reliability. Such comparisons could be:

	– valued loss
Insurance Value	– original cost
	– company accounts

A fourth comparison could be with the sale price if a company's property was sold.

If Dr Jenkins' argument for the use of insurance valuations is taken as also being valid for assessing the valuation of fixed capital in the cotton industry, given that the mills and machinery were very similar and often interchangeable, certain estimates can be arrived at. The average value of insured fixed capital has been calculated from policies issued over periods centred on the years 1800, 1810, 1820, 1830 and 1835. Table 3.1 shows the average insured value of a sample of mills for those dates.

Table 3.1

Average Value of Insured Fixed Capital of Samples of
West Riding Cotton Mills 1800–1835

Date	Number of Mills in Sample	Average Value of Insured Fixed Capital (£)
1800	96	1,735
1810	34	3,448
1820	16	2,865
1830	9	2,895
1835	10	7,575

Jenkins found that in the data on the wool and worsted industries there was a bias in his sample towards the larger mills, particularly after 1820. This bias occurred because the Sun Fire Office rarely insured the smaller mills after 1820. The firms occupying the smaller mills were attracted to the provincial insurance companies which charged lower premiums. The sample of mills in Table 3.1 is clearly unrepresentative of the total number of cotton mills in the West Riding after 1800 when the two sets of figures are compared:

Year	Number in Sample	Number of Mills Working
1800	96	232
1810	34	241

1820	16	183
1830	9	160
1835	10	158

As a way of correcting the bias in his sample Jenkins used the information available about the horsepower and number of workers employed in a number of mills in 1835. The average horse power in use in the cotton mills surveyed in 1835 was twenty and the average number of workers employed was seventy-seven. Unfortunately too few of the mills insured by the Sun Fire Office were in the official returns to provide a valid sample of all the mills in the industry. Only seven mills occur in both lists using both the 1830 and 1835 insurance details. Those seven mills used an average of twenty horse power and employed an average of sixty-three workers. On that basis the insured cotton mills, for which estimates of fixed capital are available, would appear to be fairly typical of all the mills in the region. However, seven is too low a number to take as being representative, particularly as Jenkins found that the bias in his sample necessitated reducing fixed capital stock values by 37%. In the absence of any better information the following table is presented without any adjustment for bias and therefore the figures for 1820, 1830 and 1835 may be too high.

A tentative evaluation of Table 3.1, which takes no account of the uninsured costs of site preparation etc, for which Jenkins added 5% could be that two waves of cotton mill building are indicated. A large number of small mills were built in the years up to 1810. There was then a decline, which in many areas meant a transfer of assets to the wool, worsted and flax industries. By 1835 a few of the largest early mills were still being used for cotton spinning but new mills had also come into operation and some from both groups were also being used for power-loom weaving. The new mills were larger and were more likely to have steam power.

Table 3.2
Estimated Value of Fixed Capital
in the West Riding Cotton Industry, 1800–1835

Year	Number of mills	Average Value	Value of Fixed Capital Formation
1800	232	1,735	402,520
1810	241	3,450	831,450
1820	183	2,870	525,210
1830	160	2,900	464,000
1835	158	7,575	1,196,850

CHAPTER 4

The Labour Supply
for a New Industry

The problems

THE ESTABLISHMENT OF THE COTTON INDUSTRY in Yorkshire brought two new dimensions to local industrial production. The first of these was the need to find and train a large number of people in new skills, knowledge and attitudes which would enable them to establish and sustain the growth of the industry. The second was the need to adapt to a new scale of production with mills employing hundreds of work people. These workers were often women and children who had not previously been involved in industrial production on any scale. Furthermore, the new workforce had to work at a uniform pace which was dictated by the water wheel or steam engine.

Although by 1780 there were some sizeable production units in mining, quarrying and the woollen industry, these were isolated and involved mainly male labour. The new cotton mills, on the other hand, became common throughout the area and were often the largest single source of employment in the towns, villages and hamlets where they were established. As has been explained earlier, the lack of employment for women and children in the rural areas was a contributory factor in the location and establishment of the early mills. The demand for cheap labour on a large scale led at first to the industrialisation of country areas where there was surplus female and child labour as well as water power. Further development of the industry however, resulted in serious efforts to attract workers and persuade labour to move to the mills.

With the use of large numbers of unskilled workers in one place came a degree of exploitation which was then more difficult to hide. The earliest factory legislation, the Factory Act of 1802 was brought about because of the employment and treatment of children in cotton mills. Further legislation in 1819 and 1833 was also initiated because of working conditions for cotton operatives. Cotton mill workers were the first to experience factory conditions on a large scale and therefore the first to have their conditions noticed nationally.

Despite the existence of the well established woollen and worsted industries in Yorkshire, the development of the cotton industry called for knowledge and skills in eight areas which were new. These were the ability to:

1. Design and build mills on a larger scale than previously, for a new purpose and with larger power requirements.

2. Design, build and maintain machinery which was completely new to most people and which processed a fibre which was new to the area.

3. Operate new types of machine which needed the repetition of monotonous tasks

64

over very long periods and which were operated from a central power source by dangerous shafting and belts.

4. Manage departments in mills where large numbers of people were employed and where machinery had to run continuously or production in other departments would suffer.

5. Buy raw cotton at the right price, quality and quantity at a time when supplies and markets were uncertain.

6. Sell cotton yarn and cloth at a profit at a time of economic and political uncertainty at home and abroad.

7. Account for the transactions of the new firms and partnerships.

8. Manage mills which were new, with new processes and an untrained workforce in a time of serious economic and social difficulties.

These abilities were in such demand that many firms experienced problems at one time or another because of the shortage of certain skills in workers or management or in general lack of labour.

Where early cotton mills still exist, such as at Long Preston, which was built in the 1780s, it can be seen that the design was based on contemporary corn mills but with more windows. Water powered textile mills, particularly cotton mills were usually larger than corn mills but the same materials and methods of construction were used. Most local industrial buildings of the time were built to the same pattern with thick, load bearing stone walls, small evenly spaced windows and a stone flag or slate roof.

The apprentice house at Marsden known as Throstle Nest which housed the apprentices from Haigh's Mills about 1800.

Internally, wood, but later cast iron beams and pillars supported the floors and, at first, limited the width of the buildings.

The designers of the early mills are largely unknown as are the builders. Single firms capable of undertaking the task of erecting a new mill did not exist in the area in the late seventeenth century so the task of coordinating the activities of the various craftsmen and labourers seems, in many cases, to have been carried out by the mill owners. When Clayton & Walshman of Low Mill, Keighley expanded their business and built Langcliffe Cotton Mill outside Settle in 1784 they had to supervise the construction themselves.[1] This involved finding local supplies of timber and stone of various qualities, bringing in lime, lead, iron and slates from outside the immediate area. In addition they had to rent rooms for a joiner's shop and employ specialist millwrights from Halifax. The original mill was large for it housed 150 workers in 1803 before it was extended the following year. It took eleven months to build from January to November 1784 when the first bags of cotton were brought to the mill.

Obtaining the services of craftsmen in sufficient numbers to construct a mill in rural areas would also have been a problem. When Robert Hargreaves & Co. of Addingham, who were cotton and worsted spinners, wished to build a new worsted mill on the site of a corn mill at Linton on the River Wharfe near Grassington in 1788 they had to advertise for craftsmen in a Leeds newspaper.[2] They wanted masons, joiners and carpenters to meet them at an inn in Skipton to see the plans which had been prepared.

The early millwrights had a multitude of tasks to undertake in the construction of a mill and often nothing but their own limited experience to guide them. The early Arkwright type mills could be planned to a simple formula but when other machinery was used or different combinations of machines put together with a layout on the various floors a number of detailed mechanical problems had to be solved. The power requirements of the different machines had to be calculated and also the frictional losses in the power supply. After deciding on the optimum speed at which to run the machines gear ratios had to be worked out for the transmission system between the water wheel or steam engine and the individual machines.[3] There is some evidence that estimates of power output and frictional losses were optimistic so that the planned number of spindles could not be run in some mills.[4]

Besides the building of the mill, the owners had to construct dams and goits for a water supply and supervise the construction and installation of a water wheel. If a mill was steam powered the mill owners had the responsibility of providing the engine and boiler foundations before the engine was assembled. The house to enclose the engine was also their responsibility.

By 1780 Richard Arkwright and his partners had built twenty cotton mills in England.[5] By then however, other cotton spinners were starting to ignore his patent rights, although he made strenuous efforts to uphold them. The patents were finally revoked in 1783 but Arkwright still had a dominant role for a few years afterwards. Despite the growing number of mills being built around the country and the spread of knowledge of how to construct Arkwright type machinery, there were still many local difficulties because of the commercial advantage to the successful spinners of the knowledge they had. Some of the early Yorkshire firms did not find it easy to master the techniques of operating the early spinning frames. The machinery for Low Mill in Keighley was made under the direction of Arkwright in 1780 as Clayton & Walshman had a licence from him to operate his patent. A number of other early mills in Keighley and the Yorkshire Dales were built on the Arkwright pattern which meant that the early builders could plan to a known standard.

The second cotton mill to be built in Keighley, Northbrook Mill, was built by John Greenwood about 1782. The spinning frames did not operate successfully however, as the mill and machinery were built without Arkwright's support. As a result the yarn was faulty and according to a local writer it was only after Greenwood had asked one of the young spinners working on Arkwright's machinery at Low Mill to check his spinning frames that the fault was rectified.[6]

Many of the early partnerships in the Yorkshire cotton spinning industry were made up of members with capital and members with technical skill and experience in building machines and managing mills. This latter quality was often the essential element in ensuring the success of an enterprise. When the partnership of Smith, Watson, Blakeys & Greenwood was formed in 1784 to build Greengate Mill in Keighley, only Greenwood had any experience in cotton spinning. Abraham Smith was described as a yeoman, Rowland Watson a gentleman, Joseph Blakey a wool stapler and John Blakey a stuff maker. Greenwood, on the other hand 'not having at present any capital to bring into stock' was brought into the partnership because he had

a plan in respect of other mills of a similar nature to the intended one and a genius well adapted for constructing the machines and other works to be made use of and employed in and about the said intended mill.

For the opportunity to eventually become a full partner James Greenwood was to 'employ himself in constructing and finishing machines and other works necessary to be done in and about the said intended mill, with all best skill, knowledge and judgement and with the utmost expedition he can'. To retain the advantages to the partners of Greenwood's scarce skills he was restricted from divulging the 'secrets' of the undertaking.[7]

Another Keighley partnership which was concerned about keeping to themselves the skills and knowledge of its millwright and its joiner was Craven, Brigg & Shackleton of Walk Mill. In November 1783 they employed Joseph Tempest as millwright but he would forfeit £100 if he divulged 'the secrets respecting the construction or movements of any of the machines or works'. Thomas Robinson, a joiner, was also bound to the partnership for four years, during which time he should not:

reveal or make known any secret respecting the construction or movement of any of the machines or works, now or hereafter to be carried on or set up in the said cotton mill.

Robinson was also bound for the sum of £100.[8]

Within a few years however, knowledge and skills became more widespread and by about 1790 specialist machine making firms had emerged to serve the local cotton industry as well as the expanding wool and worsted industries. One of these was the Leeds firm of Todd & Bosworth. They were prepared to provide plans and directions for the building of cotton mills as well as provide the machinery to be used in them.[9] Brass founders were also prepared to add the making of 'all kinds of cotton works' to their traditional bell casting business.[10]

Power driven worsted spinning had not been started in Keighley in 1790 when a man called William Carr set up as a mechanic so much of his work was preparing flyers and guides for the cotton spinning firms. Within a few years he started building spinning frames and bought rollers and spindles from Richard Hattersley who had moved from Sheffield to Keighley to take advantage of the growing demand for metal components for the new textile machines. Keighley became a centre for the production of textile machinery as did Halifax and Leeds. By 1790, for instance, John Jubb, a Leeds millwright, was advertising for mechanics to make and fit up cotton and worsted machinery. Two years later Andrew Lawson, a manufacturer of spinning and carding machinery in Glasgow, and Wright & Wight of Lever Street, Manchester were both advertising cotton machinery in a Leeds newspaper and the need for every cotton spinning firm to be able to look to Arkwright for guidance or produce its own machinery was over. However, firms still continued to make their own machinery using their own mechanics. In September 1803, for instance Ard Walker, in Leeds paid five shillings (25 pence) for ale for the joiners who had just finished the last two spinning frames ready for the start of spinning in his new mill.

Child labour

Although cotton mill owners in 1816 could state that the labour required of child cotton spinners was not arduous and consisted of intermittent tending of the machines,[11] the simpler but cruder machines of the early years required more expertise and effort. The children to be employed at Low Mill, Keighley in 1780 had first to be sent to Arkwright's mill at Cromford in Derbyshire to learn how to operate the carding, preparing and spinning machines.[12] As other early enterprises built spinning machines on the Arkwright pattern the skill involved with tending the frames was possibly disseminated by the early Arkwright trained operatives. Certainly the young people who returned from Cromford to Keighley then had to train the young spinners from Clayton & Walshman's other mill at Langcliffe near Settle.

A further example of the difficulties in operating early machinery successfully comes from Yore Cotton Mill near Aysgarth in Wensleydale. Again some of the young workers were sent to Winstanley's mill in Lancashire to learn the rudiments of cotton spinning. This mill, which was built in 1784/5 by William and John Birkbeck of Settle was run by Wood, Winstanley & Co. One of the partners, a Mr Sutcliffe wrote to William Birkbeck back in Settle that:

> I observed that the work of some one of the frames was almost constantly lost in changing the roving and spinning bobbins.

Most of the faults which lay behind the loss of production and low quality yarn being spun were blamed by Sutcliffe on the manager but he also wrote that: 'the children are got into bad habits which will not easily be broke'. Sutcliffe was so worried about the inability of the mill workers to spin good quality yarn that he had suggested closing the mill down for a few months and, in an earlier letter, that if better quality yarn could not be produced they should just spin low quality candle wick.[13]

The early problems in mastering the techniques of cotton spinning appear to have been overcome by about 1800 although experienced spinners from Bingley went to

Knights Mill in Great Horton, Bradford, to train the operatives there when the mill
was built in 1806.[14] The shortage which did arise to a considerable extent in the
Yorkshire cotton industry however, was not so much in terms of the skills of the
operatives , but in terms of numbers. One very important factor influencing the location
of cotton spinning mills was the availability of large numbers of potential operatives
in the immediate neighbourhood. This aspect has been dealt with in Chapter 1, but
the labour problem arose for many established firms when they wished to expand their
productive capacity or when the local towns and villages could not even meet the
original demand.

One example of a mill being started to tap a local supply of child labour was Idle
Cotton Mill near Bradford. This mill, in the Parish of Calverley, was built by John
Cromack about 1795 and leased by Robson, Edmondson & Co. who also ran an
adjoining woollen mill. Adults who worked in the mill came from other towns where
they had already been employed in the cotton industry. The situation was that the
mill 'has not taken any of the natives of the town to work it, only some few children'.
On the question of wages for the workpeople at the mill it was said that the workers
were mainly children and therefore there should be no comparison of wages between
the cotton and woollen industries where most workers were men.[15] The reason was
that the spinning of woollen yarn had not been widely mechanised and few children
were employed in the factory side of the industry.

To meet their need for spinners Clayton & Walshman of Low Mill, Keighley and
Langcliffe Mill, Settle brought extra workers from Keighley to Settle in the first instance.
In September 1785 'night spinners' were brought from Keighley.[16] However, in 1787
the following advertisement appeared in the Leeds Intelligencer:

> Notice is hereby given that Messrs Clayton & Walshman, cotton manufacturers,
> in order to accommodate workpeople are now erecting a number of convenient
> cottages at Langcliffe Place, which will be ready to enter at Monday next.
>
> Any people with large families that are desirous to have them employed, and
> can come well recommended, may be assured of meeting with every reasonable
> encouragement by applying to Messrs Clayton & Walshman at Langcliffe aforesaid,
> or at their cotton works at Keighley.
>
> April 10th 1787 [17]

A further example of the inducement to families to move and take up employment
in the cotton industry came in the Leeds Intelligencer in January 1805. This is also
worth quoting in full to indicate the type of work which was to be provided and also
to show that by 1805 looking after spinning frames was considered to be a job which
could be done by the youngest untrained children:

ENCOURAGEMENT
To the Industrious Poor

> Wanted several large families in the Cotton Manufactury where children from
> twelve years old and upwards will be taught to weave, and under that age to spin.
>
> Suitable employment will be given to their parents if not already accustomed
> to the cotton business, and comfortable habitations, at very moderate rents,

provided for their reception. Further particulars may be known by applying to Mr William Davenport, Marsden.'[18]

It appears from the above advertisement that it was the children who were required first and foremost, but of necessity their parents had to accompany them. Other possibilities of hiring children from a distance often occurred however, and this time by themselves. These were the children who had to be supported by the parishes and had been orphaned, abandoned or were destitute. The long established system of apprenticeships was used although very little training was given by the employer. Parish authorities were pleased to be able to cut the cost of the Poor Rates after advertising and disposing of groups of children on the pretext of their being put into apprenticeships. This practice became harder after 1803 when local magistrates refused to indenture parish apprentices to mills working at night or over ten hours per day.

As one of many examples a Leeds paper carried the following advertisement in November 1804:

To Manufacturers
  Several stout healthy boys and girls are ready to be put out as parish apprentices. For further particulars enquire of the churchwardens and overseers of Grantham, in the County of Lincoln.[19]

One cotton spinning firm which originally used parish apprentices extensively was Merryweather & Whitaker of Greenholme Mill, on the river Wharfe at Burley-in-Wharfedale. Thomas Brown, who was a manager at the mill gave the following information to the Factory Commissioners in 1833:

We colonised our mill originally and have several hundred apprentices from London; we now have a sufficient supply in the neighbourhood.[20]

The boys and girls were bound as apprentices to serve the firm for several years and came from the parishes in London. Besides teaching the children cotton spinning, Merryweather & Whitaker provided food, clothing and lodgings in an apprentice house. Another company which used parish apprentices was Haigh Brothers of Marsden who also built an apprentice house for the children.[21]

The industry's reasons for employing child labour were explained in some observations on the 1802 Act for the Preservation of the Health and Morals of Apprentices and others, employed in cotton and other mills.[22] Free labourers could not be obtained except on what were described as very disadvantageous terms. It was said that manufacturers, for the sake of securing one spinner, would be obliged to engage a whole family of five or six persons without any benefit of their labour. As we have seen some Yorkshire manufacturers were prepared to employ complete families but the Greenholme Mill concern preferred to use parish apprentices but then had to look after them.

In the early years of the industry some firms ran two shifts of child workers, one during the day and one at night. Clayton & Walshman of Keighley was one such firm and Garforth & Sidgwick of Skipton another. The latter firm explained in 1816 that they had given up the practice some years before because of inadequate water supplies to their mill. They had run a water wheel from a stream which allowed night and day working but then built a dam to store water to improve the day time supply but which stopped night working.[23]

A more general restriction on night working for children came with legal sanctions. West Riding magistrates passed the following resolution in May 1803:

That we will not, on any account, allow of the apprenticing of poor children to the masters or owners of cotton mills or other works of the kind, where such poor children shall be obliged to work in the night time, or for an unreasonable number of hours in the day time.

They also resolved that they would not allow the children from one parish to be bound to a master in another parish.

The passing of the 1802 Act which regulated the hours of child workers in cotton mills was opposed by West Riding mill owners who met at the Devonshire Arms in Keighley in February 1803 to consider ways of repealing the Act, or as they said, 'restoring the trade to its former freedom'.[24] The committee set up by the meeting was to organise a petition to the House of Commons to bring in a Bill to repeal the 1802 Act and also to correspond with the committee of the cotton trade in Manchester. It is perhaps significant that this committee saw itself separate from the Manchester committee and proposed to act through Yorkshire MPs and newspapers.

The members of the committee were:

Name	Location of Mill
John Greenwood	Keighley
William Willett	Hampsthwaite
Mr Clayton	Keighley and Settle
Peter Garforth Jnr	Bingley, Skipton, Bell Busk
Lister Ellis	Keighley
William Ellis	Keighley
Mr Sidgwick	Skipton
George Merryweather	Burley-in-Wharfedale
Mr Ross	Stanbury
Mr Barker	Morton
Mr Heaton	Ponden
Jonathan Barker	Morton
Mr Hollings	Stanbury
Thomas Leach	Keighley
Mr Horsefall	Keighley

How many firms employed parish apprentices in their cotton mills is difficult to say. However, there were occasional advertisements for people to manage or run the boarding houses where the children lived.[25] Usually a couple were needed and were often expected to act as instructors to the children or to be able to do some extra work themselves such as weaving.[26] In one case a man and his wife were wanted to run a boarding house for only twelve to fourteen children.[27]

When a firm had taken on apprentices they were tied to that firm for the period of their apprenticeship. Even workpeople who were 'hired servants' could not easily leave their employer. Garforth & Sidgwick, who were cotton spinners with mills in Skipton, Bingley, Bell Busk and Sedbergh, found that a millwright, a drawer and a reeler had 'absconded' in 1791 and advertised for information of their where-abouts.[28]

Two of the children employed at William Parker's cotton mill at Hebden near Grassington ran away in 1807. As Parker had hired them for a number of years he threatened to prosecute anyone who employed the children before their contract with him expired.[29] John Sutcliffe of Marsh Factory in Stansfield, Halifax, took an alternative approach when one of his apprentices ran away by offering three guineas reward for his return.[30]

The apprenticeship system was of advantage to the millowner because the young person was bound for a long time and the training which had to be given was minimal. The long established custom of apprenticeship was being used in an inappropriate context but was resorted to by millowners because of the general shortage of child labour. In addition some parents were reluctant to allow their children to work in the mills despite the value of their earnings.

Two different aspects of the life and work of child labour in a Yorkshire cotton mill come from 1802 and 1810. These illustrations show the continuing need for large number of children to work in the cotton mills and also how far they were removed from normal family life and were dependent on their employer.

In May 1802, Mr Hey, a surgeon from Leeds, and the Rev Dikes from Hull, visited the Greenholme Mill of Merryweather & Whitaker at Burley-in-Wharfedale. The mill was worked mainly by apprenticed children and Mr Hey produced a report on his visit which was to be used by those opposing the proposed Act to control the working conditions of cotton mill children.[31] The children were interviewed singly and the visitors inspected the mill, the dining rooms and lodgings. Of the 260 apprentices, 52 worked at night and they came in for close questioning. The 52 night shift children, who were all boys, worked from 7.00 p.m. to 6.00 a.m. without a break although they had a meal while they worked about midnight. At 6.00 a.m. they had another meal and then spent the time until 10.00 a.m., when they went to bed, with lessons in winter and games in summer. Five out of six of the night shift children would have preferred to work during the day. They explained that they could not sleep during the day and were woken by the other children returning to the dining room which was under where they slept. Some children complained that they could not work so well by candlelight although a minority said that they preferred night work as it gave them more time for their school work during the day.

Large mills, such as this one at Burley-in-Wharfedale, were given two years by the 1802 Act in which to prepare for the cessation of night working by young children. Despite the efforts of the local cotton masters and their committee, which included George Merryweather from this mill, night work was stopped at Greenholme Mill within a few years.

Changes in the partnership agreement between George Merryweather and Jonas Whitaker eventually led to Merryweather leaving the mill and moving his cotton spinning interests to Manchester. What was unusual was that when this move took place in 1810, one hundred and ninety of the young operatives were marched to Manchester to work there. The Halifax Journal described the event in some detail and puts a different light on the mill owners attitude to his employees.

On Tuesday last the boys and girls, consisting in number 190, belonging to the calico manufactory of Mr Merryweather, of Otley, passed through this town on their way to Manchester, to which place that gentleman is removing his extensive concern. The cleanliness, healthy appearance, and general cheerfulness of the children, attracted the notice and excited the admiration of every beholder; they were clad in blue uniform, the boys with leather caps, and the girls wearing straw hats; walked hand in hand but were relieved at times on their march by several waggons which accompanied wherein were deposited 3,000 penny loaves. By the acknowledgement of everyone who witnessed the procession, it was a sight truly gratifying: and if gratifying to the stranger, to the disinterested spectator – how much more so, how exquisite must it have been to the feelings of him, their Master and Protector, whose head devises and whose heart sanctions the superior comforts of this youthful band – this rising body of industry.[32]

The particular significance of this event is the relationship between the young workers and their employer. A new partnership was formed to run Greenholme Mill for John Greenwood and William Ellis, both cotton spinners from Keighley, joined Jonas Whitaker and continued cotton spinning at the mill until 1848. There was therefore still work at Burley but George Merryweather was able to, and presumably needed to transfer a large section of the workforce when he left the partnership.

The parish apprentices did provide cheap labour as virtually no wages were paid but many other costs were incurred by employers. If an employer carried out his responsibilities fully the apprentices from the various parishes had to be housed, clothed and fed. An element of education was provided and there was the cost of being in charge of a large number of young people. These costs, as much as the legislation, may have accounted for the decline of the apprenticeship system when other labour could be obtained.

Adult labour

In addition to problems with child labour there were also difficulties with adults. Garforth & Sidgwick had labour problems at their Castlefield Mill outside Bingley in September 1791 when twenty workers went on strike.[33] The fifteen women and five men had combined together and left the mill. They assembled outside the mill in 'a riotous and tumultuous manner' and refused to return to work. The strikers were accused of grossly insulting the different masters and overlookers at the mill and all the other workers who refused to join the strike. Garforth & Sidgwick started legal proceedings against the strikers which were only stopped after a public apology had been issued:

We do hereby publicly acknowledge the great offence which we have committed and humbly ask pardon of our said masters in pardoning us, promising never to be guilty of the like offence in future; and hope that this will be a warning to all other hired servants, and deter them from entering into similar combinations.

None of the twenty people could sign their own name.

Management and supervision

The apparent prosperity of the cotton industry led many Yorkshire men and women with some capital to spare to invest in the new mills even though they had no connection with the industry or textiles in general. The background of some of the early cotton mill proprietors, which is discussed later, accounts for the frequent lack of experience or desire to become directly involved in the industry which then led to a great demand for supervisory and managerial skills. In some cases one or more of the partners in a concern had the required skills to manage the entire operation but as mills increased in size specialist managers of carding, preparing, spinning and other processes were required. Thus there was a need for managers of entire mills, as well as a need for overlookers and managers of sections of the new cotton mills.

An example of this dual need comes from an advertisement for managers for an un-named cotton mill in Bingley in 1791 but may have been Castlefields Mill and may have been linked with the dispute mentioned above:

> Wanted immediately, a steady experienced person, who perfectly understands the nature of cotton spinning in all its branches, and is qualified to undertake the superintendence and management of a large mill in a healthy situation at Bingley near Bradford . . . A good carding and spinning master are also wanted who are capable of taking the command of a carding or spinning room.[34]

Earlier that year a manager was needed for another un-named mill:

WANTED

> To superintend a cotton spinning manufactory in the West Riding of the County of York. A sober, steady active man who thoroughly understands, and has been employed in the making of twist as great encouragement will be given, none need apply who cannot produce a proper character of his abilities and good behaviour.'[35]

When Robert Heaton built Ponden Mill in 1791 he employed local men to be foremen in two of the sections at the mill. These two, James Baldwin and John Midgley had previously been drawboy weavers but were to look after the carding and spinning departments. They therefore had no previous connection with the cotton industry or with the machines and processes they were to manage. They would also have been without the experience of working regular hours and controlling a workforce of children.[36]

If a firm running a mill wanted a manager for a particular department they were also aware that he would probably have a family. An unspecified cotton spinning firm wanted a preparing and spinning master in 1795. If he understood the management of weft carding and mule spinning that was an additional advantage but overall preference would be given to someone with a family 'as they may all be employed in the same mill'.[37]

The importance of the few, but relatively highly paid charge hands was shown when the Leeds cotton spinning firm of Beverley, Cross & Billiam sacked most of their workers in 1792. They had problems with their old pumping engine and boiler which was used to pump water back up to the dam to work the water wheel over and over again. They were waiting for a new Boulton & Watt rotary engine so the mill was at a standstill but they retained a few of their key workers.[38]

One cotton mill where there was a serious management problem was Yore Mill near Aysgarth. This was seen by one of the partners to be on two levels. Supervisors of individual rooms were not doing their job properly and the overall management was inadequate. The firm was losing money and one of the areas where the loss seemed to be occurring was in the spinning room. This loss could have been avoided, it was stated 'by the activity of a sharp and diligent superintendent'. Also 'the reeling room seemed to have no superintendent at all, and made unnecessary waste'. In general this partner at the mill felt that:

> I know so much of human nature that I despair of being able to get superintendors of these rooms, let them be even so well qualified (which I fear those we have are not) who will take the necessary care unless strictly attended to by some trusty general superintendent, who knows the business perfectly, and will dedicate his whole time to it; and who has also an absolute authority over all those that are employed under him.[39]

As this letter was written in 1786, only six years after the start of cotton spinning in Yorkshire, it does not seem surprising that there should be a shortage of men with the necessary experience and skills to supervise production in this rather remote mill. As time went on however, the situation may not have changed much. The great majority of employees in the mills were women and children so there was no obvious training for supervisors. Local firms advertised extensively in the local newspapers for managers for their mills. The Leeds and Halifax newspapers contain many examples of these advertisements which usually were for men who had to be 'well qualified' and who could give 'testimonials to their honesty and sobriety'. With the Arkwright type mill the main requirement was coordinating and managerial skill and not technical knowledge. It was this management skill which was scarce in the early years of the industry.

Book-keeping

In addition to the problems of management, owners of mills had the problem of accounting for their activities. With turnover possibly amounting to several thousand pounds a year and a multiplicity of suppliers of raw materials, machinery, equipment and labour as well as sales to a large number of people in a variety of markets, it was essential to keep a careful account of all transactions. In addition the owners had to keep an account of their profit and loss for income tax purposes to help pay for the French wars. There appeared to be a shortage of men who could act as book-keepers for the new cotton firms. Although clerks had obviously been needed in most extractive, manufacturing and merchanting firms the scale of the cotton spinners was, in many cases much larger. Many cotton firms had hundreds of employees and their purchase and sales transactions covered a wide area. Cotton spinners therefore wanted men who were 'capable of keeping the accounts and perfectly understand the business',[40] or 'wanted in a cotton factory, a book-keeper who writes a good hand and has been used to the business'.[41]

The rapid growth of the industry and its marginal nature in parts of Yorkshire, threw up problems for inexperienced companies, their owners and managers in forecasting the overall demand for yarn and cloth. The price of both raw cotton and cotton cloth were subject to violent fluctuations due, for many years, to the opening and closing of foreign markets following changing wartime conditions and strategies. The interval between the close of the French Revolutionary War in 1802 and the start of the Napoleonic Wars in 1803, for example, brought about a short term boom when foreign markets were opened. However, the resumption of hostilities and the collapse of the boom resulted in a number of bankruptcies amongst the Yorkshire spinners and manufacturers. Even after the end of the war, in 1816, there were trading problems for the cotton industry which were more severe than those for the woollen trade which did not depend so much on exports.[42] Although the general trend in the consumption of raw cotton was upwards throughout the period under discussion, there were setbacks in 1793, 1797, 1808 and 1815. These setbacks and temporary lack of markets were difficult to predict and eventually influenced many of the marginal producers who could do so to move out of the cotton industry. One of these men, Thomas Corlas, who built Hope Mill in Keighley in 1800 was, by 1812, said to be so:

> agitated by unfavourable state of the markets, he one morning went to the engine tenter and ordered him to rake out the fire and stop the engine; and this being done, he from that time ceased to be a cotton spinner.[43]

Thus after a period of only twelve years Hope Mill was changed over to worsted spinning, a decision which many other local cotton spinners copied.

CHAPTER 5

Buying and Selling
Cotton and Cloth

Despite the number of cotton mills in Yorkshire and the quantities of cotton they obviously bought, little is recorded about their purchases. It is likely however, that the Yorkshire firms followed the practice of those in Lancashire and the slight evidence available confirms this. The cotton they used was imported through London and later Liverpool, Lancaster and Hull. Although the cotton industry relied on imports and exports through the Lancashire ports it should not be forgotten that Hull was also a major port. Manchester cloth was regularly exported through Hull by the time of the French wars. This was particularly true for the heavier velvets which went to markets in Europe where nearly 250,000 were exported in 1783. In addition when raw cotton was imported from the Middle East much of it came through Hull. The route was from Smyrna, via Rotterdam, Amsterdam and Hamburg.[1] London cotton merchants were important until about 1795 and circulated price lists in the northern towns.[2] Liverpool then took first place and a system of importers, brokers, merchants and dealers grew to facilitate the marketing of the various types and qualities of cotton which were used.

Cotton purchases

The important people who were links in the chain between importers and spinners were the dealers. They bought cotton from importers, either from their warehouses or at an auction. They then sold at local auctions or by private sale. These dealers existed in all the cotton towns but were concentrated in Manchester.[3] In addition to being able to offer a range of qualities of cotton the dealers also gave credit. This enabled some spinners to process the raw cotton and sell it as yarn before they had to pay for it. In this way they were able to use the dealer's capital instead of their own.

From advertisements in the local papers it appears that there were auctions of cotton in Leeds and Hull. For example, in 1802 Fine Georgia or Sea Island cotton was for sale in Leeds in March, and New Orleans, St Lucia and Surinam in September. New Orleans, Laguira and Georgia cotton was for sale in Hull in June.[4] It would seem that there were some people who acted as cotton dealers in the Yorkshire towns but they possibly only served the smaller spinners. Benjamin Marshall, for example, was a cotton merchant in Bradford until 1804 while William Halstead and Matthew Hey were both cotton merchants in Halifax in 1803.[5]

Where there is some evidence of purchases of cotton by Yorkshire spinners their suppliers were usually in Manchester or Blackburn. By 1799 Manchester had become the great distribution centre for cotton. The cotton merchants and dealers gave credit to the smaller spinners and so bound them to continue buying. Larger firms had more freedom and eventually would buy from Liverpool as well, as they had financial independence. In some cases where firms or individuals were bankrupt, proceedings were brought by their creditors who would be suppliers of cotton which was their basic raw material. When Soloman Lumb of Temple Mill in Rishworth near Halifax was bankrupt in 1798 he owed £1,602 to James & Joseph Kershaw of Manchester. These cotton merchants had instigated the bankruptcy proceedings against Lumb and eventually forced the sale of his mills.[6] When Robert Pearson of Addingham was bankrupt in 1797 he assigned his assets to Myers, Fielding & Co. who were cotton merchants in Blackburn.[7] Similarly, when Mrs Betty Hudson of Damside Mill, Keighley and also her grandson, Thomas Parker were in financial difficulties they owed money to James Pilkington, a Blackburn cotton merchant.[8] Thomas Parker's bankruptcy hearing eventually took place in Blackburn, although his business was in Yorkshire.[9] Later the same year Jonathan and Hill Barker of Upper Mill in Morton near Keighley had to attend at Manchester for their bankruptcy hearing.[10]

The relatively short distance from the Lancashire cotton centres and the dependence of the Yorkshire firms on the market there for their yarn and cloth meant that much of the raw cotton which was used was bought there personally by the spinners. The amount purchased would depend on the size and capital resources of the spinner and his feeling for the market. There was also a necessity to check the exact quality of the cotton as this extract relating to Yore Mill in Wensleydale illustrates:

> Our stock of cotton begins to get low at the rates we are now going on what we have bought will be used up by the later end of next month. It is my present intention to set forward for the cotton market in about a fortnight it appears to me desirable to see the cotton before it is brought if we can meet with tolerable good Orleans we can manage it very fairly.[11]

Another partner in a Wensleydale mill, Joseph Driver of Askrigg Mill, also bought cotton in Manchester. Driver was a tenant at Castle Mill in Keighley from before 1800 until 1807. On his bankruptcy in 1805 he had to assign his assets to Ebenezer Thompson of Market Street, Manchester.[12]

Although the Yorkshire spinners made use of local cotton dealers the larger firms bought raw cotton every few months from the Lancashire merchants and dealers. One illustration of this practice exists in the books of Smith, Blakeys & Watson of Greengate Mill in Keighley.[13] The partnership was formed in 1784, Greengate Mill was built, and the firm started trading in October 1785. Sales of cotton twist were recorded then but no previous purchase of cotton was noted. Therefore, although records of cotton purchases exist for five years from 1786 there may be some omissions. Despite this possibility the pattern of this firm's purchases was probably typical of a medium sized cotton spinning mill in Yorkshire.

Raw cotton was bought in July 1786 from Robert Twyford, a cotton merchant in Manchester, who also supplied cotton in the following year, but then in exchange for yarn. The other supplier of cotton in 1786 was William Corlass of Blackburn, who was also a good customer for yarn. As the three entries for cotton purchases from the two suppliers mentioned were in July, September and December it would appear that stocks of cotton were bought which would then take several weeks or months to work up into yarn.

Although some of the yarn customers occasionally made payment partly in the form of raw cotton, the main supply of cotton, as can be seen from Table 5.1 was drawn from merchants in Manchester. The city was at that time the distribution and commercial centre for cotton to the entire industry.[14] In 1788 there were twenty-six merchants and twenty cotton dealers in Manchester. Those merchants became the most important source of raw cotton for Smith, Blakeys & Watson, although there were times between 1786 and 1789 when cotton was taken in part exchange for supplies of yarn.

Table 5.1
Cotton Purchases by Smith, Blakeys & Watson,
Greengate Mill, Keighley, 1786–1790

		1786	1787	1788	1789	1790
Total Purchases	(£)	298	1,580	705	2,715	2,047
Purchases from Manchester merchants	(£)	170	1,413	475	2,510	2,047
Number of Manchester Merchants		1	5	2	1	1
Percentage from Manchester merchants	(%)	60	89	67	99	100

Source: Purchase and Sales Ledger 1785–1790 Marriner MSS LUL

The cotton manufacturers in Lancashire who bought yarn and paid with raw cotton probably did so because of financial difficulties brought about by the decline in trade in 1787–8. With a reduction in the demand for cloth the manufacturers would have had difficulty in paying for their warp yarn from the water twist spinners. However, as they had a stock of cotton which they held to supply the jenny spinners for making into weft, they found that they could use this to pay off their debts. In addition, as trade was reduced, any stocks of cotton would not be needed to the same extent. A trade depression also tended to reduce the period of credit allowed so a payment in goods may have been one way round the shortage of cash.[15]

One example of this type of dealing was the Keighley firm's trade with Ridley & Crompton of Manchester. This firm bought £247 worth of twist in 1786, £1,188 worth in 1787 but only £56 worth in 1788. By that time they were in financial difficulties and sold Smith, Blakeys & Watson £481 worth of cotton in January 1788. Ridley & Crompton soon became bankrupt and were paying six shillings and eight pence in the pound in March 1789.

If the ledger account can be taken as a fair representation of Smith, Blakeys & Watson's cotton purchases it shows a change from buying small quantities of cotton from several merchants to buying large quantities from a few merchants. In 1787, the second full year of trading, cotton to the value of £1,580 was bought from eight dealers. Five of them were Manchester merchants while the other three were Blackburn calico manufacturers who also bought Greengate Mill yarn. 1787 had been a crisis year for the cotton trade after the rapid growth of the previous two years.[16] There was a certain amount of overproduction which is reflected in the firms cotton purchases for 1788. They bought only £705 worth of cotton and of that as little as £24 worth was a straight purchase from a Manchester dealer. In 1789 and 1790, the last two years for which there are records, two Manchester firms, Brocklehurst & Winterberry and Green, Mawson & Dobson, supplied all the cotton apart from a small quantity supplied by Ainsworth & Lister of Blackburn in exchange for some twist. By 1790 all the cotton bought came from one firm, Green, Mawson & Dobson and was delivered as one consignment in May. Probably, as the Keighley firm increased its turnover, acquired more capital and gained an increased knowledge of the market, the partners felt confident enough to buy cotton in bulk at what they thought was a favourable price.

It is likely that the pattern and method of purchasing cotton exemplified by Smith, Blakeys & Watson of Greengate Mill, Keighley was common amongst other firms. Robert & William Heaton of Ponden Mill and Royds House Mill, also near Keighley bought their cotton in Liverpool and Blackburn during the 1790s.

Benjamin Marriner, who became one of the partners at Greengate Mill, was still buying cotton from Manchester merchants between 1814 and 1817. The supply of raw cotton was firmly in the hands of the Lancashire agents, dealers and merchants and Yorkshire was part of the area they increasingly supplied. Thus by 1833, Green-wood & Whitaker, who were cotton spinners at Burley-in-Wharfedale, said that they were completely dependent on Liverpool for cotton and Manchester for their sales.[17]

Trading conditions

A simple division of the cotton goods market can be made by dividing it into home and foreign. Both of these were expanding during the period of this survey, but for many years the overseas section was extremely volatile. Wars and trade embargoes closed the frontiers of countries at short notice so, although trade could be profitable when foreign ports were opened to British ships, it was also hazardous. Cotton exports increased after about 1798, but the overseas market was similar to the home market in that it took mainly the cheaper yarns and cloths.[18] However, by 1815 the value of cotton exports was double that of the wool trade.

With the changing nature of foreign markets and the similarity of goods there was a great deal of movement from one market to another. This could be between Europe and the Americas or between overseas and home. Thus, although the home market was basically steadier than the overseas market, it was subject to the dumping of frustrated exports. There was also a possibility of overproduction of many counts of yarn from about 1800 when more steam power was applied to cotton spinning until the development of power-loom weaving in the 1820s.[19] Therefore trading conditions were difficult for many years and a large number of Yorkshire cotton firms changed hands because of financial problems or bankruptcy.

Home demand came mainly from the working and middle classes, particularly in women's wear. There was more profit in producing yarn of finer counts for fashion goods but much of the Yorkshire cotton industry produced for the lower end of the market. The cotton cloth made with these yarns was cheaper than woollens and linens and also easier to wash, dry and iron. In the 1790s the well established domestic worsted industry in Yorkshire also received a set back, partly because of the decrease in foreign trade and partly as a result of this increased production of cotton cloth.[20]

Sales of yarn

Within Yorkshire several changes took place in the sales pattern of the local cotton firms after the industry was established in 1780. To follow these it is important to consider the two major sections of the industry – spinning and weaving. Of these spinning came first to Yorkshire for reasons which have been explained in Chapter 1. As there was no local market for cotton yarn the early spinning firms depended on selling what they produced to the Lancashire cotton manufacturers who employed hand loom weavers to make it into cloth. This situation lasted until the 1790s when there were two new developments. Cotton weaving on the domestic system was started by independent manufacturers in Yorkshire, many who may have changed from wool or worsted weaving. Secondly, cotton spinners started having their yarn woven locally instead of sending it to the Lancashire weavers in the Blackburn area. Some cotton manufacturers then moved the other way and built spinning mills to secure their own supply of yarn. After about 1800 there was a complex system of spinners and manufacturers with some vertically integrated firms and others specialising in one branch of the industry. With the development of power-loom weaving during the 1820s and 1830s came further changes. Cotton manufacturers built weaving sheds but so did the spinners who had the advantage of existing mill sites where additional buildings could be erected. Throughout the whole period however, it was to Manchester that the Yorkshire spinners and manufacturers looked for the ultimate destination of their products.

Local examples of the trends outlined above are easy to find but difficult to quantify. One example of an early Yorkshire cotton spinning firm which sold mainly to the manufacturers in the Blackburn area was Smith, Blakeys & Watson of Greengate Mill, Keighley. They produced cotton twist on water frames built on the Arkwright principle and so were spinning warp yarn for which there was a steady demand to make cheap cotton cloth. Although they could have sold their yarn on the developing Manchester market, they chose to sell directly to the Lancashire manufacturers. They, in turn, could possibly have bought some yarn locally, but bought their warp yarn from the Yorkshire spinners. Local opposition to the building of spinning mills and the larger hand powered spinning machines meant that only weft was readily available for a while in Lancashire so this Keighley firm and other Yorkshire spinners could supply twist from a safe distance.

From the purchase and sales ledger of Smith, Blakeys & Watson a picture of their involvement with the Blackburn weaving area can be drawn.[21] This area included towns and villages such as Chipping, Darwen, Leyland, Chorley, Ribchester, Church and Samlesbury. The local 'putters-out' had been denied a source of machine spun twist following the bout of machine breaking in 1779 but the demand for their cloth was in no way reduced. The local calico manufacturers were therefore prepared to buy from as far afield as Keighley to obtain the yarn they needed to keep their hand-loom weavers occupied. These manufacturers organised their activities in much the same way as their contemporaries in the worsted trade in Yorkshire. The Blackburn area was the centre of the cotton weaving industry and a loom was considered to be an essential piece of equipment in every cottage.[22] As weaving expanded the area also became the centre for calico printing and bleaching after the development of printing by rollers and the use of improved bleaching methods.

One of the customers for Smith, Blakeys & Watson's yarn was the firm of Chippendale & Co. who took half the production of Greengate Mill during the first few months of trading to the end of 1785. This firm remained a regular customer for several years and set the pattern for the early concentration of sales to Blackburn 'putters-out'.

In 1786, the Greengate Mill concern had its first full year of trading. Sales were made to forty-three customers, some of whom had been dealt with the previous year. Total sales for the year were £4,046 with eleven customers each taking over £100 worth of yarn. Parker & Smaley of Blackburn took the largest amount – £690 worth of cotton twist. The great majority of the other customers, both large and small, were in Blackburn or the surrounding region. The exceptions were William Burnley of Gomersal who took a few shillings worth of twist in 1785 and 1786, George and James Lowe of Stockport who bought £47 and £26 worth in 1786 and 1787, Thomas Hawksworth of Ilkley with £51 and £50 worth in the same years, William Myers of Draughton and Stedal & Lonsdale of Kendal with £58 and £5 worth in 1786 and lastly Cockshott & Lister of Addingham who bought £276 worth of twist in 1786. There were no sales outside Lancashire after 1787 and, as can be seen from Table 5.2, sales to Blackburn as a proportion of total sales increased considerably.

The ledger from which the figures have been taken was left incomplete. Many accounts were not balanced and the last section of the book was left blank. Despite possible omissions the records show a rapid increase in sales from the commencement of trading which is an indication of the general market conditions. Sales doubled within the first two years from £4,046 in 1786 to £8,066 in 1788. The following year saw a decline, which can be explained by economic conditions within the industry, but 1790 brought a recovery with total sales of £8,585.

The panic in the cotton industry in 1788 resulted in many business failures. These included Blackburn manufacturers, Manchester banks and London wholesale houses.[23] The expansion of the cotton industry up to 1788 had been accompanied by a certain amount of local speculation. By then the export market had become saturated and a number of merchanting firms had become bankrupt with consequent losses to a large number of small firms.[24] It was said that 'the manufacturers calculated too high upon the foreign market, which they have unfortunately overstocked'.[25] As a result the number of bankruptcies was the highest recorded for nearly fifty years. However, the downturn in trade did not last long and in Yorkshire – 'The late shock in the cotton manufacture is in a great measure got over, and the manufacturers again employed for the foreign markets.'[26] Whatever the ultimate destination for Greengate Mill yarn, sales were to the Blackburn weavers and certain manufacturers there were starting to buy large amounts. Turner & Son, who bought a small amount of twist in 1786, bought £542 worth in 1789. Joshua Fielden & Son, also of Blackburn, who had bought large amounts in the two previous years nearly doubled their purchases to £1,586 while Henry Suddall, who had taken an eighth of the mill's output in 1788 increased his share to a fifth in 1789.

The year 1790 is the last one covered by the Keighley firm's ledger. Sales rose to

£8,616, nearly half of which was taken by one firm, that of Henry Suddall of Blackburn. He came from a Lancashire family which had prospered with the growth of the cotton industry. Coming originally from farming stock, the family took up trade as chapmen in the seventeenth century and became merchants and eventually manufacturers in the eighteenth century. By 1800 Henry Suddall was one of the most influential men in the area. He had a country estate at Mellor and other property but his success did not last for he was bankrupt in 1827.[27]

A valuable customer such as Suddall was an indication of the extent to which Smith, Blakeys & Watson had penetrated the cotton yarn market without any previous experience and as a new enterprise. The success they had in building up trade with some of the leading cotton manufacturers is shown by the size and importance of their Blackburn customers. Cardwell, Birley & Hornby had their own spinning mill at Scorton but bought yarn from Keighley to supply weavers from their warehouse in Clayton Street, Blackburn. Peel, Yates & Co. became customers of Greengate Mill in 1790 and went on to become one of the largest firms in the trade.

From the analysis of yarn sales in Table 5.2 it can be seen how an increasing percentage went to the Blackburn manufacturers. Between 56% and 66% went to Blackburn from 1785 to 1787, but from 1788, a bad year for the cotton trade, to 1790, between 82% and 90% went to Blackburn. The greatest increase came between 1787 and 1788 which shows that despite the recession in trade there was still a demand for the lower count threads which the Blackburn manufacturers used for weaving the coarser cloths.

Another Keighley firm with concentrated sales in the Blackburn weaving district was Robert & William Heaton of Ponden and Royd House Mills.[28] They sold yarn through a commission agent in Blackburn. From 1800 the agent was a man called Joseph Ainsworth. By that time Robert Heaton was running the firm himself and judging by the amount of yarn he sold the business was very prosperous for several years.

Table 5.2
Yarn sales by Smith, Blakeys & Watson, Greengate Mill, 1785–1790

		1785*	1786	1787	1788	1789	1790
Total sales	(£)	441	4,046	7,597	8,066	6,958	8,585
Sales to Blackburn	(£)	289	2,350	4,270	7,367	5,713	7,313
Blackburn Sales as %		66%	58%	56%	90%	82%	85%
No. of customers	8	43	46	40	29	21	
No. of customers in Blackburn	3	19	24	28	20	15	

* Entries date from 9th October 1785. All other years from 1st January
Source: Purchase and Sales Ledger 1785–1790. Marriner MSS. LUL

Overall the Yorkshire spinners had a number of ways of selling their yarn. The ideal situation was probably the one described above for Smith, Blakeys & Watson who sold to a few large manufacturers who bought large amounts of twist with frequent repeat orders. However, examples of this pattern of selling to the Lancashire weaving firms are hard to find after about 1800. From about that time the Yorkshire spinners either sold on the Manchester market or became involved in the hand-loom weaving market in Yorkshire and sold their pieces in Manchester.

Sales of cloth

The growth of Manchester as the centre and chief market for cotton cloth after 1780 led to a greatly increased demand for warehouses in the town.[29] At times firms shared premises in order to establish a depot when the demand for space pushed up rents. It is impossible to estimate the number of Yorkshire firms which rented space in Manchester, but there is evidence that a large number did so. These were the firms which wanted control over their own sales rather than selling through an agent. They tended to be the larger firms which could afford the rents and could finance the establishment of stocks of yarn and cloth.

Several examples of Yorkshire firms with warehouses in Manchester and elsewhere can be found. W M Willett of Rushforth Hall near Bingley was a cotton spinner, first at Castlefield Mill near Bingley, and after 1799 at Wreaks Mill in Hampsthwaite. In 1803 he insured a warehouse at Back Square, Manchester, for £2,000.[30] Rangeley & Tetley of Birkenshaw Mill insured stock and utensils in their warehouse at 24 Canon Street, Manchester for £1,500 in 1810.[31]

Two other towns where Yorkshire firms held stock were Liverpool and Nottingham. William Sidgwick of High Mill in Skipton had stock in Bancroft & Loriman's warehouse and also in Flounder's warehouse in John Street, Liverpool in 1812.[32] The south Yorkshire firm of Heathfield & Co. of Sheffield sold their yarn on the Nottingham market. Even after their bankruptcy in 1815 their assignees insured stock in the warehouse at Carlton Street, Nottingham for £2,000.[33]

Some Yorkshire merchants and manufacturers had partnerships and links with London merchants. The Haigh Brothers of Marsden were one such example and the Middletons from Sheffield another. The buyer with London finance often had advantages and a wider range of contacts.

However, it was to Manchester that the majority of firms looked for sales of yarn and cloth. John Dewhirst & Brothers, calico manufacturers from Skipton, had £2,000 worth of stock in a cellar under Rea & Thomas's warehouse in Manchester in 1817.[34] By 1833 Greenwood & Whitaker of Burley-in-Wharfedale could claim that stock in their warehouse in Manchester 'is always sufficient to supply every demand instantly'.[35] An earlier example of the links between Manchester and Yorkshire come from the activities of John Haigh & Co. of Marsden. They were cotton spinners, manufacturers and merchants. John Haigh was based in London and his two brothers, Thomas and Samuel in Manchester. However, their spinning mills were in Marsden until 1806. There they ran a complex of mills with a total spinning capacity of over 10,000 spindles. This was a case of a merchanting firm establishing spinning mills in Yorkshire and the scale of their activities can be judged by the insurance valuation of £10,000 worth of stock in their Peel Street warehouse in Manchester in 1795.[36] Some of this would have been woven at their premises in Marsden and some also at a weaving shop near Ashton in Lancashire.[37] Haigh Brothers later ran into financial difficulties and became bankrupt in 1806.

Before about 1815 some other large Yorkshire cotton firms also had a member of the family or partnership resident in the town where their goods were sold. One member of the firm of Graham Brothers, for example, lived in London in 1803 while Robert and William Graham ran the manufacturing business at Making Place near

Halifax.[38] Similarly, Edward Swaine of the merchanting and manufacturing firm of Swaines & Ramsbotham of Halifax and Bradford lived in London.[39]

One example has been found of a Lancashire firm – Peel, Yates & Co, establishing an agent in Bradford, presumably to buy cotton cloth.[40] Previously in 1811 it had been noted that there was a regular weekly market for cottons in Bradford.[41] However, these were probably attempts to circumvent the normal channel of trade which was directly to Manchester. Contemporary directories carry many pages of names of Yorkshire calico, dimity, fustian and muslin manufacturers who attended Manchester market. John Heaton of Springhead Mill in Keighley, for example 'regularly attended for the sale of his pieces, and the purchase of raw cotton'.[42] The coach which the many Bradford cotton manufacturers took to Manchester to attend the market there eventually became known as the 'Calico Coach'.[43]

In the Halifax area the increased production of cotton cloth brought about the need for a local market to match the existing one for worsted pieces. The proprietors of the rooms at the Halifax Piece Hall therefore agreed the following resolution on 9th February 1805:

> That cotton and cotton goods may be admitted into the Hall, for the purpose of being exposed for sale at the usual market hours, subject to such rules and regulations as are established for the admission and sale of worsted and woollen manufactures.[44]

Later that year when a house and warehouse were for sale in Halifax which would have been suitable for a cotton merchant it was said that 'the cotton trade has been established for some years in the neighbourhood and is rapidly increasing; and the manufacturers hall for worsted goods is becoming a mart for cottons'.[45] Halifax was even considered an important market for cotton goods produced in the Bingley area and local manufacturers were able to sell their cloth there to dealers who travelled from Manchester.[46]

With the growth in the market for cotton goods from the mid-1790s and the consequent development of a network of agents, dealers and merchants there came pressure to secure additional supplies of yarn and cloth. The Yorkshire newspapers carried advertisements from firms soliciting trade, most of them Manchester agents, but also others. A typical advertisement started in this fashion:

> To Mule and Water Twist Spinners and Calico Makers.
> A house in the habit of disposing of the above articles upon commission, in a respectable and punctual connection, in and near Manchester, would take the consignment of a further concern whose produce would be regular.[47]

Another merchant house wanted calico or wide goods to sell on the Manchester market in 1807. This concern claimed to be 'well acquainted with the purchasers resident in that town as well as those who frequent it'. The service they offered they thought to be very advantageous to 'distant manufacturers'.[48] Later the same year a Manchester agent was wanting yarn of a type many of the Yorkshire spinners produces. That was good second quality water twist from 20s to 30s quality as well as cap weft from 14s upwards. The firm in this instance was Merrick & Flintoff of Peel Street, Manchester.

In 1813 a London agent wanted local calico printers and calico, muslin and linen manufacturers to contact him if they wished to extend their trade to London. He claimed to be well acquainted with the wholesale and retail town and country drapers and also with the shipping trade.[49] Even in 1819, when many of the Yorkshire cotton mills had been converted to worsted spinning, commission agents advertised for cotton yarn. One of these would take thirty to fifty packs of cotton twist per week. He claimed to have extensive connections with 'the most respected manufacturers of cotton piece goods in Lancashire and Yorkshire'.[50]

One of the most important markets for cotton yarn produced in the Yorkshire mills must have been the local hand-loom weavers, either as individuals or through a small manufacturer. However, it is not possible to quantify the proportion which went into local cloth as against the proportion which was sold as yarn to be made into cloth elsewhere. One firm which did offer yarn to local manufacturers was William Musgrave & Co. of Simpson Fold in Leeds. They started cotton spinning about 1803 and used mules in their steam powered mill. The following year they were advertising warp yarn for sale to local manufacturers to be made up into fustians, swansdowns and toilinets.[51]

Another outlet for Yorkshire cotton yarn was to the hosiery industry of the Midlands. One instance of a Nottingham agent or dealer exploring the possibility of selling this yarn is shown by an advertisement in 1799. He wanted 'a commission in cotton twist for the hosiery market' and claimed that because of his good connections he would be able to sell a considerable quantity each week.[52] As cotton yarn was used for making hosiery as well as weaving into cloth the Midlands area was an important market for relatively fine cotton yarns. The spinners in Nottinghamshire and Derbyshire who produced these yarns sold them to both the Nottingham hosiers and on the Manchester market.[53] There were a few cotton spinning firms in the south Yorkshire area and they could have sold to the Midlands hosiers but it is only known that one of them did so. There were cotton mills in Conisborough, Stocksbridge, Ecclesfield and Sheffield as well as some smaller mills in other towns but only the large Sheffield mill is known to have had sales to the Midlands.

Within Yorkshire two examples have been found of hosiers setting up cotton spinning mills although it is not known if the yarn was used for stockings or cloth. Gervais Marshall of Leeds and William Mounsey of Otley were partners as hosiers with shops in Leeds and Otley. They built a cotton mill near the road to Menston at Ellar Gill near Otley in 1790 but only used it for cotton spinning for about ten years. The water powered mill can still be seen although there are many later additions. Similarly John Merryweather was a linen draper and hosier in Leeds who started cotton spinning at Embsay near Skipton before 1796.[54] In both these cases however, it may have been that knowledge of the industry and its products led to an investment in new mills rather than any link with hosiery manufacturing.

After about 1800 the pattern of trading for a number of Yorkshire cotton spinning firms changed as the opposition to the new mills and machinery in parts of Lancashire was overcome. The importance of the Blackburn weaving area as a direct market for yarn declined while Manchester and the looms of local weavers became more important. Yarn sold in Manchester still went to Blackburn but through the services of a merchant or agent or, for larger firms, through their own warehouse. Similarly cloth woven in Yorkshire was sold in Manchester as that town took an increasing role as the commercial hub of the industry. With the development of cotton weaving in Yorkshire there were attempts to graft the sales of this new type of cloth on to the established distribution and marketing centres for worsted cloth. This did not however detract from the importance of Manchester as it was the Manchester merchants who travelled to Yorkshire to buy and the Yorkshire manufacturers who travelled to Manchester to sell.

CHAPTER 6

Cotton Weaving

THE TECHNOLOGICAL DEVELOPMENTS in cotton spinning in the late 1790s expanded the volume of yarn which was being produced. Although Arkwright's water frame continued in use in many mills, particularly in the Keighley area, the development of the throstle with improved gearing for the rollers enabled spinning frames with more spindles to be made. Typically the water frame had forty-eight spindles by the 1790s but the new throstles could have from seventy-two to one hundred and twenty spindles. Improvements to the mule were also made with the application of power to some of the motions so the spinners were then able to operate two mules at the same time. Other improvements increased the number of spindles on each mule and yarn output increased dramatically.

Hand-loom weaving

The rare equilibrium of the early cotton industry with the supply and demand of the spinning and weaving sections being balanced was once again broken. The earliest Yorkshire cotton mills had been built to supply yarn to Lancashire hand-loom weavers at a time of local shortages. After fifteen years or so the rising output of yarn from the increasing number of mills which had been built in Yorkshire enabled a rapid growth in cotton manufacturing to take place, particulary in the West Riding. Hand-loom weaving was already very well established in the local worsted industry with a large number of manufacturers providing employment for weavers over a wide area. Weaving was often combined with a little farming in the Halifax area and the changes in agriculture had pushed many former farm workers in Craven and the Dales into working at the loom. Demand for cotton cloth continued and hand-loom weaving became a well paid occupation for several years.

It appears to have been relatively easy to learn hand-loom weaving and also for weavers to change from weaving worsted cloth to cotton cloth. An un-named woollen mill was for sale near Halifax in 1801 and included with the sale were a large number of looms. It was said that the mill could be adapted for cotton spinning and also that:

> There are looms which would make more than two hundred pair of cotton looms . . . the goodness of the weavers in the Parish of Halifax, the tradesmen in Manchester for some time have experienced.[1]

One of the leading Bradford worsted manufacturers, Richard Fawcett, said that men who had previously been woollen and worsted weavers could make more money weaving cotton cloth at that time and also that the work was easier.[2] One of the interesting features of the Yorkshire cotton industry was the part played by worsted manufacturers. Many of them in Keighley, for example William and John Haggas,

Loomshop at Chapel Lane, Addingham, used for cotton weaving.

William and Benjamin Marriner and Lodge Calvert, were also cotton spinners and sometimes manufacturers.

A number of factors make it difficult to estimate the scale and distribution of cotton hand-loom weaving in Yorkshire at this time. The looms themselves were hand made and readily available. Many weavers combined occupations, as weaving was often done in their own homes and the weaving of a piece could be interrupted when other tasks had to be undertaken. The looms could be used to weave cotton, worsted or linen without major adaptation and weavers changed from one to the other depending on the price they were paid. Most of the weavers were self employed and could work for a number of masters. The manufacturers themselves, although they needed capital for stock, often had little more than a small warehouse for yarn and cloth and very few have left more than a name in a directory as an indication of their occupation. Even then many of the smaller manufacturers had only just risen from the ranks of the ordinary weavers and could return there if trade was bad. However, the references to cotton hand-loom weaving are numerous and there seems to have been a period of about twenty years from 1795 to 1815 when cotton weaving rivalled worsted weaving in many parts of Yorkshire. The decline came in two stages. Firstly when machine spun worsted yarn started to come onto the market in large quantities and the worsted industry started to regain its pre-eminence. Secondly, when power-loom weaving became widespread from about 1825.

The references to cotton manufacturing and to the existence of large numbers of cotton weavers in particular areas are widespread. One centre for cotton weaving was the Horton area of Bradford where cotton manufacturing became well established in the late 1790s. A number of carriers were employed in taking cloth to Manchester and other places while the larger manufacturers used their own wagons. The return journey to Manchester took three days and warps were brought back to Bradford. There were a number of cotton manufacturers who employed large numbers of weavers and it is not surprising that some of them also became spinners. About 1806 John and Benjamin Knight built a cotton spinning mill in Great Horton no doubt to supply their own weavers and also other manufacturers who attended the Manchester market. John & Benjamin Knight failed in 1827 and the mill was then rebuilt for worsted spinning. However, cotton manufacturing continued in Great Horton, Little Horton and Horton Green for some years until the failure of a Manchester merchanting house brought financial disaster to most of them. Many of the cotton manufacturers such as Abraham Balme and Samuel Swaine then changed over to worsted manufacturing.

On the other side of Bradford, at Idle, another cotton spinning mill was built to supply the local weavers and both Holme Mill and Rand's Mill in Bradford originally had cotton spinning machinery before they were changed over completely to worsted spinning before 1810. John Rand was a woollen and cotton manufacturer in 1802. One man, who was described as 'perhaps the largest maker of cotton goods in this part of the country' was John Knowles. He ran Hallas Bridge Mill and Bent Mill in the Goit Stock valley near Wilsden on the outskirts of Bradford. He put-out cotton hand weaving at twenty different 'stations' extending from Skipton to Cleckheaton and made calicoes for the Manchester market. His wagons went constantly on the three day journey over Blackstone Edge to Manchester. His son John Wilkinson Knowles continued as a cotton spinner and manufacturer but was also a victim of the 1826 recession when one of his assignees was Edward Marsland, a cotton merchant in Manchester.

In the Yorkshire Dales the cotton mills provided yarn for large numbers of cotton weavers in each area where they were situated. In Settle, for example, it was said that:

> the sound of the hand-loom might be heard in every village in the district, and in almost every street. From the town of Settle and the adjacent villages there must have been a considerable output of hand-made fabrics.[3]

Further confirmation that the cotton spinners were selling yarn to local manufacturers comes from the evidence given by William Sidgwick who was a cotton spinner at High Mill in Skipton. In 1816 he was giving evidence related to the employment of children and said that he sold little yarn abroad but sold mainly to manufacturers in his own neighbourhood. He added that there were a large number of operative weavers in the area and that they were having to work long hours.

As the number of cotton hand-loom weavers increased, problems arose between the master manufacturers and the individual weavers. These complaints and disputes gave rise to an act in 1800 intended to reduce the problems by introducing a process of arbitration.[4] Evidence about the working of the Act was collected during 1802 and 1803 and illustrated the spread of cotton manufacturing.[5] One question which was asked was:

> Has not cotton weaving extended itself into Yorkshire, Derbyshire, Cheshire, Lancashire, Warwickshire and many other places within the last ten years; and

have not the weavers in those places turned from weaving linen yarn, worsted yarn and woollen yarn to cotton?

Thomas Thorpe, who was giving evidence, replied that a number of weavers had turned from woollen to cotton and the reason was that the woollen trade had suffered because of the war and so wages were better for cotton weaving. The change to cotton had not always been easy and the resulting bad work had brought many disputes with the masters.[6]

Several years later there was another example where hand-loom weavers in Pudsey changed from weaving woollen cloth to cotton cloth:

In 1824 a severe panic existed in the woollen trade and there was scarcely a cloth-loom to be heard in the village. To keep them from starving many of the people were employed in weaving cotton by hand-loom obtaining their work from a Mr Nutter or Nuttall of Bradford, whither they took their pieces on Thursdays. Mr Joseph Tordoff, of Low Moor, also put out cotton weaving at Pudsey.[7]

The nature of the organisation of the domestic hand-loom weaving trade brought a number of problems. One of these which was treated very seriously was embezzlement. Yarn was given out to weavers to make into cloth. If small amounts of yarn could be saved by the weaver these could be accumulated and made into an extra piece for his own profit. The worsted manufacturers had an organisation to combat this type of fraud which was also widespread amongst the cotton weavers. Two cotton weavers, William and Abraham Stansfield of Stansfield in Halifax worked for two fustian manufacturers, Messrs Holt & Chadwick and George Thompson. In 1792 they embezzled yarn from them which they then wove into a piece of cotton thicksett. This they then tried to sell to Turner, Varley & Co. of Mytholm but they were found out and would have been prosecuted if they had not publically admitted the fraud to deter others.[8]

One cotton weaver who was not let off so lightly was William Wadsworth of Haugh Top in Barkisland. He embezzled one cotton warp and twenty-two pounds of cotton which had been delivered to him by William Mitchell of Booth Town near Halifax to work up into a piece. Wadsworth was sent to Wakefield Prison for three months and was to be publically whipped in the centre of Halifax on his release.[9]

The cotton weavers who embezzled cotton yarn were not the only people who troubled the manufacturers. Cotton was stolen from mills and cloth from bleaching grounds. Garforth & Sidgwick of Skipton had yarn stolen in 1803 and offered twenty guineas reward for information leading to its recovery.[10] J. Smith of Horton near Bradford had seventy pieces of calico stolen in 1809 but the thief was caught and convicted.[11] Turner, Bent & Co. of Mytholm Mill in Stansfield near Halifax had six pieces of velveteen stolen from their bleaching ground in 1810.[12] Dewhirst's warehouse in Skipton was broken into in 1818 and ten dozen fents were taken.[13]

One way the weaving side of the cotton industry was organised around the turn of the century was for a spinner to give out yarn to individual weavers. An example of a spinner doing that was John Wright of Kebroyd Mill near Halifax. He had employed a number of hand-loom weavers on the putting-out system but was bankrupt in 1806 and so his machinery was put up for sale by auction. At that time he still had yarn out with the weavers but this yarn had become the property of his creditors. The following notice was therefore added to the details regarding the auction:

NB All weavers having any work in their possession are particularly desired to bring in the same on the mornings of each day's sale.[14]

An alternative way of organising the hand-loom weavers was for firms operating the spinning mills to add weaving shops to their premises and so have the weavers under their direct control. One of these was the firm of Rangeley & Tetley who ran Birkenshaw Mill from 1805. They had a weaving shop close to the mill which they insured for £150 in 1805.[15] Much of their cloth was then sold through their Manchester warehouse.

A further example of a small weaving shop was at Spring Hall Factory near Halifax. This had been run by Thomas Lee until 1807 when he was bankrupt. Lee was a cotton spinner, dealer and chapman who previously ran Bankfoot Mill for cotton spinning and Midge Hole Mill for flax spinning. Spring Hall Factory consisted of a spinning room with nine jennies, a weaving room with fourteen pairs of broad looms and ten pairs of harding looms, a dressing shop, baling room and warehouse.[16]

Another loom shop was for sale at the same time at Marsden near Huddersfield. The machinery, which consisted of fourteen pairs of fustian looms and one 204 spindle mule was also for sale.[17] These two examples are probably typical of the large number of loom shops which existed throughout the Yorkshire cotton spinning areas. They were sometimes purpose built but were more often conversions of existing buildings. No power was needed, the scale of operations was relatively small, and often the weft yarn was produced on the premises while the warps were bought in.

A number of examples of spinners with larger mills also having loom shops can be traced across the region. William Chamberlain, who was an ironmonger and cotton spinner with a warehouse near the bridge in Skipton, ran a cotton mill at Eastby from 1796 with a new mill added in 1801. In Skipton he had several buildings used for muleshops and weaving shops in 1801. He also insured a warehouse, counting house and goods in trust and on commission. Bainton, Boyes & Co. at Wansford near Driffield had a weaving shop at their mill as did Beaumonts at Crossland Factory near Huddersfield.

An intermediate stage between the small workshops and factory power-loom weaving came with the introduction in some areas of dandy-loom workshops. This compact, iron framed loom was invented in 1802 and had a taking up motion which enabled more regular, better quality cloth to be produced than with the ordinary loom.[18] They were worked only by men because the work was harder. The dandy looms were more expensive to make and usually several at a time were placed in workshops. In 1835 the use of dandy-looms was explained in this way:

Those dandy-looms are worked by persons who employ men to come to their premises to work, in a sort of factory. Those weavers have to attend, if they work at the dandy-loom, the same number of hours as they do in the factories and their toil is excessive.[19]

Examples of dandy-loom workshops in Yorkshire have been found in Kettlewell,

Todmorden, Addingham and Bradford. As the number of people turning to weaving expanded after about 1815 it became difficult to obtain raw materials on which to work.[20] There was keen competition for yarn which made it difficult for all the individual weavers to be supplied. The weavers had to accept the workshop situation as the manufacturers who converted premises could supply warp and weft so they provided employment for a weekly wage but at the loss of the weaver's independence.

Power-loom weaving

The first attempts at cotton power-loom weaving were made in Yorkshire at Doncaster by Edmund Cartwright in 1786.[21] He set up a factory there with 20 looms but they did not work well and the venture was abandoned in 1793. Although Doncaster was a distance away from the later centres of the textile industry there was a cotton spinning mill at Balby, about two miles away before 1792.[22] In Doncaster itself Charles Plummer was spinning cotton at a mill in Fishergate about 1790. It is open to conjecture that the three storey building advertised for sale in June 1805 in Doncaster was the one used by Cartwright. It was near the navigable river Don, had a number of cottages for workmen and was built of brick with the upper rooms used for weavers workshops. It was said that 'the building was erected a few years ago and are well adapted for a cotton or linen manufactury.'[23]

Following Cartwright's failure there is no record of further attempts to weave by power in the West Riding until the 1820s. The situation in Lancashire was different and there were fourteen power-loom factories there by 1818 and thirty-two by 1821. At that time there was only one firm in Yorkshire with power-looms. That was William Hegginbottom & Bros of Saddleworth who started power-loom weaving in 1821. Power-loom weaving of cotton elsewhere in Yorkshire does not appear to have started until 1825/26 when a number of cotton spinners bought looms and installed them in their mills. The installation of the looms provoked serious disturbances in 1826 when Low Mill at Addingham and Mason's Mill at Gargrave were attacked on the 26th and 27th of April.[24] Both mills were defended by soldiers who only managed to keep the rioters from damaging the new looms at Addingham. A serious downturn in trade had harmed the hand-loom weavers and the power-looms were seen as a contributory factor. A local correspondent wrote:

> Calicoes for which 5/- a piece was paid for the weaving 20 years ago, are now woven in some districts on the confines of Yorkshire at 10d a piece, and 1/- is the maximum price in Addingham, and in this part of the country.[25]

Similarly in Settle the winter of 1825/6 marked a severe decline in the local cotton trade and prices for cotton weaving were very low.

However, more manufacturers turned to power-looms and by 1835 there were twenty-four mills with cotton power-looms in the West-Riding according to the survey of power-looms used in the United Kingdom. This list is reproduced in Table 6.1 and shows a total of 3,153 cotton looms. This survey is incomplete for several other firms had power-looms by 1835. Jeremiah Horsfall of Addingham had power-looms in 1826 while John & Richard Allen were using them in Saddleworth in 1833. However, the table does make clear the dominance of the firms in Halifax Parish which were running 85% of the power-looms in the county.

Table 6.1

Cotton Power-looms in West Riding Mills, 1835

Firm	Parish	Number of Looms
Greenwood, J	Halifax	122
Crossley & Sons	Halifax	157
Bent, James	Halifax	48
Hinchliffe, W & G	Halifax	80
Horsfall, John	Halifax	46
Gill, Jonathan	Halifax	56
Hodgson & Gill	Halifax	39
Sutcliffe Bros	Halifax	130
Barker, J & W	Halifax	22
Townend, George	Halifax	13
Horsfall, Slater & Robinson	Halifax	75
Fielden Bros	Halifax	810
Whiteley, John	Halifax	30
Bold, Peter	Halifax	328
Hinchliffe, George	Halifax	80
Greenwood Bros	Halifax	40
Mallalieu & Platt	Halifax	63
Jackson & Co.	Halifax	18
Barrow, Ed	Westhouse?	10
Messrs Threlfall	Bingley	24
Greenwood & Whitaker	Otley	112
Clayton & Son	Giggleswick	52
Sidgwick & Co.	Skipton	144
Henry Bramley	Linton	36
Four Mills	Saddleworth	618
		3,153

Source: Number of Power-Looms used in Factories in the manufacture of woollen, cotton, silk and linen respectively in each county of the United Kingdom PP 1836 (24) XLV

CHAPTER 7

The First Cotton Masters

T HE RAPID GROWTH of the Yorkshire cotton industry prompted large numbers of people to enter the field of cotton spinning. As there was little cotton spinning or weaving in Yorkshire before the first mills were built in the 1780s the new cotton masters came from a variety of backgrounds. A few came from Lancashire with cotton experience but the majority came from the ranks of merchants, wool and worsted manufacturers, drapers, land owners, solicitors and bankers. The Leeds area in particular provided many cotton spinners from the merchant class. It has been estimated that fifteen of the West Riding cotton mills were built by Leeds merchants or other Leeds men with mercantile connections.[1] In Keighley, where over twenty cotton mills were built, no single group of people predominated. The men and women who were responsible for building nineteen of the mills in the Keighley area included – four Lancashire cotton manufacturers, two gentlemen, three ladies, one merchant, seven worsted manufacturers, six farmers, two solicitors, a wool stapler, a cotton manufacturer, a joiner and a machine manufacturer.[2] With the lack of experience in the industry, it is no wonder that there was a great demand for cotton mill managers or that firms were frequently rendered bankrupt.

A number of men entered the Yorkshire cotton industry with experience gained from the industry in Lancashire. Some of these people made a significant contribution to the development of local firms, but they did not overshadow their Yorkshire competitors. A number of the Lancashire men brought technical expertise, others brought capital and entrepreneurial ability. How far the Yorkshire and Lancashire elements of some enterprises were integrated it is difficult to assess. However, there were some examples where the Lancashire connection was important.

Lancashire men

The first, and possibly most significant Lancashire men to be involved in cotton spinning in Yorkshire were John, George and William Clayton and Thomas Walshman. They were in business as calico printers at Walton-le-Dale near Preston in Lancashire.[3] Walshman had also been a partner with Arkwright and others at Birkacre Mill near Chorley. He was three years older than Arkwright and had been one of his early associates in Preston.[4] The partners built Low Mill in Keighley in 1780 but Walshman retired to Preston after a few years. Claytons retained control of Low Mill and also built a large mill at Langcliffe outside Settle. William Clayton went to live at Settle near the new mill but he also had mills in Preston and was at one time considered to be one of the largest cotton spinners in Lancashire.[5]

Several partnerships which were formed to build or lease cotton mills included Lancashire men who brought a knowledge of cotton spinning, manufacturing or finishing with them. Some of these partnerships may have been entered into as an attempt to obtain a supply of yarn for weaving interests in Lancashire, but in the absence of sales data this cannot be verified. A man called Thomas Binns, who had been involved in cotton spinning in Lancashire joined a partnership to run Stubbing House Mill in Keighley in 1787.[6] One of the partners at Ripponden Mill, Ralph Holt, came from Heywood near Bury in 1801.[7] The two brothers, John and Thomas Hadwen, who were the sons of the vicar of Longwood had been trained in cotton spinning in Oldham before moving to Severhills and later Kebroyd Mills near Halifax where they established a very successful concern. The four partners who built Malham Mill in 1785 all came from Lancashire. Three of them were from Colne but the other, Richard Brayshaw, although having a local name, was described as an excise officer from Liverpool.[8] The Haigh brothers, who were cotton manufacturers near Ashton in Lancashire, ran several mills in Marsden for about ten years from 1795.

A few Yorkshiremen in turn established cotton spinning mills in Lancashire. Edmund Lodge, a wealthy Leeds merchant who built Willow Hall Mills at Skircoat near Halifax also owned Oakenrod Mill at Spotland in Lancashire.[9] Richard Paley, the Leeds soap boiler, iron master, cotton spinner and merchant ran cotton mills at Colne in 1797 as well as a mill in Leeds.[10] The movement of people and capital between Lancashire and Yorkshire emphasises the unified nature of the industry which in practice was centred on Manchester.

During the period of this survey, two families achieved prominence in cotton spinning in the Ryburn Valley near Halifax. They were the Wheelwrights of Rishworth and the Whiteleys of Soyland. A man called John Hoyle, who was spinning cotton at Temple Mill from about 1810 married Sarah Wheelwright who was a beneficiary of John Wheelwright, who was a corn miller at Rishworth Mill until his death in 1814. John Hoyle then changed his name to Wheelwright. In addition to Temple Mill, Wheelwright eventually had interests in Hollings Mill, Ripponden Mill and later Rishworth Mills. John Whiteley started cotton spinning at Stones Mill in Soyland in 1820. Later he bought a number of mills but leased them to other spinners. The mills he owned included: Stones Mill, Hazel Grove Mill, Upper and Lower Swift Place Mills, Dyson Lane Mill and later, in 1850, a new mill called Ryburn Mill.[11]

Greenwoods of Keighley

One of the most successful cotton spinners in Yorkshire was John Greenwood of Keighley. According to a recent family history his grandfather, James Greenwood, was a wool stapler and landowner in the Keighley- Haworth area.[12] His father, also called John lived at Damside in Keighley and although a worsted manufacturer built the second cotton mill in the town in 1782. This was built in defiance of Arkwright's patents and he had some problems making his own spinning machines work correctly. However, this was solved successfully and John Greenwood senior went into partnership briefly with Abraham and Christopher Hargreaves who were converting Barrowford corn mill near Colne into a cotton mill in 1783. According to Abraham Hargreave's diary John Greenwood supplied carding engines and spinning and roving frames from Keighley for their new mill. The two Greenwoods, father and son helped to fix up the first two spinning frames at Colne. However, the partnership did not last many months and after the machinery had started running Greenwood left to concentrate his attentions in Keighley.

In 1787 John Greenwood senior took room and power at Stubbing House Mill, later called Aireworth Mill, and his son was established as manager there. Following the end of the recession in the late 1780s John Greenwood built Cabbage Mill in Keighley in 1793.[13] He equipped the mill with machinery from North Brook and Stubbing House Mills and his son moved there as manager. The Greenwoods continued to operate this mill until 1844 and, together with Vale Mill it was the centre of the Greenwoods cotton interests in Keighley.

Vale Mill is in the valley between Keighley and Haworth and was run by the Greenwoods unlike many of their other mills where they had a partnership arrangement or leased them to tenants. John Greenwood senior died in 1807 and his son expanded the business both on his own account and in partnership with Lister Ellis and others. A farm was bought near Vale Mill to increase the water supply and the mill was extended. In 1804 land and property was bought from Joshua Field which included Upper and Lower Newsholme Mills and Damens Mills which were all built for cotton spinning. However, these were leased and not run by the Greenwoods. John Greenwood bought Goose Eye Mill after the partners were bankrupt about 1798 but again did not run the cotton mill himself. The real expansion came in partnership with Lister Ellis and started with the purchase of Castlefield Mill in Bingley and Wreaks Mill in Hampsthwaite following the bankruptcy of the previous owners in 1805. Wreaks Mill was on the Swarcliffe Estate and eventually John Greenwood went to live at Swarcliffe Hall. In 1807 Greenwood & Ellis bought Damside Mill in Keighley, again when the existing owners were bankrupt. In 1808 they rented Airton Mill near Kirkby Malham but possibly did not buy the mill because they only insured the contents. One of the largest mills they bought, this time in partnership with Jonas Whitaker was Greenholme Mill in Burley-in-Wharfedale. The new partners built a new mill with gas lighting and made a considerable investment with a new weaving shed within a few years. The last cotton mill John Greenwood had a share in was Ellar Carr Mill in Cullingworth. Greenwood bought the mill in 1816 but it was run by his partner Edward Craven who besides managing the mill was a civil engineer who advised Greenwood and others on water power systems. By the end of this survey John Greenwood was still running four of the largest cotton mills in the area – Vale, Cabbage, Greenholme and Castlefield. Unlike most other cotton spinners and manufacturers where he was active he did not change over to worsted spinning but retired to his country estate at Swarcliffe Hall. Similarly, Lister Ellis moved away from cotton spinning to Croft Head near Long Town in Cumberland.[14]

Heatons of Ponden

Although some early Yorkshire cotton spinners were successful in their business activities, many such as the Heatons were not. Robert Heaton, who lived at the top of the Worth Valley, was a worsted manufacturer from about 1749 to 1791. He organised a network of workers with his wool being combed, spun and woven on a domestic basis over a wide area. Heaton's two sons, Robert Junior and William also became worsted manufacturers and organised their business activities on similar lines.[15]

In 1791 both the father and the two sons decided to change over to cotton spinning. There had been a serious decline in their markets for worsted cloth so, like many local manufacturers in the Keighley area, they decided to build two water powered cotton mills. There were already nearly ten cotton mills within a few miles and, although there had been setbacks in the cotton trade, the new spinning mills must have appeared to offer more promise than the troubled worsted trade.

Robert Heaton Senior's mill was built at Pitcher Clough near the road to Colne and was called Ponden Mill. Wood planks for the mill were brought from Leeds while oak came from Gisburn and Wentworth Castle near Barnsley. Local stone and labour was used to build the walls and the local joiners made the preparing and spinning machinery. By 1791 the mill was ready so the workforce had to be recruited. James Baldwin, a weaver, was hired at 9s. per week plus lodgings, as spinning master while John Keighley, also a weaver, was set on as card master to start at 10s. 6d. per week.[16] Local children were then employed to look after the machinery at 3s. per week.

Robert Heaton's trading activities then changed. Instead of journeys with wool into Craven and Bowland to have it spun and then deliveries to local weavers and visits to Halifax Piece Hall to sell cloth the focus became Lancashire. Cotton was bought in Manchester while the twist was taken by cart from Ponden to Colne and then on to Blackburn.

Within two years of starting cotton spinning Robert Heaton died and left the concern to his two sons. Unfortunately the mill had a serious fire in 1795 but was rebuilt with the help of the money paid by the Royal Insurance Company. The cost of rebuilding the mill was £285 which confirms the insurance valuation put on the mill in 1797 of £250.[17]

At the same time as his father was building Ponden Mill Robert Heaton Junior was building Royd House Mill at Oxenhope. The land and property belonged to his father-in-law, John Murgatroyd, with whom Heaton had been in partnership as a worsted manufacturer. After the initial decision to start cotton spinning Robert Heaton's diary gives many of the details of the work that was necessary to organise the building of the mill. The two partners might have been considering spinning using mules, for Robert was given an introduction to a Mr Marshall of Manchester to see his mule spinning machinery. Another entry recorded that Richard Sagar of Burnley and his brother at Blackburn also made mules. Sagar had suggested that a room six yards square would hold two mules of about 144 to 150 spindles. Alternatively a room five yards square would hold two jennies. A jenny with 100 spindles would cost £7 and would spin one pound of cotton an hour of 24s to 30s quality. However, Royd House Mill was eventually built for spinning with water twist frames on the Arkwright pattern as were most of the local mills.

Joseph Tempest planned the mill and worked out the calculations for the water wheel and transmission.

Suppose the water wheel to be 30 feet in diameter the pitt wheel must be 23 feet 9 inches and contain 324 teeth in 18 segments, 18 teeth each. The crown wheel must be 3 feet 1 inch in diameter and contain 41 teeth. The fly wheels must contain 68 teeth each. The drum wheel 43 and 34 teeth each.[18]

During 1791, Robert Heaton Junior and Joseph Tempest were busy buying materials and superintending the building of the mill and the machinery. The axle for the water wheel was bought second-hand from a Leeds bark mill for £3 15s. od. Iron work was bought from Sturges & Co. of Bowling Iron Works near Bradford. Files, glue and nails were bought in Halifax, while some of the gearing came from Sowerby Bridge. Numerous entries were also made about the books available on mill work, eg 'Moxons Mecanic Powers. Many plates and scarce 4/−.'

In September 1791 the partners took into their employ an engineer:

'Memorandum, that John Brigg agrees with John Murgatroyd of Roydhouse and Robert Heaton of the same place to come and work for them at Roydhouse − Brass, iron and wood turning and to act as engineer for the cotton mill erected there, for one year upon this condition that he is to have his meat, drink and lodging at Roydhouse, and also forty pounds wage.[19]

After Briggs arrived work started on making the carding engines and spinning frames. Detailed notes were made of each piece of timber needed for the machines as well as if the wood was to be deal, oak or alder. Briggs was given money when he went to collect machine parts:

1792 Jan 1	Lent him when he went to Keighley for bobbings	£1 1s. od.
1792 Feb 27	Lent do. when he went to Keighley for brass wheels	5s. od.

The Sagar brothers from Burnley and Blackburn may have done some work at the mill for there is an entry for August 5th 1791: 'Paid Sagars for 4 1/2 days 0.12.0' Apart from that entry it appears that John Brigg and other local men made the cotton preparing and spinning machinery.

Cotton spinning was a successful venture at Royd House Mill for the first few years. William Heaton became the traveller for the firm with visits to Liverpool and Blackburn to buy cotton and sell yarn. By 1799 William left the firm to set up as a cotton broker in Liverpool. Robert Heaton continued cotton spinning at both Royd House and Ponden Mills and for several years was financially successful but started to lose money after about 1806 because of the trade difficulties caused by the French Wars. Royd House Mill burnt down in 1808 but unlike Ponden Mill was not covered by insurance.[20] The loss, coupled with difficult trading conditions started a decline in Robert Heaton's fortunes. Possibly as a result of the loss of half his productive capacity he turned to calico weaving in 1808 as a way of making money for cotton spinning alone was no longer profitable. Many other local spinners also started cotton weaving at that time as the demand for twist to be sent to the Blackburn weaving area was considerably reduced. However, this venture did not succeed and by 1809 he had lost £332. These losses continued to such an extent that land and property had to be mortgaged.[21] Between 1810 and 1813 further losses from cotton spinning amounted to £569. Robert Heaton struggled on as a cotton spinner at Ponden Mill but died in 1817 surrounded by debt.

PART 2

CHAPTER 8

Introduction

IN 1835, Baines, in his *History of the Cotton Manufacture in Great Britain*, described the location of the industry in Yorkshire at that time. He said that thousands of workers were employed in the industry and that weaving was carried out in most of the areas where there were spinning mills. The mills were to be found on the: 'Calder, the Aire and the Wharfe, in Saddleworth, the valley of Todmorden, Halifax, Skipton, Keighley, Bingley, Addingham, etc.'[1]

Baines was describing the situation in 1835, fifty-five years after the first cotton mill was built in Yorkshire. By then many of the early mills had been changed from spinning cotton to other textile uses or were being used for a variety of other purposes or had been abandoned. The mills which Baines did know about, and there were about one hundred and fifty, or less than half of those which had been built, were usually much larger than the earlier mills and many of them continued to be used for spinning and weaving cotton until at least the end of the century.

One of the questions which is often asked about the early cotton industry in Yorkshire is, 'where were the mills?'. There is then some surprise to find that, although there were concentrations in certain areas, the mills were spread across the three Ridings which made up Yorkshire and that a large number were in what are now called the Yorkshire Dales. Indeed the Yorkshire Dales National Park has the remains of a large number of mills, many of which, at a distance, look little different from when they were built two hundred years ago. The purpose of the following chapters therefore, is to describe all the mills which have been traced, to give details of the firms which owned or ran them and to give information regarding a few surviving mills which might be of interest to those interested in industrial archaeology or local history.

The stress of a survey such as this, on what was happening in Yorkshire, can appear to give the county boundary great importance. This has not been the intention as the Yorkshire cotton industry did not exist in its own right but was part of an industry based on Liverpool and Manchester. However, most of the people involved, the capital and the after effects were part of the economic growth and development of Yorkshire. To help with an understanding of the scale of the local industry and of what it must have meant when these large industrial buildings came to dominate small hamlets and villages, the whole of the three Ridings have been divided into ten areas. Again, the boundaries are artificial and have only been drawn for convenience so that mills with similar characteristics could be grouped together. They also relate to modern centres and boundaries.

The first areas chosen are based on modern towns and cities, the geographic areas around Leeds, Bradford, Huddersfield, Halifax and Keighley. All these are interesting in their different ways. There was the strong influence of merchants in Leeds. Bradford, later to become the centre of the wool textile industry, had few mills in the centre because there was little water power, but a good number in the valleys around. Huddersfield again had few mills but the Colne valley has been included. Halifax, and westwards to the Lancashire border, became dominated by cotton after a slow start. Keighley was a leading centre of the cotton trade for thirty years or so from 1780 but the worsted industry re-established itself once worsted spinning became mechanised.

The Saddleworth area was lost to West Yorkshire some years ago, perhaps because it faced west and was seen as part of Lancashire. So, for this survey, it has been treated as a separate area. Another region where there have been strong links with Lancashire and where the cotton industry grew in importance was Craven and the area to the west of Skipton. Two other areas where cotton spinning flourished for a while were east and south Yorkshire and they have been treated separately. The last area, and perhaps the most interesting, is the Yorkshire Dales. Here are the remains of far more Arkwright type mills than the whole of Lancashire and here the enterprise of the early cotton spinners can be seen in the form of multi-storeyed mills standing at the side of the Wharfe, the Aire, the Ribble and numerous smaller streams.

A list of all the mills has been provided with the references to this section together with a grid reference where it is known. The aim in the following chapters has been to describe the mills with information based on original sources wherever possible. A date has been given for when each mill was first known to be operating. That will be the first accurate date that has been traced but on occasions the mill may have been in use for cotton spinning for some years earlier. That was obviously true when a mill was for sale after the owners or tenants were bankrupt but only the sale notice is available.

The name given for each mill may have changed as some mills on the same site had several names. This came about when they were rebuilt or changed owners. Where possible all known names have been given. The main body of information for each mill relates to:

1. the size of the mill

2. the power source

3. the firms which occupied the mill during this period

4. when firms ran more than one mill

5. the type of spinning machinery used and the total spindle capacity at various dates

6. the mill's earliest insurance valuation

7. any significant new building on the site between 1780 and 1835

8. if and when cotton spinning ceased

9. changes to other textile use

10. if the earliest buildings can still be seen

CHAPTER 9

The Leeds Area

Aᴸᴛʜᴏᴜɢʜ the cotton industry was taken up with enthusiasm throughout the West Riding, some areas stood out as having particularly large investments in the new industry. One of those areas was Leeds where about a dozen cotton spinning mills were built or converted before 1800. Of particular significance in Leeds was the involvement of a group of wealthy merchants who built a number of large steam powered mills in the 1790s. Fortunately for them many of the early mills seemed remarkably adaptable in producing either wool, worsted or cotton yarn as comparative profitability shifted and ownership changed. This input of merchant capital resulted in a number of large mills being built although there were a number of smaller concerns in Leeds which relied on water power, or at times, horse power. Thompson & Naylor had a horse mill in Hunslet where they used mules and jennies before 1805.[1] Richard Shores insured all his cotton utensils, stock and machinery for the low sum of £70 in 1793,[2] while Hanah Brooks insured her cotton machinery and stock for £100 in 1808.[3] Somewhat larger was the firm of Brannands & Hinchliffe who insured their cotton machinery, stock and utensils in Nether Mills for £600 in 1792 and their utensils and stock in a counting house and shop nearby for a further £400.

Despite the early investment in mills, machinery and steam engines, the cotton industry declined rapidly in the Leeds area after about 1815. Some of the larger firms such as Beverley, Cross & Billiam and Markland, Cookson & Fawcett went out of cotton spinning by 1800.[4] By 1831 there were only eighty people engaged in the cotton industry in Leeds out of 8,674 workers in all textile industries.[5]

Besides cotton spinning there is also evidence of cotton weaving in Leeds and the surrounding districts. Oates & Keighley, who were merchants in Leeds, had eight pairs of cotton looms for sale in 1804. Thomson & Naylor were fustian manufacturers in Hunslet up to 1805 when they were also spinning their own yarn on jennies and mules.[6] William Gatcliffe was a cotton manufacturer in Woodhouse Lane before and after being in partnership as a mule spinner at Simpson Fold in 1807. Gatcliffe insured his stock for large amounts of money and may have been one of the more important cotton manufacturers in the area. In 1803 his insurance cover was:

	£
Stiffening Room	
Washing Room	300
Putting-out Room	
Drying Store Room	
Stock and utensils therein	1,200
Stock and Utensils in Twist	
Warehouse	1,000
	2,500[7]

The contribution of the cotton industry to the industrialisation of the Leeds area can be judged in terms of factory size and steam power. Several of the Leeds mills were large, particularly those where Boulton & Watt engines were installed. From the details of cotton mills in Leeds which

Drawing of front elevation of Bank Mill, Leeds, 1791 by John Sutcliffe.

follow, it can be seen that the majority of the early mills had steam power which was a complete contrast with other areas of Yorkshire at the time. The use of steam spread from Leeds to the rest of Yorkshire in the cotton industry and later in the woollen and worsted industries. Although cotton spinning developed in Leeds at the same time as in other parts of Yorkshire it was distinguished by the input of merchant capital and the use of steam power. However, the expected yarn sales to other parts of the country did not materialise and the local cotton manufacturing industry was not large enough to absorb the output from all the Leeds mills. After a few years the mills were changed to spin wool or worsted yarn and assisted the development of those industries by the transfer of capital assets.

The mills

Scotland Mills consisted of two small mills on Adel Beck which were used for cotton spinning and wool scribbling in 1792. The tenant at one of the mills was James Whitely and at the other George Baron. James Whiteley & Sons were dyers in 1801 and were wishing to find a subtenant for their cotton spinning section. They had four frames with seventy-two spindles each and used power from an overshot water wheel.

Scotland Mill was still partially used for cotton spinning in 1808 when some cotton spinning frames were offered for sale. The mill at that time had three water wheels – thirty-seven feet by five feet, twenty-one feet by four feet and twenty feet by two feet. No further references to cotton spinning at this mill have been found.

Waterloo Mill had belonged to John Storey but it was later occupied by his son-in-law Ard Walker, who was one of Storey's executors. The mill was water powered but had a 'fire-engine' for raising water back to the dam. In 1787 Walker and the other executors insured the mill and contents for the following sums:

	£
Oil and cotton mill	1,000
Utensils and stock	900
Fire engine	100
	2,000

The water wheel, which measured twenty-four feet by four feet was taken out in 1802 and a new steam engine made by Fenton, Murray & Wood was installed and the mill enlarged. By 1804 the cotton machinery was insured for £2,500 out of a total insurance cover of £5,000.

By 1815 the mill was being run by Richard, William and Ard Walker Junior for cotton spinning but they left in 1816 when the mill was leased to Thomas and

Benjamin Ingham for woollen manufacturing.

At Hillhouse Bank. This mill was built for cotton spinning in 1790 and was possibly run by J & J Whitaker & Co. That firm was a partnership between John Whitaker Senior, John Whitaker Junior, Jonas Whitaker and Richard Paley who owned the mill. Jonas Whitaker was a partner at a cotton mill in Burley-in-Wharfedale from 1792 and Richard Paley, although a soap boiler and iron merchant in Leeds, also had interests in other cotton mills. The partnership was dissolved in May 1795 and the lease of the mill was advertised for several months the following year. The mill was described as 'a large cotton mill . . . in which is a fire engine, twenty-two mules, ten water frames and other suitable machinery . . . Also a large and good warehouse adjoining to the above cotton mill.' Application to lease the mill had to be made to Richard Paley. Towards the end of 1796 a partnership was formed between Richard Paley, Thomas Wilkinson and John Holdforth to run the mill for cotton spinning. They insured the mill and contents for the following sums:

	Royal Exchange	Phoenix
	£	£
Cotton mill	500	500
Machinery	500	500
	1,000	1,000

Thomas Wilkinson had been a partner at Settle Bridge Cotton Mill until 1785 and was also a partner in another Leeds mill. Paley was a partner at Bowling Iron Works near Bradford and installed one of their steam engines in this mill. The engine infringed Boulton & Watt patents and in 1796 James Watt Junior visited the mill and later threatened legal action over the use of the 45 hp Bowling engine.

According to replies made to the West Riding magistrates in 1803 the mill employed about four hundred workers. Although that seems a large number the mill and contents were insured for £9,000 in 1804 which confirms the size of the enterprise. The mill was still being used for mule spinning in 1803 as the following advertisement shows:

> To Carders etc.
> Wanted, a person capable of undertaking the management of a card-room. A sober steady man who understands the business may have constant employment and good wages by applying to J. Hutton, at Mr Paley's mill, Bank, Leeds.
> A person accustomed to making up twist, and a few good mule spinners will meet suitable encouragement by applying as above.

Richard Paley was bankrupt by the end of 1803 and the creditors eventually forced the sale of this mill in 1804. The four storey mill measured one hundred and twenty-three feet by thirty-two feet within the walls and was powered by a steam engine with two boilers. The spinning machinery consisted of:

15 spinning frames with 848 spindles
20 mules × 192 spindles each

The mill was then run by the remaining partners trading as Wilkinson & Holdforth until Wilkinson died in 1807 after which time John Holdforth continued alone.

According to Crompton in 1811, the mill held thirty throstles with seventy-two spindles each, so the mules appear to have been removed. John Holdforth was followed by James Holdforth who added silk spinning in 1816 and built a new mill in 1822. The mill was still being used for spinning cotton on throstles in 1833.

At Hunslet. The firm of Halstead, Coupland & Wilkinson were distillers in Leeds up to 1779 when Halstead died. Wilkinson gained experience of cotton spinning in Settle and the two remaining partners started cotton spinning in Leeds in 1791. Their mill and contents had the following insurance cover in 1791:

	£
Cotton mill	500
Machinery	1,000
Utensils	1,000
Blacksmiths shop	100
Utensils in	100
Engine house and water wheel	150
Five cottages	150
	3,000

By 1801 a new Boulton & Watt steam engine had been added and judging by the increased insurance cover the mill had been enlarged. In 1803 the firm advertised for a card master, a picking master and a man accustomed to working a Boulton & Watt double powered patent steam engine. The position of manager of the carding department may not have been filled satisfactorily because Thomas Coupland & Sons were advertising for a man again in 1805. The person to be taken on then had to be a 'perfect master of the business'.

Coupland's cotton mill was five storeys high in 1807 and had a floor area of 1,807 square yards.

Although Crompton noted that Wilkinson & Co. had twenty throstles with seventy-two spindles each in 1811 the mill was also being used for mule spinning in 1818. Three years later Couplands were running two cotton mills in the complex of buildings. One was a four or five storey mill with a 40 hp engine. The other consisted of two interlinking buildings forty-three and a half feet by twenty-five and a half feet and fifty-three and a half feet by twenty-seven and a half feet which were equipped with gas lighting. These had the use of a 16 hp steam engine. All the buildings were for sale when the Couplands

were bankrupt but Richard Coupland continued using the larger building until at least 1830 when it was advertised for sale again. The five storey mill measured one hundred and twenty-three feet by twenty-nine feet within the walls and power was still provided by the 40 hp engine in a separate engine house. Also for sale were ' a quantity of throstles and other machinery therein well calculated for spinning cotton twist for exportation'. The mill was then taken by Dearlove & Fenton. The cotton mill was part of a complex of buildings including a paper mill and fulling mill. They were all for sale again in 1835. Cotton spinning was probably given up after the mill was sold.

At Hunslet Moor. This mill was run in 1792 by a partnership between John Beverley, a pawnbroker, John Cross, a cotton manufacturer and former partner of Arkwright, and John Billiam, a doctor. Their mill was eventually powered by a 22 hp Boulton & Watt engine with a twenty-five and three quarter inch by five feet cylinder. However, the mill originally had an atmospheric engine which was found to be most unsatisfactory. From letters to Boulton & Watt it appears that the mill was four storeys high and measured twenty-five yards by ten yards inside the walls. It was expected to hold thirty to forty spinning frames with eighty-four spindles each which would have given the mill a possible total of 3,360 spindles but the total only reached 1,800 in 1793. In 1795 the mill and contents were insured for the following sums:

	£
Cotton mill	600
Steam engine	800
Mill work	500
Machinery	1,000
Stock	600
	3,500

The amount of machinery in the mill increased later that year but the mill burnt down in 1796 and was not rebuilt.

At Far Bank Mills. Edward Markland, John Cookson and Joseph Fawcett, who were merchants in Leeds and Manchester, built this mill in 1792. The millwright was John Sutcliffe from Halifax who designed the mill and installed the water wheel and atmospheric engine for pumping water back to the dam. A 30 hp steam engine had to be ordered later that year from Boulton & Watt as the pumping engine did not perform as expected. The Boulton & Watt engine had a cylinder twenty-eight inches by six feet and was used to run both cotton and worsted spinning machinery. They probably made their own spinning machinery at the mill for they were advertising for mechanics and iron and wood turners at the mill in April 1792.

The mill was large and was further extended in 1796. the insurance cover was then:

	Sun £	Phoenix £
Mill and new building under one roof	1,500	1,500
Steam engine and boilers	750	750
Mill work	500	500
Machinery	1,750	1,750
Stock	400	400
Wool combing warehouse and stock	1,600	1,600
	6,500	6,500

When the mill was extended, the water wheel, which measured fourteen feet by seventeen feet, was offered for sale. Presumably this wheel had been installed in conjunction with the pumping engine but was no longer needed when the Boulton & Watt engine was installed. The firm expanded and advertised for wool combers, cotton pickers and cotton spinners.

Although it was said that the cotton workers would 'meet with constant employ', cotton spinning was given up in 1797.

At Far Bank Mills. This mill was possibly an extension of Far Bank Mills but with its own power and in the occupation of a different firm. Benjamin Gowland and Richard Clark, cotton manufacturers of Chapel Town in Leeds, ordered a 20 hp steam engine from Boulton & Watt in 1795. The engine had a twenty-four inch by five foot cylinder and was used to power their cotton spinning and wool scribbling machinery. Clark was replaced by James Bayne by 1797 when the cotton and woollen machinery was insured for £600 and their stock for £800.

There were then some other changes in the partnership. Up to 30th December 1797 the partners were Benjamin Gowland, James Bayne and Henry Cooper, but Bayne left at that time and the firm continued as Gowland & Cooper.

Their premises were to let in 1799 when they consisted of three rooms thirty yards one foot long by ten yards fourteen inches wide. There was also a ground floor which was shorter and contained the counting house. The machinery included:

16 frames – 1,024 spindles
10 carding machines

It is likely that Gowland & Cooper sold their cotton machinery about 1800 and then concentrated on the woollen side of their business.

Mabgate Mill or Sheepscar Mill. This mill was established by 1793 when Samuel Blagborough, John Holroyd and Joseph Holroyd insured the building and contents for the following sums:

	£
Cotton mill near Mabgate	1,800
Utensils, stock and goods	1,700

Steam engine	100
Warehouse, stables, offices	400
	4,000

The first steam engine was replaced with a 20 hp Boulton & watt engine in 1796. That engine had a twenty-three and three-quarter inch by five foot cylinder and cost £697 but was insured for £1,000 when it was installed. The old water wheel was sold at the same time.

By 1800 the insurance cover had risen to £6,600 with much of the increase being accounted for by the increased valuation of the machinery, steam engine and stock. Blagborough had left the firm by 1806 but Holroyds continued as cotton spinners and manufacturers until 1810 when the mill was leased to Crabtree & Green. In 1811 Crompton noted that this mill held twenty frames or throstles with seventy-two spindles each. Cotton spinning stopped in 1816 and the machinery then consisted of:

    28 water frames × 72 spindles
    4 water frames × 48 spindles
    2 throstles × 96 spindles
    1 throstle unfinished

Nether Mills. These mills were used for a variety of purposes – fulling, scribbling, dyeing and also cotton spinning. The partnership of James Brennand, Joseph Bowling and John Hinchliffe rented part of the mills in 1792 and insured their property for these sums:

	£
Utensils, stock and machinery	600
Utensils and stock in counting house and shop near	400
	1,000

The section of the mills occupied for cotton spinning consisted of three rooms, thirty-two yards by twelve yards, over the eastern end of the mill and driven by one of the five water wheels. Nether Mills were to let in 1799 when it was stated

that the cotton spinning section would work six hundred spindles. By 1814 there was no longer any cotton spinning at this mill but part was still used for cotton carding.

Steander Mills. Richard Lobley was a cotton and woollen manufacturer who moved from a section of Nether Mills in Leeds to this mill at Steander in 1803. The mill was steam powered and was used to produce yarn for Lobley's weaving shops in Kirkgate. Cotton spinning was discontinued in 1806.

At Simpson Fold. The partnership between William Musgrave and William Gatcliffe was in existence by 1803 when they traded as Musgrave & Co. and were advertising for workers.

> To Cotton Carders
>   Wanted immediately, a person that has been used to the modern way of carding cotton on the working roller engines. Liberal wages will be given to a sober man who answers to the above. A few mule spinners are also wanted.

In 1806 they insured their property as tenants in the mill in Simpson Fold, for the following sums:

	Sun	Royal Exchange
	£	£
Mill work	150	—
Machinery	150	2,200
Stock	600	500
	900	2,700

The four-storey mill measured eighty-three feet by thirty-four feet in 1807 when it was for sale. Power was supplied by a 10 hp steam engine and it was said to have been 'erected at very great expense' as the mill was constructed on 'fire-proof' construction principles. It had iron pillars, brick arches and iron window frames.

The mill, which adjoined an iron foundry, was probably sold in 1808 and then used for other purposes although William Gatcliffe was still described as a cotton manufacturer in 1811 when he lived in Woodhouse Lane.

Lowfold Mill, Hunslet. Cotton machinery was advertised for sale in 1811 at a cotton mill in the Hunslet area. It included four double spinning frames and the owner of the mill at the time was Joseph Hudson. This mill may have been taken by John Howard in 1823 and used for spinning cotton and woollen yarn. A new mill was added in that year and production continued until the 1830s. The buildings at Low Fold were for sale in 1832 and still included one which was being used for cotton spinning.

At Morley. This early mill was run by Samuel, John and Joseph Wetherill together with John Jubb and Allen Edmondson from at least 1785 when they insured the mill and contents for the following sums:

	£
Cotton mill and water wheel	300
Utensils and stock	500
Joiner's shop, Smith's shop, Counting house, Warehouse and Store House	100
Utensils and goods	100
	1,000

Wetherills & Co. were cotton manufacturers at Mitershaw and Churwell near Leeds. John Jubb was a millwright from Leeds and Allen Edmondson was later a partner with William Baynes at a cotton mill at Embsay near Skipton. No further references have been found and this mill was probably converted to wool scribbling and carding by 1805.

At Wortley. M & J Bateson ordered a 20 hp steam engine from Boulton & Watt

in 1796 for their cotton mill at Wortley. The engine was a double acting type with a twenty-three and three-quarter inch by five foot cylinder. The mill was only used for a short time, if at all, as a cotton mill but was then used as a woollen mill.

At Sandal near Wakefield. In 1804 a partner was wanted by the person running this mill, who may have been William Young. The mill held six mules with 1,320 spindles and was powered by a 10 hp steam engine. There was also a vacant room, seventy-two feet by thirty-four feet, which, it was suggested, could also be equipped with machinery. Any prospective partner had to bring £1,000 to £1,200 to the business which was in an area where plenty of hands were available. It is not clear how long cotton spinning was carried on at this mill as no further references have been found.

At Thorner. Henry Tarboton and Thomas Carr insured their cotton mill and contents at Thorner for these sums in 1794:

	£
Cotton mill and steam engine	200
Machinery	150
Moveable utensils	350
Stock	150
Utensils, stock and goods in Mr Tarboton's house	50
Utensils, stock and goods in Mr Carr's house	100
	1,000

The mill was powered by a steam engine which was not supplied by Boulton & Watt so presumably was of local manufacture.

No further references to this firm have been found but in 1818 a cotton mill still existed in Thorner so it may have been this mill under different ownership. It was insured then by George Thompson who had been established in the area as a fustian manufacturer since 1805 or before so he may have taken over this mill from Tarboton & Carr. George Thompson was possibly a relative of Benjamin Thompson who had been spinning cotton at a horse mill in Hunslet until 1805.

This mill and contents were insured for the following sums in 1818:

	£
Mill	800
Mill work	150
Machinery	1,300
Stock	100
Steam engine	350
	2,700

In 1831 the mill was for sale or to let as George Thompson wanted to give up the business. The spinning machinery included fourteen throstles with one hundred and sixty-eight spindles each. There were also some mules and machinery for spinning candlewick. Thompson did not manage to sell the mill so it was put up for auction in 1835. It was then three storeys high and measured seventy-one feet by thirty feet with a 16 hp steam engine. The mill was being used to spin 28, 30 and 32 hanks twist. In 1839 this mill was still used for cotton spinning, had a 14 hp steam engine and employed forty-six people.

CHAPTER 10

The Bradford Area

THE GROWTH of Bradford to be the 'Worstedopolis' of the nineteenth century has obscured the small but significant contribution made by the cotton industry to the industrialisation of the area. Two early incidents have often been referred to when desciblng the growth of the mechanised worsted industry in Bradford. One was the opposition to John Buckley's proposal to build a steam powered mill near Manchester Road in 1793, and the second was the use of the first spinning machines in the town by James Garnett in the Paper Hall in 1794. John Buckley was a cotton manufacturer so the mill he proposed building was most likely a cotton mill as it was a cotton mill he ran in Todmorden after being turned away from Bradford. Cotton machinery was also probably used in Bradford before 1794 as this advertisement of 1792 indicates:

To Be Sold
The engines for the working of cotton which were lately the property of Mr Thomas White of Brick Lane near Bradford deceased, consisting of one scribbling machine, one carding machine, one drawing and roving frame, with about ninety tin cans, a ratching frame and a reel all nearly new.
Apply to Mr John Jarrat of Bradford the owner.'[1]

Few mills for spinning cotton were built near the township of Bradford so this area covers a number of towns and villages near Bradford where cotton mills were built in the 1790s. It includes Bingley, Wilsden, Baildon and Birkenshaw as well as Bradford itself. However, cotton spinning never gained much foothold in the township of Bradford. Holme Mill and Rand's Mill were used for cotton spinning as well as worsted spinning for a few years after 1800 and Knight's Mill in Great Horton and Idle Cotton Mill were successful for several years but it was cotton weaving which had the major impact.

One of the partners who built Holme Mill was Nathanial Murgatroyd. He was a cotton and worsted manufacturer who gave evidence in 1806 to the Parliamentary Committee investigating the woollen industry.[2] At that time Murgatroyd employed fewer than ten worsted weavers but between thirty and forty cotton weavers. He felt that it was as difficult to weave cotton pieces as it was worsted, but weavers could, if they wished, change from one to the other. In previous years he had employed a hundred cotton weavers but had found that he could not make a profit so he had not given them any more yarn. The weavers did not wish to take worsted yarn and so had gone to other masters for cotton yarn to weave. The situation appeared to be that both masters and weavers in the area switched their efforts between cotton and worsted depending on the profit to be made.

By 1810 cotton weaving had become so well established in Bradford that a Colne merchant was wanting cotton pieces made in the area. The proposal was

Providence Mill in Bingley in the 1840s when it was used for worsted spinning.

that from one hundred to four hundred pieces per week would be required on a commission basis. Security was needed while the cloth was being made up, but the Colne merchant would pay the cost of transport.[3]

Cotton weaving thus, for a few years, became of great importance, particularly in the areas round Little and Great Horton and Allerton. Some of the local cotton manufacturers even extended their activities to other areas round Bradford and employed agents to 'put-out' yarn to weavers.[4]

The cotton manufacturers needed premises for warehousing and for sizing the yarn. Nathanial Murgatroyd, who has been mentioned above, was a cotton manufacturer with premises at the Tyrls in the centre of Bradford. There he sized and dried his yarn before it was given out to the weavers.[5] Another firm was Ingham & Fox who had premises at High Street and Barkerend Road in Bradford. On their bankruptcy in 1810, their assets included 1,600 pieces of super calico as well as large quantities of warp and weft. At their warehouse in Barkerend Road they had wind-

ing engines and warping mills, boilers for twist and paste as well as sizing bowls and drying rails.[6] A number of other Bradford cotton manufacturers were also rendered bankrupt at that time including five in the Horton area.

The greatest concentration of cotton mills in the Bradford area was around Wilsden, Culllingworth and Bingley. Most of the mills were built in the 1790s but four of them were still spinning cotton in 1835. Castlefield Mill, built just outside Bingley in 1791, was one of the largest in the area with 2,232 spindles in 1805.[7] Providence Mill followed in 1801 and was the first mill to be powered by steam in the parish. Two other cotton mills may also have been built in the town but no corroborating evidence has been found. They were said to be at the corner of Park Road and Main Street, run by a Mr Whiteley and another started by Messrs Gott, but turned into cottages in 1825.[8] Cottingley Mill was also said to have originally been a cotton mill.

The Harden Beck valley provided sites for several mills, some on the Wilsden side and some on the Bingley side. As there has been little further development in the valley it is still possible to see the sites and some of the mill buildings.

The Mills

Castlefield Mill, Bingley. The construction of this large cotton mill was started in 1790 when iron was first bought from Kirkstall Forge and the mill was finished in 1791. It was owned by a partnership between Johnson Atkinson Busfield, Joseph Wood, James Sidgwick and Wilmer Mackett Willett who were described as cotton merchants. They insured the mill and contents for the following sums in 1792:

	£
Water cotton mill and machinery	2,500
Utensils, stock and goods	2,500
	5,000

The partnership traded as Sidgwick, Wood & Co. but these two men left the firm in 1794 so the name was changed to Wilmer Mackett Willett & Co. Willett left Castlefield Mill by 1797 and soon afterwards bought Wreaks Mill in Hampsthwaite near Harrogate. William Busfield joined J A Busfield in 1797, by which time the insurance valuation had increased to £7,000 so perhaps the mill had been enlarged. Soon afterwards Peter Garforth Junior bought the mill but he died in 1804 and the mill was for sale in 1805. At that time the four storey mill measured forty yards by eleven yards and was powered by an eighteen foot by eighteen foot water wheel which was said to be quite new. As well as the normal preparing machinery the mill held thirty-one spinning frames with seventy-two spindles each.

Peter Garforth Junior had been one of the most important Yorkshire cotton spinners with mills at Skipton, Bell Busk and Sedbergh. On his death Castlefield Mill was bought by another of the large York-shire firms, Greenwood & Ellis who already owned several other mills.

In 1808 John Greenwood and Lister Ellis, cotton merchants of Keighley, insured this mill and contents for the following amounts:

	Sun £	Royal Exchange £	Phoenix £
Mill	900	600	500
Mill work	420	280	100
Machinery	1,440	960	1,000
Stock	240	160	400
	3,000	2,000	2,000

By 1811 the thirty-one spinning frames had been increased to thirty-six with a total of 2,598 spindles.

William Ellis took over sole control of Castlefield Mill in 1817 and continued cotton spinning there until after the end of this survey. By 1833 the mill was still water powered with a 40 hp wheel and at that time one hundred and thirty people worked there.

The complex of mill buildings is still there but with a number of later additions. As the downhill flow of the river Aire is not very prominent at Castlefield the fall to the wheel of nine and a half feet was achieved by digging a tail goit in a tunnel under the river to rejoin the main stream about half a mile away. The roof of the tunnel can still be seen where it crosses the river.

Providence Mill, Bingley was built in 1801 by three brothers, William, Charles and Thomas Hartley. It was the first steam powered mill in the parish. Thomas Hartley left the partnership in 1809. In 1811 Crompton noted that the mill held twenty throstles with eighty-four spindles each. The three storey mill was for sale in 1814 and changed over to worsted spinning.

At Bingley. A small mill at the corner

of Park Road and Main Street run by a Mr Whiteley.

Elmtree Mill was originally built for worsted spinning by Messrs Gott about 1811 but was later used for cotton spinning before being converted into cottages.

Cullingworth Mill was used briefly for cotton spinning from about 1830.

Hewenden Mill. This mill may have been built near an old corn mill about 1780 and used for both worsted and cotton spinning. It was owned by a Mrs Mary Morvill. Up to 1800 the mill was run by a partnership of William Nichols, Jonathan Barker the younger, John Barker and Mary Morvill who was the widow of Moses Morvill. When their partnership expired in 1800 they were described as cotton and worsted spinners and Jonathon Barker was a partner in a firm of cotton spinners at Upper Mill in Morton until 1808. However, it appears that from about 1800 Hewenden Mill was used solely for worsted spinning.

Eller Carr Mill, Cullingworth. According to a local writer, Eller Carr Mill was built by Joseph Harrison for cotton spinning but no date is given although he says that Harrison ran the mill for some years. The mill was probably built in the 1790s but Harrison, a worsted and cotton spinner, dealer and chapman was bankrupt in 1808. The mill was for sale in August 1808 when it was said to be 'lately erected'. It was three storeys high and measured thirteen and a half yards by nine and a half yards with a water wheel which measured twenty-seven feet six inches by two feet six inches. The machinery included:

3 cotton spinning frames – 64 spindles
3 cotton spinning frames – 48 spindles

2 worsted spinning frames – 48 spindles

After Harrison's bankruptcy, the mill was taken by Edward Craven and John Haggas of Oakworth Hall near Keighley and used for cotton spinning. They enlarged the mill and built cottages nearby. Unfortunately they were also bankrupt eight years later in 1816. John Greenwood of Keighley then took the mill in partnership with Edward Craven and they continued cotton spinning at Eller Carr until after 1835. Craven acted as architect and hydraulic engineer for the Greenwoods of Keighley at their various mills.

Bent Mill, Wilsden. According to replies made to the Factory Enquiry Commissioners in 1833 Bent Mill was built in 1799. The first tenants may have been John Knowles and John Smith up to 1809. The mill may then have been enlarged by William Wilkinson who was John Knowles' father-in-law. John Knowles, and later the firm of J W & W Knowles occupied this mill from 1809 until 1833 or later. In 1811 Crompton noted that John Knowles had twenty throstles with seventy-two spindles each, giving a total of 1,440 spindles.

According to a local writer, John Knowles also ran Hallas Bridge Mill which was only about two hundred yards away down the valley. He also wrote that:

'John Knowles made calicoes for the Manchester market, and his waggons went to and fro over Blackstone Edge, a three day's journey, constantly. He was, perhaps the largest maker of cotton goods in this part of the country, 'putting out' hand weaving at twenty different stations extending from Skipton to Cleckheaton. Hallas Mills were afterwards carried on for a short time by his son, John Wilkinson Knowles, in the cotton trade.'

The firm of John Knowles & Co. was

hit by the financial and trade crisis of 1826 and the assignees, Edward Marsland, cotton merchant of Manchester, and William Ellis, cotton spinner of Castlefield Mill in Bingley tried to sell the cotton machinery from Bent Mill that year. It consisted of:

22 throstles × 96 spindles

2 throstles × 100 spindles

8 mules × 216 spindles

12 mules × 210 spindles

16 carding engines

The firm had therefore been running 2,312 throstle spindles and 4,248 mule spindles which justifies the comments made above about the size of the firm.

The Knowles family must have survived their financial problems for they continued cotton spinning at Bent Mill into the 1830s although possibly as tenants. In 1833 they employed sixty-one people although they were reduced to throstle spinning only. The mill at that time had a steam engine and water wheel, both of which generated about 12 hp.

Bent Mill was later turned over to worsted spinning and the mill buildings still stand today.

Hallas Bridge Mill, Wilsden

was built some time before 1802 when the partnership which ran the mill was dissolved. The partners were Jonas Foster, Denholm, Joseph Foster, Wilsden and Abraham Foster of Denholm and they had been engaged in cotton twist spinning. Joseph Foster left the partnership and the other two men were bankrupt by 1805. However, the bankruptcy may have been overcome for Crompton noted that Fosters had twelve frames with fifty-two spindles each in 1811.

Soon afterwards this mill was taken by John Knowles from Bent Mill and in 1818 he was trying to sell the small 4 hp steam engine which had been made by Murray & Co. of Leeds. The mill, with twenty-four cottages, was to let in 1826 and still had the 4 hp engine. At that time the mill was three storeys high and measured twenty and a half yards by ten and a quarter yards. The water wheel was eighteen feet by five feet and the machinery included six pairs of mules. Hallas Bridge Mill was changed over to spin worsted by 1829 and has since been demolished.

Goit Stock Mill, Harden

was built by Benjamin Ferrand Esq for Timothy Horsfall who was a tanner until he turned to cotton spinning. Horsfall's partners were Richard Holdsworth from Otley and James Anderton from Bradford. The partners insured the mill and contents for the following amounts in 1792:

	£
Cotton mill	500
Utensils and machinery	1,500
Warehouse	30
Stock	470
	2,500

By 1797 Anderton had left the partnership but Thomas Colbeck from Keighley and John Horsfall from Bradford had joined the firm at Goit Stock. By 1811 Crompton noted that Horsfall & Co. had ten frames with ninety-six spindles each and ten with seventy-two spindles each. They were also cotton manufacturers and had a warehouse in Bridge Street, Bradford in 1816. Timothy Horsfall died in 1811 and his sons took over the mill but then established themselves as woolstaplers and worsted manufacturers in Bradford. By 1830 this mill had been taken by Thomas Sleddon & Co. for cotton spinning but it was then taken by James Upton, a cotton spinner with a mill near Sedbergh. Upton insured Goit Stock Mill for the following amounts in 1834:

	£
Goit Stock Mill with gas, water and steam power	500

Steam engine and water wheel	250
Mill work	250
Machinery	2,400
Stock	250
	3,650

Wilsden Mills, Wilsden were built next to a corn mill by Messrs Barraclough, Smith & Tetley, the corn mill being owned by Mr Barraclough senior. In January 1792 Messrs Tetley & Co. were advertising for a well qualified card master:

Wanted Immediately

A card master to a cotton factory. Any person well qualified for that business may meet with good encouragement by applying to Messrs Tetley & Co. of Wilsden in the Parish of Bradford.

None need apply but such as are perfectly master of their business, and can be well recommended.

It is not clear if the mill had just been built but that may have been the case. In 1803 the partners were John Smith, Jonas Tetley, William Briggs and Joseph Barraclough and they employed fifty-three people.

Crompton noted that Smith, Tetley & Co. had the following machinery in 1811:

12 mules × 252 spindles

4 mules × 264 spindles

6 throstles × 52 spindles

In 1816 John Smith was insane and the partnership was dissolved although it was not announced until 1822. By 1820 a worsted mill was built next to the original cotton mill which was then four storeys high and measured seventy-four feet by twenty-three feet within the walls. This mill was driven with a 10 hp steam engine and a water wheel. The machinery included eight mules with 1,812 spindles and twelve throstles with 1,200 spindles all made by F Sleddon of Preston. Besides the two mills there were two warehouses, a counting house and five cottages. It is possible that cotton spinning was given up after 1822 and both mills were then used for worsted spinning.

Old Mill, Wilsden. George Tweedy was spinning cotton at this mill from about 1793. In 1810 he was joined by his son of the same name and John Anderson with his son Christopher. They then increased the amount of machinery at the mill and Crompton noted that they had sixteen frames with forty-eight spindles each in 1811. Shortly afterwards the partners changed the mill over to worsted spinning.

Over Mill, Hawksworth. Land for two cotton mills was for sale at Hawksworth in 1791. It was said that a stream ran through the land which would give a fall of sixteen feet to the wheels at the two mills. Two mills were built, possibly with both being used for cotton spinning but few details are available. One changed over to worsted spinning about 1812 but Thomas & James Holroyd were still spinning cotton at the other mill up to 1835.

Low Mill, Hawksworth. This mill, or the one above, may have been occupied by William Mawson.

Clifton Mill, Baildon. The same advertisement for the land above mentioned that there was a cotton mill driven by the same stream further down the valley. This may have been another name for Gill Mill which was used as a woollen mill about 1778 but had several changes of use and ownership.

Idle Cotton Mill. Two mills were built on the site of a corn mill by John Cromack. One of the mills was used for wool scribbling and carding but the other was used for cotton spinning. The partnership running both mills consisted of Thomas Denbigh, Christopher Edmondson, Robert Stansfield and Jasper Robson who traded as Robson, Edmondson & Co. This firm employed twenty-seven people in the mill in 1803. In 1806 they insured the contents of the five storey mill for the following sums:

	£
Mill work	300
Machinery	1,500
Stock	200
	2,000

John Cromack tried to sell the two mills in 1809. The cotton mill then measured thirty-six feet by twenty-seven feet within the walls and the two mills were driven by a 25 hp steam engine. Much of the cotton yarn produced at this mill was used locally by the large number of hand loom weavers in Idle.

Robert Stansfield died in 1810 but Crompton noted that Robson & Co. had eighteen throstles with eighty-four spindles each in 1811. This mill continued to be used for cotton spinning for a few more years. Later tenants included Isaac Barrow and David Green of Leeds and later, William Lister and Samuel Brook. Cotton spinning finished in 1826 and the mill was converted to woollen manufacturing.

Knight's Mill, Great Horton, Bradford. This steam powered mill was built by John Knight & Co. in 1806. John Knight was a local cotton manufacturer on a large scale with agents in several areas of Bradford who distributed yarn and collected the woven pieces to be taken to Manchester. This mill gave him his own supply of yarn and in 1811 he had twenty mules, each with two hundred and sixteen spindles. Besides selling cloth in Manchester John and Benjamin Knight linked with Edward Scholefield of Watling Street,

London, as merchants and commission agents.

Knights continued spinning cotton at this mill and also at a mill in Morton until their bankruptcy in 1827. At that time the four storey mill measured thirty yards by eleven and a half yards and was driven by a 15 hp steam engine. The machinery in the mill included:

18 throstles with 108 spindles each
1 pair of mules with 19 dozen spindles
1 pair of mules with 20 dozen spindles

Alongside the mill there was a five storey warehouse which measured seventeen and a half yards by eleven and a half yards and a large number of cottages.

This mill was converted to spin worsted following Knight's bankruptcy.

At Bradford. Thomas White was spinning cotton at a mill in Bradford prior to 1792 when his machinery was for sale.

Holme Mill, Bradford. Holme Mill, according to local historians of the last century, was built as a steam powered worsted mill by Henry Ramsbotham, who was a worsted manufacturer, and it was the first steam powered mill in Bradford. Prior to the building of this mill in 1800 Ramsbotham had spun his yarn at a horse mill in the centre of Bradford. He had been in partnership with Nathanial Murgatroyd and George Walker but Murgatroyd left the partnership in 1801. Murgatroyd was a cotton manufacturer, which was a flourishing trade in the Bradford area at that time. The demand for cotton yarn by the local domestic cotton manufacturers may explain that possibly from 1803 until 1807 Holme Mill was also used for cotton spinning.

Henry Ramsbotham was joined in the partnership by some of the Swaine family

who were bankers, merchants and manufacturers in Halifax and elsewhere. However, that might have been after the mill
burnt down and was rebuilt in 1803.

When Holme Mill was for sale in 1807
as a result of the bankruptcy it was four
storeys high and measured thirty-two and
a half yards by eleven yards. The machinery appears to have been worsted and
cotton in the ratio 2:1 for there were
eighteen worsted spinning frames with
1,044 spindles and nine cotton throstles.
It is likely however, that the mill was taken
over completely for worsted spinning after
1807.

Rand's Mill, Bradford. This mill was
built by John Rand in 1803 for spinning
worsted and cotton yarn. Despite the rise
of Bradford as the centre of the worsted
industry some of the local manufacturers
and spinners such as John Rand were also
involved in the cotton trade. It is doubtful
however, if the cotton side of his activities
was very extensive or carried on after
about 1815. This was a steam powered
mill and was driven by a 10 hp Fenton,
Murray & Wood engine until 1813.

At Birstall. Another mill which was partially used for cotton spinning was at Little
Town near Birstall. From 1811 to 1817 a
room measuring twenty-five and a half
yards by twelve and a half yards had been
used for cotton spinning. Power was provided by the steam engine at the mill.
Cotton spinning possibly stopped in 1817
when the room and power were to let
and a woollen manufacturer was wanted
as a tenant.

Clough Mill, Birstall. This mill at
Clough in Hightown near Liversedge was
used for wire drawing and cotton spinning
from 1824 until 1833 or later. The mill
had a 10 hp steam engine, employed

twelve people and was occupied by William Emmett and others in 1833.

Gomersal Hall Mill. Gomersal Cloth
Hall was built as a place where local clothiers could exhibit and sell their woollen
cloth. Unfortunately it did not prove
popular and was therefore sold for manufacturing purposes. The layout of the
building is explained in the following
advertisement and is relevant to its later
use as a cotton spinning mill:

LEEDS MERCURY
14th March 1786

[To Woollen and Cotton]
[Manufacturers]
 [To be Lett]
Gomersall Hall well adapted for a
manufactory. A large strong building
consisting of two wings and a place
called the Headland which joins the
wings together. The wings contain
four rooms 62 yards long each 20½
feet wide and 14 feet high. The
Headland is 61 yards long 10½ feet
wide and 14 feet high with a chamber
over it of the same size. There is
excellent water all under the building
at 12 feet deep and coal within half
a mile at 2s. 6d. per ton at the pit.
The building stands in a populous
country and near the centre of the
following market towns viz Leeds
eight miles, Wakefield nine,
Huddersfield eight, Halifax eight and
Bradford six.

The hall had been built on land belonging to Sir James Ibbotson and later reverted
to his possession. Thomas Carr was Ibbotson's agent and after trying unsuccessfully
to sell the building in 1789 took a lease
on the property himself. The exact date
when Thomas Carr first installed cotton
spinning machinery in the hall is not clear
but was probably about 1788. Carr took

as his partner Richard Paley, a soap boiler and cotton spinner from Leeds, who had an interest in many cotton mills and other enterprises in Yorkshire and Lancashire. Their partnership lasted until 1796. However, in 1795 Thomas Carr, who was then described as a cotton manufacturer, insured the hall and other property for the following sums:

	£
Dwelling house and offices	100
Furniture	100
Cottage	25
Drying room for woollen cloth	50
Stable and barn	50
Utensils and stock in above	100
Cotton mill north end of building	200
Cotton machinery	700
Stock	300
Scribbling mill – East Wing	50
5 Cottages – rest of East Wing	50
	1,725

By 1796 the insured value of the cotton machinery had been increased to £1,500 which indicates considerable expansion.

The power for driving the machinery was not described in the early insurance details but in 1801, when the cotton machinery was insured for £2,500, a steam engine was also entered with cover for £500. The entire range of buildings and their contents were then insured for a total of £5,330.

Thomas Carr was bankrupt at the end of 1803 and his personal effects as well as his cotton machinery were for sale. Besides the usual preparing machinery he had twelve mules and sixteen throstles. The machinery was driven by a 16 hp steam engine and one third part of a 24 hp engine. A firm trading as Crowther & Hirst then appear to have taken the mill after Carr's bankruptcy. They were probably tenants of the Swaines of Halifax to whom the mill was sold. In 1806 the mill was for sale again when it was said to contain 2,000 throstle spindles and 2,000 mule spindles driven by a 17 hp steam engine. Enquiries had to be made to Mr Ramsbotham who was a partner of the Swaine brothers.

Swaines' Commercial Bank collapsed in 1807 and so their various manufacturing interests had to be sold. Gomersal Hall Cotton Mill was one of them so the machinery was offered for sale. The mill was still for sale in 1808 by the assignees but it appears that two of the brothers, Edward and Joseph Swaine, later managed to retain control of the mill and convert it to a woollen mill. All the evidence suggests that Thomas Carr gave up his interest in this mill in 1803 but he may have been retained as manager for in 1811 Crompton noted that 'Carr, Gomersal' had the following machinery for cotton spinning:

20 throstles × 96 spindles = 1,920 spindles

16 mules × 252 spindles = 4,032 spindles

The change-over to wool must have come shortly afterwards.

Heckmondwyke Cotton Mill. Half of this steam powered mill in Heckmondwyke was being used for cotton spinning prior to 1802 when the cotton machinery was for sale. The machinery consisted of cards, drawing frames, roving frames and nine mules. It was said that 'The purchaser may be accommodated with all or part of the machinery, also with the place to work them'. The rest of the three-storey mill was used for wool scribbling but could be converted to cotton spinning. The mill was not running in 1805 although still described as a cotton mill. It is likely that it was then changed over completely to woollen manufacturing.

Birkenshaw Mill. This was originally a wool scribbling mill which was occupied by John Ellison and William Tetley in 1793. However, by 1805 the mill was occupied by John Rangely of Leeds and George Tetley of Birkenshaw with the help of a mortgage from Elizabeth Marrow of Chapel Allerton, Leeds. The mill, then used for cotton spinning, was insured with the contents for the following sums in 1805:

	£
Cotton mill and warehouse	350
Mill work	100
Machinery	750
Stock	100
	1,300
Weaving shop	150
Tenement and warehouse	100
Stock in warehouse	200
	450

By 1806 a steam engine had been added and the mill was enlarged with the machinery more than doubling in insurance valuation. One section burnt down in 1807 and the machinery was destroyed.

Rangely & Tetley were cotton manu-facturers as well as spinners and sold much of their cloth from their warehouse at 24 Cannon Street in Manchester. In 1811 part of the new mill had been rebuilt. One section contained sixteen throstles with seventy-two spindles each. The other section had fourteen mules with 2,952 spindles. Eight pairs of those mules were for sale when Rangely & Tetley were selling some of their cotton machinery at the end of 1811. The whole mill was for sale in that year when several other firms rented sections. A man called Abraham Sharp rented part of the mill from about 1816 to 1826. Rangely & Tetley had cut down their activities and were bankrupt by the beginning of 1819. Abraham Sharp may then have taken over more of the mill but he left the mill in 1826 whereupon it was left empty for six years before being converted to spin worsted. In 1830 one mill was five storeys, forty yards by twelve yards and the other was two storeys, forty-five yards by ten yards. Power was supplied by a 40 hp steam engine.

The Huddersfield Area

HUDDERSFIELD's early connection with the cotton trade is recorded in its coat of arms where a ram holds a sprig of cotton in its mouth. Few cotton mills were built in the town however. Most were built to the south and south-west in the Colne and Holme valleys. The majority of mills were in Slaithwaite and Marsden and built in a twenty year period from about 1790. In 1803, when Waterside Mill was rebuilt after a fire, it was said that 'the spirit of erecting cotton mills being high at the time' it was to be rebuilt as a cotton mill. This mill was one of the few mills which was still being used for cotton spinning at the end of this survey. All the rest of the mills in Marsden and Slaithwaite, apart from Shaw Carr Wood Mill and Holme Mill, had been converted to woollen mills by 1815. Many had been converted even earlier so their use as cotton mills was often very short.

Two mills where cotton spinning was retained were at Meltham and Colne Bridge. Meltham Mill in Almondbury Parish had been built as a woollen mill but was converted to spin cotton in 1805 by the successful firm of Jonas Brook. Brook was joined by his brothers and they started specialising in producing cotton warps and sewing cotton by 1822. They went on to become one of the largest cotton spinning firms in Yorkshire and by 1833 had three steam engines to work their mills.[1]

Colne Bridge Mill in Kirkheaton Parish was one of several Yorkshire mills started by Leeds merchants. In this case the firm of Rawstorne & Co. occupied this mill which, together with the stock, had an insured value of £5,000 in 1795.[2] There were several changes of ownership up to 1835 and also a possible lack of investment as this mill was still dependent on water power and jennies were still being used alongside the spinning frames in 1833.[3]

By the mid-1830s a number of mills in the Huddersfield area were being used for cotton spinning alongside other activities. Two were combined with silk and two with wool processing. Another mill was being used for manufacturing fancy goods with a mixture of cotton and worsted. This was the start of the movement towards light mixed fabrics which was to give the Yorkshire cotton spinners a new market during the rest of the century.

The Mills

The Factory, Marsden. This mill was run by John Haigh from 1795, or before, until his bankruptcy in 1806. He, and his brothers, were cotton spinners, manufacturers and merchants in London and Manchester and they owned several mills in Marsden where they used pauper children from London. In 1795 this mill was insured for the following amounts:

	£
Cotton factory	1,000
Machinery	1,500
Stock	120
	2,620

In 1808 the mill was four storeys high plus a garret and measured ninety feet by thirty feet. Within the premises were a counting house and a weaving shop with a winding house adjoining. The 24 hp steam engine drove both this mill and the adjacent Upper End Mill. In 1808 J Haigh & Co. had the following spinning machinery and looms in their various mills in Marsden:

    32 mules × 252 spindles

    2 mules × 216 spindles

    20 throstles × 120 to 128 spindles

    12 twisting jennies 76 to 87 spindles each

    34 pairs of looms

No further references to cotton spinning at this mill have been found after 1808.

Upper End Mill, Marsden. This mill was leased by John Haigh until 1806 when he and his brothers were bankrupt. The mill may have been built about 1792. In 1806 it was four storeys high with the following dimensions:

Ground floor	70ft × 30ft
Second floor	108ft × 30ft
Third and fourth floors	129ft × 30ft

The water wheel measured sixteen feet by twelve feet and had a twelve foot fall of water to it. There was also a nearly new steam engine with a twenty-seven and a half inch cylinder.

Upper End Mill was for sale again in 1808 together with the other mills leased by John Haigh – New Mill, Franks Mill and The Factory. Some time after 1808 Upper End Mill was converted into a woollen mill.

New Mill, Marsden. John Haigh of 7 Old Jewry, London, a cotton spinner, manufacturer and merchant, converted the old corn mill next to his other mills in Marsden into a cotton mill in 1796. It was insured in that year for the following amounts:

	£
Cotton mill (Old corn mill)	1,000
Mill work	600
Machinery	1,000
Stock	200
	2,800

In 1806 this mill was for sale following John Haigh's bankruptcy. The mill was three storeys high plus a garret and measured sixty-four feet by thirty-three feet. Powering the mill was a sixteen foot by twelve foot water wheel operating from a fourteen foot fall. New Mill was for sale in 1821 by which time it had a new water wheel measuring sixteen feet by eleven feet and still being used for cotton spinning in 1830.

Franks Mill, Marsden. This mill was being used for cotton spinning in 1804 when John Haigh registered it with the West Riding magistrates. Following Haigh's bankruptcy in 1806 the mill was offered for sale in 1808. At that time the three storey mill measured thirty-five feet by twenty-three feet six inches within the walls and was powered by a thirty foot by two foot six inch water wheel. It is possible that this mill was bought and converted to wool processing about this time.

Crow Hill Mill, Marsden. John, Thomas and Samuel Haigh insured this cotton mill and its contents for the following sums in 1795:

	£
Crow Hill Mill	500
Mill work	100
Machinery	500
Stock	50
	1,150

Haigh Brothers were cotton spinners, manufacturers and merchants with other mills in Marsden. It is not clear what happened to this mill after 1806 when the Haigh brothers were bankrupt but it was for sale in 1810. At that time it was used for mule spinning and held five mules with two hundred and fifty-two spindles each. There were also five jennies and other cotton machinery for sale at the mill. It was suggested that the mill could be used for dressing woollen cloth so cotton spinning may have stopped in 1810.

Smithy Holme Mill, Marsden. This mill was built about 1795 and occupied by Haigh Brothers who also occupied other mills in Marsden. In 1795 they insured this mill and contents for the following sums:

	£
Smithy Holme Mill plus smith's shop adjoining	1,000
Mill work	300
Machinery	1,000
Stock	200
	2,500

John, Thomas and Samuel Haigh were cotton manufacturers and merchants with weaving shops near Ashton in Lancashire and a warehouse in Manchester. No further references have been found to cotton spinning at this mill and Haigh Brothers did not appear to occupy it in 1806 when they were bankrupt.

Lingards Wood Bottom Mill, Marsden. This mill was occupied as a wool scribbling, carding and fulling mill by Daniel Haigh in 1797. However, by 1806, part of the mill was being used for cotton spinning on six throstles. This section of the mill, or possibly all of it, was occupied by John Haigh, who was bankrupt and therefore the mill was for sale. It is doubtful if cotton spinning then continued.

At Marsden. This mill was probably built by George Woodhead of Hollin Edge near Elland. Woodhead was a cotton manufacturer and merchant and he insured this mill and contents in 1793 for the following amounts:

	Sun £	Royal Exchange £
Mill	300	700
Utensils, stock and goods	1,400	1,400
	1,700	2,100

By 1796 a steam engine had been added which had an insurance value of £400 with £100 for the engine house. George Woodhead died in 1799 and the mill was then offered for sale together with the steam engine and machinery. Joseph Cartledge of Elland, who was also involved in cotton spinning, was Woodhead's executor and continued to run this mill until it was again offered for sale in 1802. The three storey mill had a garret and measured ninety-two feet by twenty-eight feet within the walls. There were two dams which provided water for the twenty-five foot wheel and also the steam engine. Besides the mill there were workshops, a counting house and ten cottages. The mill was used for spinning cotton twist and held fourteen frames with six hundred and seventy-two spindles as well as the necessary preparing machinery.

The mill was on the turnpike road from Huddersfield to Manchester on a site which adjoined the Huddersfield Canal and was said to be 'in the centre of the cotton trade and in a good neighbourhood for the supply of hands'.

The mill was for sale again in 1808 and may have then been converted into a woollen mill.

Jumble or Cloughlee Mill, Marsden. This mill was owned by two gentlemen from Marsden, John Gartside and John Parkin and may have first been used for cotton spinning in the 1790s when Lawrence Stansfield occupied the mill for cotton spinning. They insured the mill for £600 in 1805 but their next tenant,

William Heywood, was bankrupt in the same year. The mill finally came onto the market in 1807 together with the dams and warehouse. The 'patent steam engine' which was 'new erected' had a sixteen and a quarter inch cylinder although there was also a water wheel at the mill. Heywood's cotton spinning machinery included:

6 mules × 300 spindles
4 mules × 240 spindles

The mill was for sale or to let again in 1810 but no further references have been found for cotton spinning at this mill.

At Marsden. According to returns made to the West Riding magistrates in 1803 Thomas Gill occupied a small cotton mill in this area. It may have been a temporary conversion of one of the small fulling mills on the river Colne. Thomas Gill was in partnership with William Ashton as cotton spinners but this partnership was dissolved in 1805.

At Marsden. This mill in Marsden may have been a new wool scribbling mill which was for sale because of Francis Davenport's bankruptcy in 1795. It was said that the mill could be used for cotton spinning but it is not certain when cotton spinning commenced. By 1805 William Davenport was advertising for large families where the children would be employed in spinning and weaving. However, the scale of his activities at this mill is not known and no further references have been found. William Davenport was also a publican at the Red Lion so cotton spinning may have been a short lived venture.

Holme Mill, Marsden. Scholes, Varley & Co. were carding and spinning cotton at this mill from 1812 until after 1835. The mill may have been one of the earlier cotton mills which was taken over by this firm. The water powered mill derived 15

hp from the river Colne and employed seventy-three people for mule spinning in 1833.

Steps Mill, Honley. Several mill buildings were constructed at different times on this site. There were two separate fulling mills by 1788 but by 1803, or earlier, one building was being used for cotton spinning. William Bailey & Co. was the firm which occupied Upper Steps Mill until 6th March 1803. The partners were:

William Bailey, Senior – Batty Mill, parish of Kirkheaton

Richard Brown – Hill House, parish of Huddersfield

William Bailey, Junior – Honley

William Bailey Junior left the partnership in 1803, while the other two partners were selling their cotton machinery in 1805. They still owned the mill and the purchaser of the machinery could rent the mill if he wished. The partners had been involved in mule spinning for they had ten mules to sell, together with their other machinery. Further references do not indicate that cotton spinning continued after 1805.

Crossland Factory, South Crossland. This mill was built as a woollen mill in 1792 but part or all was converted to cotton spinning for a few years between about 1793 and 1812. In 1805, George and Walter Beaumont, who had previously been woollen manufacturers, insured the mill, machinery and stock for the following sums:

	£
Five storey cotton mill.	
1,760 sq yds	733
Mill work	133
Machinery	1,334
Stock	100
	2,300

The Beaumonts appear to have continued

with the woollen business in some part of the mill between 1805 and 1812, for when Walter Beaumont was rendered bankrupt in 1811 he was described as a woollen and cotton manufacturer. There were also three wool scribblers and three wool cards in the mill.

By 1811 the mill complex was extensive. One mill was four storeys high plus an attic. It measured thirty-two yards by eleven and a half yards and was powered by a thirty-four foot by four foot water wheel. A second four storey mill measured seventy-six feet by seventeen feet and had a water wheel fifteen feet by three feet. There was a three storey building used for storing bales of cotton, a further building with a packing shop, a counting house, overlooker's house and store room with a three storey workshop at the back. There was also a weaving shop, two rooms adjoining for depositing weft and twist, a mechanic's shop, smith's shop and dyehouse. At the same time nineteen cottages occupied by work people, a barn, stable, cowhouse, garden and a mansion, Crossland Hall, were also for sale.

The cotton spinning machinery consisted of:

6 water frames × 144 spindles
18 mules × 252 spindles
6 twist frames × 144 spindles

The mill and other premises had belonged to Richard Henry Beaumont of Whitely Hall who had recently died. In 1812 the mill reverted to woollen manufacturing.

Shaw Carr Wood Mill, Slaithwaite.
According to the Dartmouth Terrier of 1805 this mill was built in 1787 by John Varley. In 1793 J & F Varley offered the mill to let and indicated that it could be used for spinning cotton or worsted. In 1803 it was one of the two cotton mills in the area being run by Thomas Varley.

In 1811 Crompton noted that Varley had twenty-four mules with two hundred and sixteen spindles each.

The Varley family continued cotton spinning at this mill until at least 1833 when Dyson & Varley made returns to the Factories Enquiry Commissioners. By that time they employed eighty-five people and worked the mill with water and steam power. Water from the river Colne provided 20 hp and the steam engine 12 hp.

Black Moor Holme Mill, Slaithwaite.
This mill was built by William Townend of Pontefract and John Varley of Lingards in 1796 and used as a cotton spinning and wool scribbling mill. In 1800 Townend & Varley insured the mill for £200 and their tenant, Thomas Haigh, insured his cotton machinery in the top part of the mill for £350 and his stock for £50.

In 1811 Crompton noted that Thomas Varley had five mules with two hundred and sixteen spindles each and two throstles with one hundred and twenty spindles each in a mill which may have been this one. No further references have been found so the mill may have been taken over completely for woollen manufacturing about 1815.

Phoenix Mill or Waterside Mill, Slaithwaite. This cotton mill was built on the site of a previous mill which was burnt down in 1802. The previous mill had not been used for cotton but this mill was built specifically for cotton spinning on land which belonged to Lord Dartmouth.

In 1804 the three partners, Edmund Eastwood, John Varley and John Schofield insured their machinery and stock in this mill for the following sums:

	Sun	Royal Exchange
	£	£
Machinery	1,000	?
Stock	300	?
	1,300	1,800

The mill at that time was four storeys high excluding the basement and attic, and had 1,152 square yards of floor area.

The firm traded as Eastwood & Co. but the partnership was dissolved on the 3rd April 1804. John Schofield, a cotton manufacturer, took over part of the mill separately until his bankruptcy in 1810 while George Scholes from Manchester joined the remaining partners, Eastwood and Varley.

In 1833 Scholes & Varley replied to the Factory Enquiry Commissioners stating that the mill had been used for cotton spinning since it was built, that they used water power from the river Colne and employed one hundred and seventeen persons.

Meltham Mill, Meltham. Meltham Mill was built between 1791 and 1793 as a woollen mill. It was converted for cotton spinning in 1805. Jonas Brook, who built and ran the mill, evidently decided that cotton spinning would be more profitable than woollen manufacturing. Brook was joined by his brothers, James, William and Joseph and they built up one of the largest cotton, and later silk firms, in the area. One mishap was a fire in October 1827 but the mills were insured with the Leeds Fire Office.

The firm was making sewing cotton and cotton warps by 1822 and had added silk thread by 1833. The yarn was spun on mules and throstles, powered by steam engines with 88 hp and water wheels with 30 hp. In 1835 the mills employed six hundred and fifty people. The firm of Jonas Brook & Brothers went on to exhibit cotton thread at the Great Exhibition of 1851.

At New Street, Huddersfield. This large five storey mill was built about 1800. It was the second steam powered mill in the area and had a 'patent' steam engine with a nineteen inch cylinder. It was occupied by Samuel and John Sutcliffe Senior for cotton spinning. In 1811 Crompton noted that their machinery included:

4 mules × 300 spindles
6 throstles × 120 spindles

In 1815 the five storey mill, including the basement and attic, measured thirty-three feet by thirty-two feet ten inches within the walls. There was also a new card room, packing rooms and counting house. By then the spinning machinery had increased to:

6 mules × 252 spindles
2 mules × 216 spindles
11 small throstles with 436 spindles between them

This mill was possibly occupied during the 1820s by Joshua Lockwood & Co. but was burnt down in 1828 and was not covered with insurance.

At Huddersfield. The firm of Beresford & Whitmore ran a mill in Huddersfield in 1833 where eight-eight people were employed in cotton spinning.

Colne Bridge Mill, Colne Bridge. Rawstorne & Co, who were Leeds merchants, were occupying this mill for cotton spinning from 1793 or before. The mill and contents were insured for the following sums in that year:

	Sun	Phoenix
	£	£
Water cotton mill	750	750
Utensils, goods and machinery	1,250	1,250
	2,000	2,000

The insurance valuations increased

slightly in 1795 but were drastically reduced in 1798 to £750 with both the Sun and Phoenix fire offices. However they increased again when a new mill was built in 1800. The original partnership between Thomas Hewetson and Atherton, Richard and James Rawstorne was dissolved in February 1800 as Hewetson had died. The firm was then continued by Atherton Rawstorne.

Colne Bridge Mill was one of the largest cotton spinning mills in the area according to the insurance records. In 1803 the mills and their contents were insured for £8,000. The old mill was five storeys high and the new mill four storeys with gas added in 1814.

By 1818 the mill had been taken by Thomas Atkinson but was still used for cotton spinning. On February 14th that year there was a fire at the mill and seventeen girls aged from nine to eighteen, who were working on the night shift, were killed.

The mill, which had been burnt down, was rebuilt and the lease for both mills taken by Thomas Haigh. Sir Thomas Pilkington owned both mills but they were later bought by the Rev George Augustus Dawson of Suffolk. Thomas Haigh & Sons occupied the mills for cotton warp manufacturing until after the end of this survey. In 1833 they were engaged in cotton carding, spinning and doubling with frames and jennies. They employed one hundred and sixty-two people.

Sheppard's Factory, Ossett Street Side, Dewsbury. Two rooms were let in Sheppard's Factory in 1804 for cotton spinning, together with one third of the power of the 20 hp steam engine and some out buildings. The rooms measured eighteen yards by ten yards and were 'furnished with new machinery for cotton weft and twist spinning'. Application had to be made to Messrs Sheppard & Wild of Making Place near Halifax. The cotton spinning machinery consisted of:

2 mules with 3 sets of rollers × 216 spindles
4 mules with 2 sets of rollers × 228 spindles
1 throstle × 120 spindles
2 water frames

The next firm to occupy the mill was probably Taylor & Co. with the following partners:

John Taylor, Purlewell Hall, Batley
William Hanson, Street Side, Dewsbury
Joseph Brook, Gawthorpe, Dewsbury

They traded as clothiers, scribbling millers and cotton spinners. After Taylor's death about 1808 the other two partners were to continue the business at this mill. William Hanson was later joined by Robert Blakey when they were woollen and cotton manufacturers, but this partnership was dissolved on October 20th 1814. William Hanson & Sons continued cotton spinning at Ossett Street Side until after 1835 when the mill had a 36 hp steam engine and employed eighty people.

CHAPTER 12

The Halifax Area

HALIFAX PARISH was very large and at its western border adjoined Lancashire. It was from the Lancashire boundary that the cotton industry spread into Halifax Parish and it was the close geographic connection with Lancashire which sustained the growth of the industry and ensured its firm establishment during the nineteenth century. The upper reaches of the river Calder and its various tributaries had provided power and sites for over a hundred cotton mills by 1835 although not all survived until then. The upper Ryburn Valley, for instance, was one of the few areas in Yorkshire where the cotton industry took root to such an extent that eventually it replaced the traditional woollen and worsted industries.[1] The poor soil conditions and high rainfall meant that small holders needed other sources of income. The 'putting-out' of cotton warp and weft spread over from Lancashire in the 1780s with the cotton pieces delivered to Manchester every Thursday. Robert, John and James Lees of Oldham, for example, used a warehouse in Ripponden to 'put-out' warp and weft to weavers in the local villages. Cotton weaving was first introduced into the Todmorden area when warp and weft were brought over from Lancashire. The growth of this weaving trade prompted local men to build small water powered spinning mills at any point where there was a useful fall of water. Higgin Chamber Mill in Sowerby was built by 1788 and occupied by Elias Fletcher. He spun his cotton on mules which were to be the common type of

spinning machine in use in the area. Many of the men who built or leased the mills in Calderdale, such as Elias Fletcher started as 'putters-out' and moved into cotton spinning to secure their own supplies of cotton yarn.

Trade with Lancashire was facilitated by the turnpike roads with the first Turnpike Act for Yorkshire bringing about improvements to the road from Ripponden to Littleborough and Rochdale. This was completed in 1772 while the second link through Rishworth to Oldham and Manchester was opened in 1803.[2] Further transport links were improved when the Rochdale Canal was completed in 1802.

By 1800 the best water power sites in the Calder Valley were already occupied by converted fulling mills or the early cotton mills. Later developments were therefore scattered along the more remote or less satisfactory sites before the widespread use of steam power. Some mills were even built at over one thousand feet above sea level. However the use of side valleys meant that the fall to the wheel was shorter, a mill could have sole use of the water and it was easier to build small dams.[3]

One of the first and most successful ventures was Joshua Fielden's workshop at Laneside in Langfield.[4] He was a cotton manufacturer who adapted three cottages into a workshop and later added powered carding and other machinery. Fielden Brothers went on to establish one of the most successful cotton firms in the area and by 1829 had over eight hundred cotton power looms in the new mill and shed which incorporated the original Laneside premises.[5]

With the industrial development of Halifax Parish came the development of new towns as industry moved from the upland hamlets, where it was based on hand-loom weaving, down to the valleys for water power and access to the roads and canal. Most townships had a few mills by 1800 when there were about forty-five mills in the entire parish. Contemporary directories recorded this growth:

1793. Here are also lately erected many mills for the cotton manufactury, which is rapidly increasing.[6]

1811. The spinning of cotton forms no inconsiderable branch of business in the neighbourhood.[7]

The majority of the early mills were built in the steep cloughs or side valleys which run down to the river Calder. Unlike other areas of Yorkshire few of these early mills appeared to be built on the Arkwright pattern. They were usually a little later than the mills built in the Dales or the Keighley areas and were often built for mule spinning. Many were termed 'factories' in contemporary records and power may just have been used for the carding process. Some of the mills were small, about the size of two or three cottages, and outlived their purpose within a few years. The occupants then moved and built larger mills with a better supply of water or turned to steam and relied on coal being brought in by canal. The steep valleys gave a good fall of water to most water wheels but made the construction of very large reservoirs difficult. Several were elaborate structures built into the hillside but by their nature could not hold large supplies of water so in the 1820s there were a number of schemes to build large dams on the moors to serve a number of mills.

The Halifax area was one of the two areas in Yorkshire where there was a significant rise in the number of cotton mills from 1800 and contrasted with other areas where numbers fell. The firms in the area were eventually able to take advantage of the large growth in demand for cotton warps for use in mixed fabrics by the West Riding worsted industry. Although this trade did not develop on any scale until the late 1830s a number of Halifax spinners were producing cotton warps for combining with worsted weft by 1835. For all the firms which had grown with the expansion of the Lancashire cotton industry in the first decades of the century this provided an additional local market and also extra trade for the increasing number of manufacturing firms which were equipped with power looms.

The Mills

Todmorden

Pudsey Mill. Stansfield. Pudsey Mill was mentioned in the Land Tax Return for 1795 but the first reference to this mill was by Crompton in 1811 who called it 'Putsey Mill'. He noted that there were five mules with two hundred and sixteen spindles each and one throstle with one hundred and forty-four spindles. Other references to this mill have not been found although a firm trading as Heap & Co. may have been tenants. Later, this mill became Glen Dyeworks.

Frieldhurst Mill. Stansfield. This mill was originally a cotton mill first recorded in 1795. It was changed to corn milling and then reverted to cotton. Crompton noted the following machinery in 1811:

6 mules × 240 spindles
4 throstles × 120 spindles

Whether or not cotton spinning was always carried out is not clear but John Law was using the mill for that purpose in 1837 while later the mill became a 'room and power' factory and eventually a dyeworks.

Barewise Mill. This mill was being used for cotton spinning by 1808 but no other information has been found.

Kitson Wood Mill or Naylor Mill. Stansfield. This mill was owned and occupied by Richard Naylor from about 1804 and in 1811 the mill held six mules with two hundred and forty spindles each. According to Crompton. Naylor was still occupying this mill in 1822.

Lydgate Factory or Low Mill in Stansfield was built about 1804 and was about one mile from Todmorden on the Burnley road. In 1818 it was occupied by James Eastwood when it was for sale. The mill then was then four storeys high and measured twenty-four yards by ten yards although a large part of the ground floor was taken up with ten cottages. The two water wheels had iron axle trees. A later tenant may have been Richard Naylor.

Ewood Mill or Malt Kiln Mill in Stansfield was first used for cotton spinning about 1800. In 1811 Crompton noted that the Malt Kiln held six mules with two hundred and sixteen spindles each and six mules with one hundred spindles each.

Holme Mill in Stansfield was built about 1795 and Crompton noted that a building at Holme held two mules with two hundred and sixteen spindles each in 1811. It was then owned by Joshua Fielden. The railway was later built across the site.

Cross Lee Mill in Stansfield was built about 1804 and occupied by Abraham Barker. The only detailed reference found has

been from Crompton's survey when he noted that there were three mules with two hundred and sixteen spindles at this mill. Later occupants were possibly John Barker and then John & Joseph Barker.

Greenhurst Mill. Stansfield. This small mill in Wickenberry Clough was built about 1797 and occupied by Michael Helliwell and later John Helliwell. It is now a row of cottages.

Haugh Stone Mill in Stansfield was built about 1796 for cotton spinning. Crompton noted that there were three mules with two hundred and sixteen spindles each at Haugh Stone in 1811. The firm of T & J Helliwell may have occupied the mill at that time.

Hole Bottom Mill. Stansfield. This mill was built about 1790 and occupied by George Stansfield in 1796. In 1811 Crompton noted that there were six mules with two hundred and forty spindles each at Hole Bottom. Mule spinning continued at this mill and there were twelve mules in 1818 when the mill was for sale. The owner then was still George Stansfield and he was offering the cotton machinery for sale as well as the mill. Power was provided by two water wheels and a steam engine. The mill, which was part two storeys and part three storeys, had overall dimensions of thirty six yards by twelve yards.

The mill may have been leased to Abraham Stansfield who was bankrupt in 1819. However, despite his bankruptcy, Stansfield is recorded as having continued cotton spinning at Hole Bottom Mill until at least 1834. The mill was then taken by G & W Hinchliffe but still for cotton spinning.

York Field or Oak Hill Clough Mill. Stansfield. This mill was in operation by 1790 and was being used by Robert

Greenwood in 1801 when he insured the mill for £150 and his stock for £50. In 1811 Crompton noted a mill at Yorkfield held six mules with two hundred and sixteen spindles each.

Ridgefoot Mill in Todmorden Township was built by Anthony Crossley of Todmorden Hall in the early 1790s. The mill was first run by Buckley & Sanderson and later John Buckley & Sons. John Buckley had not been allowed to build a cotton mill in Bradford and possibly started spinning cotton at this mill about 1794. In 1811 Crompton had noted that Todmorden Water Mill, which was probably this mill, held twelve mules with two hundred and forty spindles each.

Waterside Mill. Langfield. Joshua Fielden was a domestic cotton manufacturer who established a simple jenny workshop in three cottages at Laneside in 1782. Later carding engines were added together with other powered machinery and eventually, about 1829, the Laneside buildings were incorporated into the new Waterside Mill.

In 1811 Joshua Fielden died and the firm became known as Fielden Brothers. Crompton's survey that year indicated that they were running mules and throstles:

  10 mules × 228 spindles

  6 throstles × 144 spindles

Fielden Brothers went on to become one of the largest and most successful firms in the area. They developed power-loom weaving with an eight hundred loom shed in 1824 and another, a few years later, with one thousand looms.

Steam Mill, Todmorden Township. This was possibly the first steam powered mill to be built in the area and was sometimes called Salford Mill. Crompton noted that there were twelve mules with two hundred and sixteen spindles each and four

throstles with one hundred and twenty spindles each at Todmorden Steam Mill in 1811. This mill was later converted into shops and houses.

Swineshead Mill in Langfield was a small mill, possibly just used for carding, which was run by Joshua Fielden for a few years from about 1803.

Todmorden to Hebden Bridge

Lumbutts Mill in Langfield was originally a corn mill and then occupied by the Fielden family for cotton spinning from about 1794 until early in the twentieth century but the date when cotton spinning started is not clear. In 1811, when the mill had been running a few years there were ten mules with two hundred and forty spindles each.

About 1830 a new water supply was provided to the mill to turn a unique arrangement of water wheels. A tower was built which held three water wheels vertically above each other so that the run-off water turned the wheel below. The wheels were thirty feet in diameter and six feet wide. Together they could produce about 54 hp.

Jumb Mills. Langfield. There were two mills on this site very close to Lumbutts Mill. One had been a fulling mill and was then converted to use hand powered cotton machinery but re-built about 1803 for cotton spinning and used by A & W Uttley who were later joined by Samuel Greenwood. The other was built in 1801 and used by Samuel Hollinrake and James Byfield. In 1811 Crompton noted that one mill held ten mules with two hundred and forty spindles each. The first mill was run from about 1825 by Thomas Uttley. The other mill was run by Samuel Greenwood & Co. and later by John and James Greenwood. In 1833 this mill was powered with a 4 hp wheel with water from Lumbutts Clough. At some time a new mill called Jumb Mill was built which incorporated the two earlier mills.

Causey or Midgehole Mill. Langfield. This mill was built and occupied by Firth & Haworth about 1803 on land belonging to Samuel Hanson. In 1811 there were ten mules with two hundred and forty spindles each at this mill according to Crompton. The Fieldens bought the mill about 1813.

Causeway Wood Mill in Langfield was just below Jumb Mill down Lumbutts Clough and was rebuilt in 1826 by Firth & Haworth who had moved down from Causey Mill about 1813 and taken over an existing small mill on this site. In 1833 the firm of Firth, Haworth & Firth occupied the mill for cotton twist spinning and employed fifty-four people. The mill had

a 5 hp water wheel and an 8 hp steam engine.

Folly Mill. Langfield. This was the fourth mill down Lumbutts Clough and was built before 1811 when Crompton noted that there were four mules with eight hundred and sixty-four spindles. In 1833 the firm of Firth, Haworth & Firth occupied the mill for cotton twist spinning end employed fifty-four people. The mill had a 5 hp water wheel and an 8 hp steam engine.

Oldroyd Mill. Langfield. Oldroyd Mill was established by 1794 for the owner, Robert Atkinson, and the tenant, John Haworth, both supported the Bill for the Rochdale Canal that year. John Haworth was later joined by James Hollinrake and Thomas Knowles as partners at Oldroyd Mill where they were engaged in cotton spinning in 1801. The partnership also ran nearby Lob Mill for worsted spinning but

Woodhouse Mill, Langfield, Todmorden. This burnt-out mill was steam-powered and built beside the Rochdale Canal in 1832 for cotton spinning.

had bought Oldroyd Mill from Atkinson. They insured the mill and contents in 1801 for:

	£
Oldroyd Mill	2,000
Mill work	100
Machinery	400
Stock	150
	2,650

The partnership of William Ingham, James Hollinrake, Thomas Knowles and John Haworth, trading as Ingham, Hollinrake & Co, worsted and cotton spinners, Lob & Oldroyd Mills was dissolved on 27th April 1805. Hollinrake may have died or left the partnership for other reasons. Haworth may also have left the partnership for he was later the tenant at Cinderhills Mill as a cotton spinner. By 1822 William Hollinrake was occupying this mill and as he was also a cotton manufacturer he sold his cloth in Manchester. Sometime before 1845 the mill was taken by Messrs Firth & Haworth from Causeway Wood Mill.

Woodhouse Mill was built in 1832 by Richard Ingham on the side of the canal.

Castle Clough Mill in Stansfield was a small mill built at the side of Castle Hill about 1796 and today is two cottages with traces of the dam behind. The small size of the mill and the restricted nature of the water supply must have limited its use. Richard Ingham, who was a cotton manufacturer, insured the mill and his stock in it for £300 in 1801. He may have given up this mill when he built Cinderhills Mill about 1804.

Cinderhills Mill. Stansfield. This mill was built about 1804. However, the mill was described as 'newly erected' when it was advertised for sale in January 1818 so it may have been enlarged. The mill was five storeys high and measured twenty-four yards by twelve yards. It was well

sited for ease of transport for it adjoined both the Halifax and Burnley turnpike road and the Rochdale Canal. The tenant at the time was John Hauworth. Richard Ingham of Castle Lodge in Stansfield bought land and property, including Cinderhills Factory, for £1,300 in 1820. In 1824 additions were made to the mill, the most important being a 26 hp steam engine from Peel, Williams & Co. of Manchester. The firm of Richard Ingham & Sons continued cotton spinning and later cotton weaving at Cinderhills Mill until the end of the century.

Millsteads Mill in Stansfield. The mill was first recorded in 1805 but Richard Ingham built this mill in 1811. The mill remained in his possession and in 1833 was powered by a 10 hp steam engine and was used for spinning 'twenty-eight hanks weft and twist'. Ingham also ran Cinderhills Mill which was nearby.

Clough Hole Mill. This was a small mill in Rodwell Clough which was possibly built as a carding mill about 1805. Today it is a cottage with limited access from the main road.

Lobb Mill in Langfield was in use from 1796 when Ingham, Hollinrake & Co. bought screws and iron rods from Butlers at Kirkstall Forge but may have been a worsted mill until about 1809 when Abraham Hollinrake converted part of it to spin cotton and it was then occupied by Ingham, Knowles & Co. In later years Samuel Hollinrake was also using it for cotton spinning.

Stoodley Bridge Mill in Langfield was built as a cotton mill about 1808 by Thomas Sutcliffe and the Sutcliffe family continued spinning cotton at this mill, despite financial difficulties in 1825, until at least 1837. Other members of the family joined Thomas, and the firm was described as being cotton spinners, manufacturers and fustian manufacturers in contemporary directories. They were also listed as selling their goods on the Manchester market. A good part of the mill was rebuilt in 1813 after a fire and there was a further fire in 1829 when the old mill was burnt down but replaced with a new spinning mill.

The partnership of John, Thomas & James Sutcliffe was bankrupt in November 1825. Their assignees offered the mill for sale on a site which adjoined the river Calder and the Rochdale Canal. The mill, which now also had a steam engine, held the following machinery:

18 mules
14 throstles
5 power looms

Sutcliffs appear to have been able to overcome their financial problems and continued cotton spinning, if not power-loom weaving. In 1833 they were using the mill for mule and throstle spinning and employed one hundred and fourteen people. The steam engine then generated about 18 hp.

Guteroyd Mill. Langfield. This mill belonged to John Knowles until 1803 when he died. His tenant was John Sutcliffe but it is not clear when the mill was built or for how long it was used. The building is now a cottage although the present owner has cleared out the dam.

Eastwood Mill. Stansfield. There were two mills on this site, one of them originally a corn mill. In 1785 a Thomas Eastwood conveyed a 'twining mill' and in 1805 a Thomas Eastwood, who was a cotton manufacturer, was married. By 1810 one of the mills at Eastwood was being used by Thomas Eastwood and in 1811 Crompton noted that it held four mules with two hundred and forty spindles

each. The Eastwood family continued using the mill for cotton spinning until about 1835.

Cockden Mill. Stansfield. The site for the lower mill was for sale in 1791 when Abraham Barker, a millwright from Todmorden, would show people the premises. Barker probably built the mill about 1797 and occupied it himself for cotton spinning. When he was bankrupt in 1805 and Cockden Mill was for sale, his cotton machinery included three mules with two hundred and six spindles each. The mill was then taken by Hiram Hay but he had to assign his estate to trustees in 1808.

Burnt Acres Mill. Erringden. The only reference found to this mill is Crompton's note that there were three mules, each with two-hundred and forty spindles, at the mill in 1811.

Staups Mill. Stansfield. John Horsfall was a cotton manufacturer at Staups in Stansfield in 1805 and used this mill for spinning his yarn in that year. In 1811 Crompton recorded that the mill had two mules with two hundred and sixteen spindles each which meant that it was very small or only partially filled with machinery. Contemporary directories indicate that John Horsfall continued cotton spinning at Staups Mill until about 1834 when it was taken by James Bent. Two walls of the mill are still standing in the narrow wooded valley.

Cowbridge Mill or Underbank Mill. Stansfield. In 1794 this mill was used by Richard Horsefall for carding. In 1819 it was for sale and application had to be made to Christopher and James Rawdon. The mill then five storeys high and measured sixteen yards by fourteen yards with a sixty foot fall of water available. From 1822, until after 1837, the mill was used by William Horsefall, who was a sateen and velveteen manufacturer. It is not clear if he was a descendant of the Richard Horsefall who ran the mill in 1794.

Spa Mill in Stansfield was the third mill down Jumble Hole Clough and the first reference for this mill establishes that it was built about 1788 and by 1815 it was owned by Christopher Rawdon. From 1825 until at least 1837 it was occupied by the firm of Mallalieu & Platt for cotton spinning. This firm also ran Nutclough Mill.

Jumble Hole Mill or Low Underbank Mill in Stansfield. This mill in Jumble Hole Clough was very near to the one

above and established by 1788. It was leased by George Ashworth by 1815 when it was owned by Christopher Rawdon. The Ashworths were also cotton manufacturers and in 1825, for instance, William Ashworth was manufacturing dimity, jean and jeanette cloths and selling them on the Manchester market.

Winters Mill or Marsh Factory in Stansfield. John Sutcliffe was spinning cotton at a mill in Stansfield from about 1809. In 1818 it was called Marsh Factory when it was for sale. The four storey mill at Lower Winters measured fifteen yards by twelve yards and had water wheels as well as a 5 hp steam engine. John Sutcliffe would show prospective purchasers round

The remains of Staups Mill.

the premises. If it was the same mill, Thomas Ramsden was spinning cotton at Winters in Stansfield in 1822. When this mill was for sale in 1834 the mill had five floors plus an attic and measured twelve yards by eleven yards. There was also a weaving shop of two storeys plus an attic which measured twelve yards by twelve yards with shafting for power looms. There was then an 18 hp steam engine and two water wheels.

Near Hebden Bridge

Calderside Mill. Stansfield. John Whitely was spinning cotton and manufacturing cloth at Calderside Mill from 1824 until 1864. The mill had been converted from part of a dyeworks on the same site.

Rodmer Clough Mill or Lister Mill. Stansfield. This mill was described as newly erected when it was offered to let in 1793. There was a thirty foot fall of water to the mill and it was described as having 'a new water wheel, with principal machinery calculated for working frames for spinning cotton or worsted are already fixed'. There was a good house at one end of the mill and more land available if required. The owner was Thomas Lister of Halifax.

Lister did not let the mill but took a partner, William Dewhurst, and they used the mill for cotton and worsted spinning. In 1801 the valuation for insurance purposes was:

	£
Lister Mill at Clough in Stansfield	250
Mill work	50
Worsted machinery	300
Cotton machinery	590
Stock	400
	1,590

The worsted spinning frames and other worsted machinery were for sale in 1802

when prospective purchasers had to see William Dewhirst at Rodmer Clough.

The mill may then have been used solely for cotton spinning for in 1811 it held ten mules with two hundred and forty spindles each. It is not clear which firm occupied this mill after 1802 but in 1814 it was William Ingham and from about 1830 to 1837 it was occupied by William Gledhill & Sons.

Salt Pie, Edge or New Edge Mill, Heptonstall. This mill was built about 1807 by Henry Sutcliffe.

Land Mill. Stansfield. This mill was first used for cotton spinning in 1796 by John Greenwood but a partnership between John Marshall and Harry Riley as cotton spinners at this mill was dissolved on 28th November 1808. In 1811 Crompton found that the mill was being used for mule spinning with four mules of two hundred and forty spindles each running in Land Mill. During the 1820s John Greenwood occupied this mill.

Hudson Mill in Stansfield was another corn mill which was converted or added to for the purpose of cotton spinning about 1786 so it was one of the earliest cotton mills in the valley. A partnership between George Widdop of Stansfield, William Sutcliffe of Fieldhead and Michael Heaton of Robertshaw was established to spin cotton at this mill before 1799 and traded as George Widdop & Co. In that year the partnership was dissolved but Widdop may have continued alone.

In 1804 Turner, Bent & Co. of Mytholm Mill, who were cotton spinners and merchants, agreed to rent the new section of Hudson Mill which had been built for cotton spinning after a fire. The lease was to be for thirteen years. In 1811 Crompton noted that Hudson Mill held three mules with two hundred and forty spindles each.

By 1825 Enoch Barker was occupying this mill as a cotton spinner and manufacturer and he continued there for several years.

Slater Ing Mill or Bob Mill. Heptonstall. This mill was built some time before 1805 when the mill was to let and the cotton machinery was for sale. The three storey mill on Colden Water measured sixty feet by thirty feet with the top floor extending to sixty-nine feet as it was built over the water wheel. This measured twenty-four feet by four feet and was supplied from a 'good dam and reservoir'. The mill was owned by William Greenwood of Little Lear Ings, Heptonstall. The machinery in the mill included the usual cards and preparing machinery together with four throstles. William Greenwood tried to auction the mill in June 1809 when the tenant was Robert Sutcliffe. In September Sutcliffe was selling his machinery which included seven throstles. This mill was then possibly taken by Gamaliel Sutcliffe for cotton spinning as it was very near Lumb Mill.

Lumb Mill. Heptonstall. This mill was built about 1801 by Gamaliel Sutcliffe and occupied by him until at least 1835. In 1811 Crompton noted that this mill was running the following machinery:

13 mules × 276 spindles

39 throstles × 120 spindles

There was also an Upper Lumb Mill which was occupied by the Sutcliffe family

Mytholm Mill in Stansfield was built by a partnership between four men about 1789. James King was probably the most important of them and he had been a woollen manufacturer before he started cotton spinning. The other partners were Alexander Turner, a Leeds merchant, Richard Paley, who was a soap boiler in Leeds and who had interests in cotton mills in Leeds, Gomersal and Colne, and Richard Varley. Within a few years Hamlet Bent, the firm's salesman, and Paul Lister, who was involved in cotton spinning at Morton near Keighley, also joined the partnership.

This cotton mill was built alongside an older mill which was used as a rasping and chipping mill. This older mill may also have been used for cotton spinning for a short time. The partnership was evidently wealthy for they insured their property and goods for £11,000 in 1792:

	£
Mill	700
Utensils, stock and machinery	1,500
New Mill adjoining and communicating	500
Utensils stock and machinery	700
Workshops, counting house, warehouse, dwelling houses and offices	500
Utensils and goods	6,000
Joiners and smiths shop	50
Utensils and stock	50
'Scowering' house and chamber	50
Utensils and stock	150
Dyehouse	100
Drying house	100
Utensils and stock	600
	11,000

On March 1st 1796 the partnership between King, Turner, Varley, Bent and Paley as merchants and cotton manufacturers was dissolved. Varley and Paley withdrew so the new partnership traded as Turner, Bent & Co.

By 1805 Mytholm Mill was five storeys high and contained 1,193 square yards of floor space. The machinery in 1811 was a combination of mules and throstles with:

12 mules × 240 spindles

12 throstles × 120 spindles

By 1825 the firm was trading as Hamlet Bent & Co. and were cotton spinners and fustian manufacturers. About 1834 the mill was acquired by Binns & Wright for cotton and silk spinning and manufacturing.

Mytholm Mill had one of the largest water wheels in the valley which measured fifty-two feet six inches by nine feet six inches with the water put on to the wheel at two levels.

Bankfoot Mills. Heptonstall. There were two, if not three mills on this site and although they were sometimes referred to as Bankfoot Upper or Bankfoot Lower there is some confusion about the occupants at particular times. Both mills may have been corn mills before being converted or added to and some of the original mill buildings may have remained. One of the mills was run by Thomas Lee until his bankruptcy in 1807. As he occupied Midge Hole Mill for spinning flax and fulling, Spring Hall for jenny spinning and hand-loom weaving, he was involved in several ventures. Lee had used Bankfoot Mill for mule and throstle spinning and the following machinery was for sale, with the mill, in 1808:

8 mules × 216 spindles

2 mules × 240 spindles

4 throstles × 120 spindles

There was, of course, all the preparing machinery as well as a steam engine for power.

Crompton's survey, which did not mention the firm, listed Lower Bankfoot Mill as having:

8 mules × 216 spindles

2 throstles × 100 spindles

In 1813 the machinery in the mill was for sale again when there were nearly four thousand mule spindles on mules with from 216 to 252 spindles and one throstle.

It is not clear which firm occupied the mill until 1825 when James Bent, a cotton and fustian manufacturer, took the mill for cotton spinning. He continued there until his bankruptcy in 1833.

The other mill at Bankfoot, which was probably the Upper Mill, was probably used for cotton spinning by Abraham Hollinrake, who succeeded Messrs Firth & Parkinson. However, Crompton included this mill in his survey as Bankfoot Upper Mill when there were:

13 mules × 228 spindles

Clough Mill at Hebden Bridge Lanes in Heptonstall. The first mention of this mill occurs in the Royal Exchange Registers when Thomas Sutcliffe insured the mill and contents in 1801:

	£
Cotton mill at Hebden Bridge Lanes	150
Machinery	200
Stock	150
	500

Crompton's survey in 1811 indicates that this mill was used for mule spinning with three mules running 648 spindles between them. Thomas Sutcliffe was still spinning cotton at Clough Mill in 1825 and the mill may have the one occupied by John Sutcliffe in 1833 when returns were made to the Factories Enquiry Commissioners. It was said that the mill had always been used for mule or throstle spinning, that it had a 5 hp water wheel and employed seventeen people for throstle spinning.

Gibson Mill or Lord Holme Mill. Heptonstall. This mill was given the name of its owners, the Gibson family, who occupied the mill for cotton spinning for the first and last thirds of its working existence. In the intervening time, the Gaukrogers of nearby Bridge Mill were tenants. According to Gaukroger's replies to the Factories Enquiry Commissioners in 1833, the mill had been built in 1803, although it was not filled with machinery until 1805.

The Gibsons lived at Greenwood Lee,

a house on the valley side above the mill. They were cotton manufacturers and started cotton spinning at the house. This was probably inadequate for their needs, so Abraham Gibson Junior built this mill on Hebden Water. Unfortunately he fell from his horse and was killed in February 1805. The new mill was therefore offered to let. It was said to be a 'Capital Factory', with three storeys, each eleven feet high, plus a garret. The mill measured forty feet by thirty-six feet within the walls, and was powered by a new water wheel measuring sixteen feet by six feet. There was no separate dam as the river itself provided a suitable reservoir. It was said that the mill could be applied to cotton, worsted or woollen manufacture. It would seem that neither the mill nor the family house was let, for the family continued to live at Greenwood Lee, and Abraham Gibson's father and nephew, both also called Abraham, took the mill themselves for cotton spinning. They were also cotton manufacturers and sold their goods on the Manchester market in common with most of the other manufacturers in the area. In 1811 they were running six throstles with one hundred and twenty spindles each.

Gibsons leased the mill to Titus and James Gaukroger in 1833 and they continued to use the mill to spin throstle twist. The mill had a five horse power water wheel at that time and twenty-one people were employed there.

In later years the mill was extended and steam power added. It was not used for textile purposes after about 1900.

Greenwood Lee. Heptonstall. Greenwood Lee was, and still is, a house. It was bought by Abraham Gibson in 1760 and used as the family house while he and his son carried on their business as cotton manufacturers using hand spinners and hand-loom weavers to make their cloth

towards the end of the century. Some time, about 1802 or before, it was decided to add a power source at the house. The reason may have been that part of the house was being used for carding cotton and then spinning it on jennies. As production increased it would have been difficult to increase the productivity of the carding section so a twenty-four foot by eighteen inch water wheel was installed which was used to drive a picker, a billy and a thirty inch double carding engine. Part of the upper floor had to be cut away to accommodate the wheel and water was brought to it down the yard by a conduit.

New Bridge Mill in Heptonstall was built by John Foster about 1796 and occupied by Foster & Sugden. John Foster insured the mill, which was his property, for £600 in 1801. Crompton's survey in 1811 showed that the mill was being used for cotton spinning with twelve frames of forty-eight spindles each. Foster & Sugden continued cotton spinning at New Bridge Mill until 1819. From 1825 the mill was taken by Titus & James Gaukroger who also occupied Lord Holme and Upper Lumb Mills. By 1833 the mill was still water powered with a 15 hp wheel drawing from Hebden Water. Sixty-three people were employed of whom twenty worked at night. Even at that date the firm had to admit that they did not know when the mill was built.

At Midgehole. Heptonstall. This mill was established prior to 1803 when the partnership which ran the mill and traded as Lawrence Moorehouse & Co, cotton spinners and manufacturers, was dissolved. The partners had been:

Lawrence Moorhouse, Birch Cliffe
William Redman, Midge Hole
John Pickles, Crimsworth Dean
William Riley, Hebden Bridge
William Riley had died so changes had to

be made within the firm. In 1811 Crompton noted that there were four mules with two hundred and sixteen spindles each in the mill. Lawrence Moorhouse was spinning cotton at this mill until 1825 when the firm was restyled Joseph Moorhouse & Brothers. The firm was still engaged in cotton spinning and manufacturing and sold their goods in Manchester.

At Midgehole. According to replies made to the Factories Enquiry Commissioners in 1833 this mill was built in 1783 and had always been used for mule or throstle spinning. John Sutcliffe occupied the mill in 1833 when it had a 5 hp water wheel and he employed seventeen people for throstle spinning.

Lee Mill in Heptonstall was originally a fulling mill dating back to the sixteenth century. Some time before 1820, Abraham Cockroft and William Thomas occupied this mill for cotton spinning but their partnership was dissolved in that year. By 1825 it was being used by William Thomas for spinning cotton but it was rebuilt in 1832 for manufacturing heavy fustian cloth. By 1836 it held thirty-nine power looms and employed thirty-seven people.

Stubbing Mill or Hebble End Mill. Heptonstall. Stubbing Mill would have been worked before 1806 for the tenant, Samuel Chatburn, a cotton spinner, dealer and chapman, was bankrupt in January that year. His machinery was put up for auction and included four new throstles with one hundred and twenty spindles each. In March 1806 the mill was to let and was described as a cotton spinning factory on both the river Calder and the Rochdale Canal. The mill may have been owned by William Sutcliffe.

A later occupant of the mill was John Wright, who was there from the 1820s until about 1834. The mill by then was called Hebble End Mill. In his replies to the Factories Enquiry Commissioners in 1834 Wright said that the mill was used for cotton spinning with the river Calder generating about ten horse power and he employed about forty-six people.

Kershaw Mill. Heptonstall. This mill, which was named after its owner, was built in 1834 and destroyed in a fire in 1843. James Kershaw was a cotton spinner and manufacturer.

Foster Mill in Wadsworth was originally a corn mill but prior to 1808 it was converted into a worsted spinning mill. It was for sale in 1808 and may then have been converted to spin cotton for the Ramsbottom family occupied the mill for that purpose from 1816 or before until the mid-1830s. Henry Ramsbottom also ran Ewood Mill in 1825.

Nutclough Mill. Wadsworth. This mill was said to be newly built when it was offered for sale in 1797. It was water powered and said to be suitable for carding and spinning cotton. Prospective purchasers had to apply to Mrs Sutcliffe or her son.

In 1811 Crompton noted that there were ten mules with two hundred and sixty-four spindles each at Nutclough Mill. It is not known which firm ran the mill at that time or indeed until 1822 when James Mallalieu, a warp manufacturer, occupied the mill. The firm of Mallalieu & Platt then occupied the mill for cotton spinning between 1825 and 1837 or later.

Mayroyd Mill or Gemland Mill. Wadsworth. This was originally a corn mill and continued to be used for corn milling well into the nineteenth century. However, part of the mill was used by James Law for carding cotton in 1794 when the mill was owned by Robert Atkinson. Law was bankrupt before March 1804 but in the meantime Lambert Barnes, a cotton manufacturer, had occupied part of the mill for cotton spinning for he had cotton machinery and stock insured for £80 and £20 in the mill in 1798. This cover was raised to £150 and £50 in 1801.

No further references to cotton spinning at Mayroyd Mill have been found after William Wheelhouse, a corn miller, took over the mill in 1802.

At Ibbotroyd. Wadsworth. This mill at Ibbot Royd was described as newly erected when it was to let in December 1798. Within the mill were two water wheels with fifteen yards fall and power to turn '6 cotton engines'. This mill may have had another name as no further details have been found until 1845 when Blackburn & Pickles were using it for cotton spinning.

Hawksclough Mill. Wadsworth. The first reference to this mill indicates that John Riley used it for dimity, fustian, jean and jeannette manufacturing in 1825. Later directories give substantially the same information but say nothing about the mill.

Mytholmroyd and Cragg Vale

Rudclough Mill in Erringden was advertised for sale in June 1801 without a specific use being given. A purchaser must have soon been found for the cotton mill was then let on a forty-year lease. Thirty-seven of the years were available when the mill was to let again in 1804, after it had been used for cotton spinning, possibly by Sarah Waddington. The mill was for sale in 1810 although it was then occupied by John Stansfield for cotton spinning. By 1822 William Hinchliffe had taken the mill and the Hinchliffes continued cotton spinning there until it burnt down in 1897.

Marshaw Bridge Mill in Erringden was advertised for sale in 1794 with eleven

Lumb Mill, Warley, Halifax. Built about 1803 for cotton spinning, but later heightened and used for worsted spinning.

years of the lease still to run, but no mention of when the lease commenced. The mill was said to be on a good stream of water with an eighteen foot fall. Land and a house were included in the sale. In the mill were three spinning frames together with cards and other machinery for preparing cotton. The machinery only took up half of the mill for it would contain another three spinning frames and preparing machinery. Six frames was not a large number, so it could not have been a very large mill at that time. After the sale of the mill in 1795 it was turned over to wool scribbling.

Marshaw Bridge Mill was converted back to cotton spinning by 1822 when it was occupied by R & A Ingham. Inghams continued at the mill for several years and, in 1833, Richard and Joseph Ingham were spinning 14s to 40s quality yarn there for their hand-loom weavers. The mill burnt down in 1843.

Castle Mill. Sowerby. This mill appears to have been in the hands of the Greenwood family for the whole of its existence. It was first mentioned by Baines in 1822, when it was occupied by John Greenwood for cotton spinning and, possibly also for paper making and wool card manufacture. John Greenwood continued cotton spinning at the mill until 1860.

Cragg Mill in Erringden had been used for wool scribbling until 1798 when part or the whole of the mill may have changed over to cotton spinning. William Currer died in 1808 and his mills at Luddenden and Cragg were offered for sale. It would seem that Cragg Mill had been enlarged, possibly after 1798. There was a large dam and a fall of twenty-four feet to the water wheel. Up to 1821 Cragg Mill was occupied by Henry Briggs who also ran Lud-

denden Foot Mill. His machinery at the two mills was then for sale and included:

22 mules × 300 spindles
4 mules × 144 spindles
25 throstles × 132 spindles
15 throstles × 84 − 120 spindles

Cragg Mill was then taken by G & I Hinchliffe for cotton spinning and they continued there until at least 1850.

Luddenden Valley

Spring Mill. Warley. This mill was built at a point 1,150 feet above sea level in a bleak and isolated position and was the highest mill at Cold Edge. It was built by the Emmett family in 1800 and sold in the same year to Samuel Schorfield who used it for cotton spinning. In 1811 Crompton noted that 'Scawfields' had:

10 mules × 252 spindles
5 throstles × 120 spindles

Schorfield sold the mill and fourteen cottages to Robert Abbott in 1817 but may have continued cotton spinning at this mill as a tenant until his death in 1825 when it was advertised to let. The mill was then three storeys high plus an attic and measured twelve yards by thirteen yards within the walls. There was a large diameter water wheel which measured thirty-six feet by two feet eight inches. Spring Mill was then used for worsted spinning.

Hoyle Bottom mill or Square Mill. Warley. This mill was used for cotton spinning in the early 1800s and may have been used by Samuel Schorfield from Spring Mill at some time. However, when it was to let in 1825 it was said to be 'newly erected'. It was then four storeys plus an attic with an iron wheel which measured thirty-three feet by four feet two inches. It was converted to worsted spinning around that time and rebuilt in 1872.

Lumb Mill or Stones Mill. Warley. This mill was built by John Garforth of Mixenden about 1803. David and John Wright later went into partnership with John Garforth and Samuel Garforth but this was dissolved in September 1818 when David Wright sold his share back to John Garforth for £390. The mill at that time measured seventeen yards by thirteen yards two feet. In 1811 this was probably the mill referred to as 'Garfits' by Crompton when he noted that there was the following spinning machinery:

8 mules × 252 spindles
5 throstles × 120 spindles

Samuel and John Garforth, who were father and son, were bankrupt in 1826, as were many other cotton spinners, and Lumb Mill was advertised for sale along with a house, a warehouse and the machinery. This included:

4 mules × 264 spindles
2 throstles × 120 spindles
6 throstles × 120 spindles

The bankruptcy and the sale of the mill led to it being converted to spin worsted in 1828 by Thomas and Matthew Murgatroyd. The mill and dam can still be seen.

Wainstalls Mill. Warley. In the early days of cotton spinning a number of mills were built on this site as there was a good supply of water. The first mill was built about 1804 or before by Jonas Tillotson. By 1811, according to Crompton, he had:

8 mules × 252 spindles
4 throstles × 120 spindles

Jonathon Calvert was taken into partnership with Tillotson by 1815 when Wainstalls mill was for sale. The cotton mill was for sale again in 1820 when there was also a warehouse with cottages underneath. Michael and Jonas Tillotson, who were running the mill, had three pairs of large mules and six throstles. The reason for the sale was that Jonas Tillotson was

bankrupt. The mills were then sold to William Appleyard in 1821 and both were used for cotton spinning.

Jowler Mill or Holme House Mill. Warley. This mill was built by Jonathon Bracken for his daughter Agnes and appears to have just had power for the preparing machinery. In 1811 Crompton recorded that the mill held five jennies with one hundred and twenty-six spindles each. The mill was occupied by John Garnett who changed over to worsted spinning in 1818 or possibly wire drawing before that date.

Dean Mills in Midgley were originally fulling mills but one was converted to a paper mill in 1769 and the other to a cotton mill in 1792. Sometime later the buildings were joined together to form one mill. For at least the first forty years of its existence the cotton mill was occupied by Jonathon Bracken. In 1804 the mill was for sale for some reason and at that time it was part paper and part cotton. Contemporary trade directories continued to record Jonathon Bracken's occupation of Dean Mill, his attendance at Manchester to sell his cotton goods and also his activities as a paper maker. According to Crompton the section of the mill used for cotton spinning held the following machinery in 1811:

8 mules × 240 spindles
6 throstles × 120 spindles

By 1850 the mill was used exclusively for paper manufacture.

Luddenden Mills. Warley. These two mills were originally corn mills dating back to the thirteenth century. By 1805 Thomas Farrer was using the lower mill for cotton spinning for he was bankrupt in that year. His machinery was for sale in 1806 and included mules, throstles and fourteen pairs of looms. In 1809 the mill was to let. At that time it was three storeys high and measured forty-two feet by thirty-six feet with a twenty-one foot fall of water. in 1811 Crompton noted that there four mules with two hundred and forty spindles each at this mill. A later occupant may have been Samuel Riley.

Luddenden Foot Mill. Sowerby. William Currer was a carpet manufacturer and cotton spinner who owned several mills. This mill was next to an old corn mill called Boy Mill and there was also a woollen mill on the site or nearby. At Currer's death in 1807 a one-third share of the cotton mill was offered for sale and it may have been used as a cotton mill for some years. The mill was large, five storeys high, and said to be capable of holding six thousand mule spindles and two thousand throstle spindles. These statements were verified by Crompton who noted, in 1811 that 'Luddenden Factory' had the following spinning machines:

  20 mules × 300 spindles = 6,000 mule spindles
  14 throstles × 120 spindles = 1,680 throstle spindles

A contemporary trade directory noted that the occupant of the mill was William Currer, Son & Co. but shortly afterwards this mill was turned over to worsted spinning.

Brearley Mill in Midgley was originally a wool scribbling mill but a section of it was used for preparing cotton and spinning on jennies by 1803. The person who had leased the mill was Samuel Winpenny and he was bankrupt then so his machinery was for sale. It is possible that this mill was not used again for cotton spinning as it was used for wool carding and scribbling in 1807 and corn milling in 1812.

Cooper House Mill. Warley. A small part of this mill was probably used for cotton spinning from about 1794 by Joshua

Crowther who was a woollen manufacturer. In 1797, Thomas and Henry Lodge, who were cotton spinners at Willow Hall Mills, took part of Cooper House Mill and insured the mill work for £200, machinery for £650 and their stock for £200. The mill continued to be used for wool processing and some cotton spinning until it was changed to worsted spinning in 1832.

Longbotton Mill. Warley. This was originally a fulling mill which was changed to cotton by 1792. Samuel Milne insured the mill and contents for the following sums in that year:

	£
Mill	1,000
Utensils and machinery	1,000
Two cottages adjoining	120
Utensils etc	330
	2,450

Cotton spinning ended by 1797 and the machinery was sold to Ard Walker in Leeds in 1800 when Walker bought 10 spinning frames and other machinery for £168.

Higgin Chamber or Ing Head Mill. Sowerby. The lease for Ing Head Mill at Higgin Chamber was taken by Elias Fletcher in 1788. He used it for mule spinning and was also engaged in cotton weaving when he was bankrupt in 1804. At that time he had the following machinery:

  3 mules × 204 spindles
  3 mules × 144 spindles
  1 mule × 166 spindles
  1 twist frame × 84 spindles for making double warps
  3 pairs of cotton looms

The cotton machinery was in Higgin Chamber Cotton Mill but the lease of Ing Head Cotton Mill was available. The two mills were contiguous with each other with one possibly having been an earlier corn mill.

One of the mills was then taken by John Pickles and Samuel Jennings who dissolved their partnership in 1805. Crompton noted that Higgin Chamber Mill had eight mules with one hundred and eighty spindles each in 1811.

By 1822 Higgin Chamber Mill had a steam engine to supplement the water wheel and was occupied by James and George Mitchell for cotton spinning with the second mill occupied by Dewhirst, Halstead & Co. However, in 1825, the owner, Thomas Horsefall offered the mill or mills for sale. The details of the two buildings were:

First Mill had three rooms plus attic.
Measured 16 yards × 12 yards.
10 hp steam engine
Second Mill had three rooms
Measured 16 yards by 10 yards
10 hp steam engine

Horsefall was bankrupt in 1833 when James Mitchell was still a tenant until he was rendered bankrupt in the following year. The mill was later turned over to worsted spinning and burnt down in 1856.

Boulder Clough or Swamp Mills. Sowerby. There were two mills on this site which were later used for different purposes, not always cotton spinning. John Ramsden from Halifax and John Ashford from Elland built one of these mills about 1790. They insured it for £150 in 1792, with the machinery and utensils inside the mill insured for a further £500. In 1801 Thomas Smith leased a small cotton mill at Boulder Clough and then insured the mill and contents for:

	£
Boulder Clough Mill	200
Mill work	50
Machinery	200
Stock	50
	500

The mill was to let in 1804 with seven years of the lease still to run. The mill was then four storeys high and measured twelve yards by ten yards. The machinery included:

2 mules × 240 spindles
2 mules × 228 spindles
2 mules × 216 spindles

In 1808 both mills were to let but neither had recently been used for cotton. One had been used for carding and spinning sheep's wool, which was probably the former cotton mill, while the other was used for wire drawing. In 1815 both were used for wire drawing but by 1819 Upper Swamp Mill, as it was then called, was used for worsted spinning and Lower Swamp Mill was used for cotton spinning. It was four storeys high and had a twenty-three foot water wheel. The tenants were Thomas Mitchell & Son who ran both mills. No further references have been found.

Sowerby Bridge to Halifax

Jumples Mill in Ovenden was first a corn mill but was changed to a woollen mill about 1785 by Charles Hudson and to a cotton mill about 1816 by James Hegginbottom. Besides water power the mill had an early pumping engine, made by Emmetts of Birkenshaw, which was used to pump water back up to the water wheel. When the mill was advertised for sale in 1815 it was three storeys high and measured ninety-one feet by thirty-four feet outside the walls. The water wheel measured thirty-four feet by five feet and the pumping engine was still in place. Alterations were made to the mill in 1829 and an 8 hp steam engine replaced the earlier pumping engine. In 1833 the mill was used for preparing and spinning cotton and James Hegginbottom employed sixty-two people at the mill.

Hebble Mill in Ovenden was an early steam powered woollen mill but was for sale in 1804. The three storey mill measured eighty-six feet by thirty-two feet and had a twenty foot by three foot six inch water wheel as well as the crank steam engine which had a forty-two inch cylinder. It is likely that the mill was taken by Thomas and James Greenwood and converted to cotton spinning. In 1811 they had twelve mules with two hundred and forty spindles each and were also involved in domestic cotton manufacturing. By 1830 Greenwoods were also woollen manufacturers but continued spinning cotton at this mill until after the end of this survey.

Holme Field Mill in Ovenden was built by John Watkinson about 1797 when he bought iron and screws from Kirkstall Forge. He insured the mill and contents for the following sums in 1800:

	£
Cotton mill	130
Machinery	300
Mill work	10
Stock	60
	500

In 1811 Crompton noted that Watkinson had ten mules with two thousand four hundred spindles. Watkinson was bankrupt in 1812 and the mill was later converted to worsted spinning.

Shaw Lane Mill. Ovenden. In their replies to the Factories Enquiry Commissioners in 1833, W & L Dewhirst & Co. stated that this mill was built, or applied to cotton spinning, in 1793. It is not clear what firm occupied the mill at that time or for the next few years although in 1811 it may have been 'Mathewsons'. Crompton noted that a firm of that name in this district were running six mules with three hundred spindles each and two with two

hundred and sixteen. About 1816 James Kershaw may have been spinning cotton there but from 1822 or earlier the firm of William Dewhirst occupied the mill. This firm manufactured calico cloth and sold it on the Manchester market. In 1833 the mill was powered by a 14 hp steam engine and a 14 hp water wheel with water drawn from Ovenden Beck. Dewhirst employed seventy-six people.

Grove Mill in Ovenden was established by 1818, when Peter Bold was using the mill to spin cotton yarn to make into fustian cloth. He continued at the mill until about 1840 when he was succeeded by John Bold. The mill was about half a mile from a stream of water so a water wheel, and later a hydraulic ram, had to be used to pump water up to the dam. In 1834 Peter Bold insured part of the mill and his property for the following sums:

	Sun Yorks Fire	
		Office
	£	£
Two spinning and carding rooms with gas and engine house	400	
Steam engine and boiler	150	
Machinery	800	
Mill work	50	
Stock	100	
	1,500	80

Old Lane Mills. Ovenden. One part of the mill was being used for cotton spinning by January 1793 when the premises were to let. There were several buildings, including one under construction which was just at foundation level 'but may yet be adapted to the convenience of the taker or purchaser'. Power was supplied by a water wheel but there was also a pumping engine built by Booth & Co. which was 'calculated to throw up five hundred gallons of water at each stroke'. Potential tenants had to apply to Messrs William Mitchell & Co. at the mill. Until 1797 the partnership had consisted of, Luke Stavely of Halifax, John Mitchell the younger from Ovenden, William Mitchell from Boothtown and William Whitfield from Northowram. This partnership was then dissolved and possibly William Mitchell and William Whitfield continued.

By 1799 the buildings had been completed and two mills were for sale or to let. These were:

Old Mill – recently erected: three storeys 54ft × 27ft, 12 ft high ground floor room with 6 chambers above. 26 foot fall to the water wheel.

New Mill – five storeys 89ft × 31ft within the walls. Steam engine.

40 sets of cards

20 frames with 1,200 spindles

8 mules of 228 to 250 spindles each

The New Mill was offered for sale or to let again in 1812 together with the cotton machinery. Further details supplied in the advertisement were that the mill was powered by a twenty-eight foot by five foot water wheel as well as a 16 hp Boulton & Watt engine. The mill had possibly been occupied by Thomas Smith but application had to be made to William Mitchell. Crompton had noted that Thomas Smith was running two thousand five hundred mule spindles and one thousand and eighty throstle spindles in 1811.

Old Lane Mill was converted to worsted spinning by 1827.

New House Mill in Ovenden was built by George Moss. The mill was steam powered from the start and was used for mule and throstle spinning. In 1811 Crompton noted:

4 mules × 204 spindles

3 throstles × 120 spindles

The mill was dismantled in 1820 and the steam engine transferred to Dean Clough Mill in Halifax but George Moss continued in business as a cotton manufacturer.

North Bridge Mill. Halifax Township. Samuel Dyson insured part of North Bridge Mill and his property for £250 in 1798 when he commenced a nine year lease.

	£
Cotton Mill	80
Machinery	90
Stock	80
	250

The mill, which had three water wheels, was partly used for paper manufacture and partly for cotton preparing. The cotton section held 'two small cotton engines and one cylinder' in 1799 when it was to let for £33 per annum. Thomas Emmett & Co. then took the mill and insured their cotton spinning machinery for £150 in 1801. By 1808 the part of the mill used for paper manufacturing had been converted for woollen manufacturing and Samuel Dyson was offering the mill for lease or sale. The cotton machinery in the mill in 1809 consisted of eight mules, three throstles and two jennies, as well as the usual carding and preparing machines.

Samuel Dyson must have been successful in leasing or selling the mill for George Haigh & Co. were spinning cotton there by 1811. Crompton noted that George Haigh was running eight mules at the Paper Mill and he was still there in 1816.

New Bank Mill in Halifax Township appears to have been built as a cotton and worsted spinning mill. In 1832 the mill and contents were insured for the following sums:

	Sun	Globe	Leeds & Yorks
	£	£	£
Cotton and worsted mill	1,500	1,500	
Steam engine	250	250	
Mill work	500	500	
Machinery	1,500	1,500	
Stock	250	250	
Boiler-house	125	125	
Boilers	50	50	
	4,175	4,175	2,000

John Haigh & Brothers were cotton and worsted spinners, manufacturers and merchants. They appeared to have the resources to expand this mill, for in 1836 the insurance valuation, particularly for the machinery, had increased considerably.

Lee Bridge Mill. Halifax Township. This mill was about a mile from the centre of Halifax and was used for cotton spinning before 1803. In that year, William Ramsden, who occupied the mill was advertising for a manager of the card room. At the same time the following machinery was for sale:

> 6 cotton twist frames × 48 spindles
>
> 1 mule × 216 spindles
>
> 1 mule × 204 spindles Both mules newly repaired

In the following year William Ramsden was advertising for a carding engine and a carder. The mill was to let in 1809 complete with the water powered machinery. Application had to be made to Messrs Ramsdens at the mill or Messrs Boulton & Ramsden in Manchester. Lee Bridge Mill was then changed over to worsted spinning.

Shibden Mill. Northowram. Between about 1803 and 1806 Thomas Moat and George Panter were cotton spinners and manufacturers at Shibden Mill which had previously been a wool and worsted mill.

Unfortunately the partners were bankrupt in 1806 and their machinery was for sale by their assignees. The machinery included seven mules with from eighty-eight to two hundred and seventy-five spindles each as well as an unspecified number of jennies.

The mill was then taken by Messrs Holdsworth & Co. who continued with cotton spinning for a few years. Crompton noted that they had five mules with two hundred and forty spindles each in 1811. It is likely that G & J Holdsworth changed Shibden Mill back to a woollen mill about 1815.

Bowling Dyke Mill in Northowram was a wool scribbling mill but part was let off for cotton spinning from about 1800. Samuel Emmett, a carpet manufacturer, owned the mill, which had a steam engine, and let part to Thomas Ramsden who insured his cotton machinery in the mill for £320 in 1801. In the following year Samuel Emmett was advertising land for sale at Bowling Dyke which would be suitable for building a mill to be used for mule or water twist spinning. A separate mill may have been built for, in 1811, Crompton noted that Emmetts were running eight thousand three hundred mule spindles. Thomas Emmett Junior continued cotton spinning at Bowling Dyke until 1815 when the mill was taken by Ackroyds for worsted spinning.

Hoyle House Mill in Warley was another small mill which was eventually turned into cottages. However, the first person to occupy the mill was Joseph Charles Gautier, a cotton manufacturer, who insured the property, machinery and stock for £500 in 1792. Three years later Elkanah Hoyle & Son took the mill. In 1801 they insured the mill and contents for the following sums:

	£
Hoyle House Mill	400
Mill work	50
Machinery	550
Stock	50
	1,050

Holroyds continued to occupy the mill until it was converted into cottages in 1810.

In 1804 the mill measured thirty yards by ten yards and had a thirty foot water wheel.

Willow Hall Mills in Skircoat were built by the same firm and on adjoining sites but at different times. Edmund Lodge (1721–99), a wealthy Leeds merchant, bought Lower Willow Hall about 1774 to add to land he already owned in the area and, about the same time, moved from Leeds to Halifax. He had extensive manufacturing and merchanting interests with cotton mills in Lancashire and Scotland. Upper Willow Hall Estate was for sale in 1782 and may also have been bought by Lodge.

Lower Willow Hall Mill was built about 1783 in brick, which was an unusual material for the neighbourhood, although not for a person from Leeds where there were many brick built mills. The mill became known as 'The Brick Factory' and was used for cotton spinning with another, possibly older mill, used for fulling and cotton spinning. By 1797 the insurance valuations were:

	£
Cotton factory	750
Machinery	1,250
Stock	1,000
	3,000
Low Mill (part cotton, part fulling)	400
Mill work and fulling stocks	100
Cotton machinery	150
Cotton stock	50
	700

Later that year another policy was issued on a stone building, which was in the course of erection and was intended to be a cotton mill. The section of the mill already built was insured for £600.

Edmund Lodge's two sons, Thomas and Henry, were to take over their father's interests and insurance policies were issued in their names and that of his widow in 1800. The new stone mill was nearing completion in 1798 when the insurance value was:

	£
Cotton mill at Willow Hall	1,500
Mill work	400
Machinery	1,200
Stock	250
	3,350

All the machinery may not have been installed for in 1799 the insurance value of the machinery was raised to £2,000.

The brick factory was equipped with a 30 hp Boulton & Watt engine in 1802. It was a twenty-eight and one eighth inch by six foot cylinder engine with sun and planet motion. That was bought new from Boulton & Watt but a second-hand engine from the same makers was bought for the stone mill from Howard & Houghton's paper mill in Hull. This second engine was rated at 10 hp, was also a sun and planet engine and had a seventeen inch by four foot cylinder. In 1805 gas lighting was introduced which made this the first textile mill in the country to use the new form of illumination.

About 1810, Thomas and Henry Lodge gave up cotton spinning themselves and rented the two mills. The stone factory was let to William Huntriss, who equipped the mill for mule spinning. According to Cromptons survey of 1811 he had eighteen mules with two hundred and forty spindles each. Huntriss left this mill in 1813, possibly to start worsted spinning in Halifax, and the mill was offered to let. It was equipped with the 10 hp Boulton & Watt engine, a forty-two foot water wheel and contained 2,200 yards of floor space. There were also cottages and a blacksmith's shop. The stone mill was then possibly taken by Thomas and John Sutcliffe, who had rented the brick factory from Thomas and Henry Lodge since 1810. In 1811, according to Crompton, they had been running forty-eight mules with one hundred and twenty spindles each.

The brick mill was also offered for sale in 1813 when it was still tenanted by the Sutcliffes. The sale notice described it as being 'substantial' and 'the timbers uncommonly strong'. Power was provided by the 30 hp Boulton & Watt engine, but no water wheel, and the mill contained 3,000 square yards of floor space of fire proof construction. Included in the sale were several houses, including one for an overlooker.

Neither of the sale notices mentioned gas lighting, although Samuel Clegg, who was previously employed by Boulton & Watt is supposed to have installed gas lighting in Henry Lodge's cotton mill in 1805.

In 1833 the brick mill had a 32 hp engine and the stone mill a 25 hp engine. The mills were used for both ccotton spinning and weaving and employed three hundred and twenty-six people.

The two mills continued to be run by the Sutcliffes until after the end of this survey.

Wharfe Mill. Skircoat. This mill may have had other names and may have been used for cotton spinning from the early 1780s. There was a fire in 1806 when the firm of R & W Wainhouse occupied the mill which was often called Wainhouse Mill. In 1811 Crompton noted that 'Wainhursts' were running twenty-four mules with three hundred spindles each, which,

with seven thousand two hundred spindles, made this the largest mill, in terms of spindle capacity, in the area. Thirty years later when the mill was occupied by Robert & William Wainhouse, the number of mule spindles had only increased to ten thousand. Cotton spinning continued at this mill until it was turned into an engineering works.

Stern Mills in Skircoat consisted of a corn mill and a fulling mill, to which cotton spinning was added for a number of years. Thomas and John Sutcliffe occupied part of the mills in 1802 but moved to Willow Hall Mill in 1810 so it is not clear who was running the mill in 1811 when Crompton noted that the following spinning machines were in use there:

8 mules × 240 spindles
1 mule × 120 spindles
5 mules × 126 spindles
jennies with 630 spindles

The small size of some of the mules and the use of jennies indicates that much of the machinery was hand operated and when the mill was to let in 1818 it was said that it was mainly a corn mill but part was used for cotton spinning, possibly by John Swallow. Cotton spinning continued in this section of Sterne Mills until at least 1832.

Copley Mill on the river Calder was a woollen mill but for a few years around 1810 was used for cotton spinning. Crompton noted that in 1811 Copley Mill held the following spinning machinery:

12 mules × 228 spindles
8 mules × 252 spindles
8 throstles × 120 spindles

Copley mill was to let in 1835 and could be rebuilt but no use was given.

Brow Mill. Skircoat. The partnership between James Milnes and Watts Wrigley, who were cotton spinners at Bolton Brow in Skircoat, was dissolved on 12th December, 1803 so it would seem that the firm had been running this mill before that time. Wrigley was to pay any debts of the partnership and continued cotton spinning at the mill until his bankruptcy in 1806. He must have been able to continue in business after 1806 for Crompton noted that he had four mules with two hundred and fifty-two spindles each in 1811. The mill was later occupied by Binns & Wrigley.

In 1832 two factories occupied by Messrs Watts Wrigley & Sons were to let. The first measured thirty-one yards by twelve yards and was four storeys high with a 12 hp steam engine. The second was sixteen yards by nineteen feet, three storeys high and had an 8 hp engine.

Marshall Hall Mill. Elland. This mill had been run by Samuel Broadbent & Co. up to 1807 when the partnership was dissolved. The former partners had been Samuel Broadbent, John Milner, Joseph Rushforth, Benjamin Rushforth and Charles Broadbent and besides cotton spinning they had also traded in cotton goods. In 1811 Crompton noted that this mill held twelve mules each with two hundred and forty spindles. Marshall Hall Mill was changed over to woollen manufacturing by 1822.

Thornhill Briggs Mill. Hipperholme. The Thornhill Briggs Estate was for sale in 1791 and it was mentioned in the advertisement that it was a suitable location to build cotton mills. In 1797 Joseph Cartledge bought screws, rods and 'turning iron' from Kirkstall Forge when this mill was built. In 1811 Crompton noted that Joseph Cartledge was running twenty throstles each with one hundred and twenty spindles. James Cartledge was bankrupt in 1825 but his assignees

carried on running the two mills at Brow Bridge and Thornhill Briggs. The firm of Joseph Cartledge & Sons were recorded as cotton spinners in contemporary trade directories until after 1830 by which time the mill was still being run by the firm's assignees. In 1826 the mill had a 30 hp steam engine as well as a water wheel for the twenty-three foot six inch fall. The mill was on the side of the Calder & Hebble Navigation and also for sale was a mansion and twenty-four cottages. The mill continued to be used for cotton spinning until after the end of this survey.

At Hartshead. In 1803 this mill, or part of it, was occupied by a man called John Webster for cotton spinning and he had six employees. It is likely that this venture into cotton spinning was only a temporary measure as this was probably a woollen mill.

Ryburn Valley

Temple Mill. Rishworth. This mill was built by Soloman Lumb who unfortunately, over extended his financial resources in building the mill and was rendered bankrupt as soon as he started trading. He insured the mill and contents for the following sums in 1799:

	£
Cotton mill	1,000
Mill work	200
Machinery	600
Stock	200
	2,000

Temple Mill was seven storeys high with 'each storey consisting of one light, spacious room, capable of containing eight mules of three hundred spindles each'. The mill measured sixty-nine feet by forty-two feet inside the walls, and had a fall of water of thirty-four feet to drive the wheel.

Soloman Lumb, who was a cotton

manufacturer, dealer and chapman, had also envisaged weaving cotton cloth at the mill, for he had built a loom house adjacent to the mill which could hold fifty pairs of looms.

Temple Mill was run by Lumb's assignees for several years until they found a purchaser. Lumb's bankruptcy dragged on after the initial proceedings and he was shot and dangerously wounded in 1800. One of the first tenants, Henry Knowles, was also bankrupt in 1804. He had used the mill for mule spinning and some of his stock and machinery were for sale in 1804 including:

9 mules × 300 spindles

2 mules × 240 spindles

The mill was unoccupied in 1808 but by 1811 had been taken by John Hoyle who, according to Crompton, had sixteen mules with three hundred spindles each. Hoyle continued until 1832 when the mill was taken by William Stead. It is likely that some other firms took room and power at Temple Mill during the same period. About 1837 steam power was added and the mill burnt down in 1860.

Booth Bridge Mill in Rishworth was built in 1794 by John Pickup from Jumples Mill in Ovenden. The first tenants were John Bottomley of Longwood and John Dyson of Scammonden who took the mill for cotton spinning. The mill was very small at that time and only insured for £100 in 1795, with the machinery valued at £200. In 1802, when Booth Bridge Mill was for sale, together with the cotton machinery, it was only three storeys high and twenty-seven feet by twenty-four feet inside. However, a new four storey mill had been built alongside which measured thirty-nine feet by thirty-eight feet. Although this building was not occupied in 1802, future references are confusing because of the two buildings.

Tenants in the first decade of the century included John Dyson, John Hoyle, Robert Lewis, Henry Binns and Richard Holt. In 1811 Crompton noted that there were twelve mules with two hundred and forty spindles each at 'Booth Factory'. The tenant from about 1818 until 1882 was John Stansfield Wells, while the owner was John Dyson. Both mills had been for sale in 1818 when there was a twenty-four foot wheel

Booth Wood Mill in Rishworth was mainly used for paper manufacture, possibly by John Garside who bought spindles and an axle tree from Kirkstall Forge in 1785. However, for a period of at least five years it was partially occupied by James Garside for preparing and spinning cotton. Garside used at least three large rooms on three floors of the mill, which appear to have measured nineteen yards by eight yards. Besides the usual preparing machinery, Garside had three mules with two hundred and twenty-eight spindles each when the lease became available after his death in 1805.

Spring Mill in Rishworth was built about 1800 for cotton spinning. The tenants from 1807, until well after the end of this survey, were James, and later, Benjamin Mallalieu.

Slithero Mills in Rishworth were fulling mills built in the sixteenth century. One of the mills was possibly changed to cotton spinning in 1783 when it was leased to Michael and John Richardson who were cotton manufacturers. Both activities were carried on at the mills however, for in 1798 they were described as being used for 'carding and spinning cotton and flax, and fulling woollen cloth. John

Richardson insured the mills and contents for the following sums in 1798:

	£
Slithero Mill	300
Mill work	350
Machinery	400
Stock	50
Warehouse, counting house, and cottages	300
Two cottages	100
	1,500

While John Richardson continued fulling woollen cloth and spinning cotton in one mill, John Edwards used the other mill for spinning and weaving woollen cloth.

In 1811 Crompton noted that John Richardson had the following machinery:

12 mules × 204 spindles

4 throstles × 120 spindles

Later extensions possibly linked all the buildings together and between 1826 and 1841 the Edwards family occupied Slithero Mill for cotton spinning. Firstly a partnership between Henry Lees Edwards and Thomas Grove Edwards and then, from 1882, John Edwards & Son. This family also ran Dyson Lane Mill.

Thrum Hall Mill or Beestonhirst Mill or Beeston Mill. Soyland. The machinery at Beeston Mill belonging to C & E Kenworthy was for sale in 1808 together with their household goods by order of their trustees. This included:

6 throstles × 120 spindles

3 throstles × 132 spindles

6 throstles × 108 spindles

1 mule × 228 spindles

12 patent looms

By 1811 the mill had been taken by George Hartley and Crompton noted that he had the following machinery in 1811:

12 mules × 240 spindles

In 1815 George Stansfield Wells may have taken this mill for cotton spinning but by the early 1820s Lawton & Mallalieu oc-cupied the mill and they were followed by Joseph Lawton.

Hanging Lee Mill in Soyland was built in 1788 and used for wool scribbling until 1802 when it was sold to Thomas Stead for £4,609 and changed to cotton spinning. The mill was three storeys high and measured forty-four feet by thirty-four feet within the walls. There was a thirty foot fall of water to the wheel which measured thirty feet by four feet.

The mill was for sale in 1821 but without the machinery for it was tenanted by James Mason from 1812 onwards. Thomas Stead, the owner, was bankrupt by 1822 and James Mason, who was also a cotton manufacturer, bought the mill from him.

Hazel Grove or Hutch Clough or Cockroft Mill in Rishworth had a number of other names but was built about 1792 by John Holroyd senior and John Holroyd junior and sold to Robert Brear of Middleton, Manchester for £1,050. It was described as newly erected when it was offered for sale in 1794. The four storey mill was twelve yards square and had a 'large' water wheel to drive the cards only, as it was a jenny mill at that time.

This mill was sold to Soloman Lumb of Temple Mill in Rishworth for £1,200 in 1796. Lumb purchased the mill with the help of a mortgage from W Brisco of Devonshire Place, Marylebone, Middlesex. For insurance purposes the mill was valued for the following sums:

	£
Cotton mill	300
Mill work	30
Machinery	120
Stock	50
	500

By 1799 Soloman Lumb was in financial difficulties and had to assign his property to James and Joseph Kershaw, who were cotton merchants in Manchester. In all, he owed them £1,602, and Lumb's two mills were put up for sale to pay off the debt. This mill had been leased to John Moore, and in 1800 was said to be 'in full employ for country work, and capable of working five or six carding engines'. When the mill was offered for sale again in 1802 there were two tenants, John Hartley and William Thomas junior of Soyland, who were cotton manufacturers.

Robert Holroyd, a merchant from Kebroyd, bought the mill in 1803 for £2,000 which indicates that it might have been extended. He allowed the existing tenants, Hartley & Thomas to continue but John Firth took part of the mill. Holroyd sold the mill to William Thomas in 1809 and it was sold again to John Whitely of Soyland who was to own a number of mills in the area. The mill was for sale again in 1821. It had not been enlarged and was still four storeys high and eleven yards square and said to be well supplied with water. Levi Lumb would show prospective purchasers around the mill. By the 1830s a steam engine had been added.

Swift Place Mill. Soyland. There were two fulling mills on this site until 1803 when this one was let to John Knight Senior, John Knight Junior and William Bamber, cotton spinners from Manchester. These partners bought the mill in 1805. Later John Whiteley, who was 'one of the most successful spinners of his time in the valley' bought the mill and Knights & Bamber continued as tenants until 1858. Whiteley & Clayton may have taken part of the mill about 1820. John Whiteley also acquired Stones Mill, Hazel Grove Mill, Upper and Lower Swift Place Mills, Dyson Lane Mill and later, Ryburn Mill.

Lower Swift Place Mill in Soyland was a fulling mill up to 1803 when it was for

sale after Samuel Dyson's bankruptcy. For a short period it became a paper mill but then cotton spinning was carried on there by Joseph Barrett, Joseph Cocker, Benjamin Meller and later, John Whitely.

Stones Mill in Soyland was built about 1800 for cotton spinning and run by Fenton, Lambert & Co. In 1811 Crompton noted that 'Lambert Factory' held the following machinery:

14 mules × 204 spindles
8 throstles × 120 spindles

Robert and Fenton Lambert occupied the mill until a little time after they dissolved their partnership in August 1819. The firm at that time were cotton spinners and manufacturers and sold in the Manchester market. Thomas Stead, the owner of the mill, who may also have had an interest in a cotton mill at Morton near Keighley, sold the premises to John Whiteley who started cotton spinning there. Stones Mill at that time was four storeys high and measured sixty feet by thirty feet within the walls. Power was provided by a twenty-seven foot by five foot water wheel. Besides the mill there was a warehouse, two large rooms for bleaching or dyeing, two dry houses and a sizing room. Steam power was added in 1833 and the mill was enlarged in later years.

Dyson Lane Mill. Soyland. There were two mills at Dyson Lane. One was a woollen mill but the cotton mill was built about 1803 by John Haugh, a blacksmith, and occupied by Henry Binns for cotton spinning until about 1845. He was a cotton spinner and fustian manufacturer and sold his cloth on the Manchester market. At some time the mill was sold to John Whitely but Binns continued as tenant.

Lower Dyson Lane Mill. Soyland. This mill was originally a fulling mill to which cotton spinning was added by 1822. John

Hoyle, who was also a cotton manufacturer, ran the mill from 1822, or before, until about 1830.

Hollings Mill. Soyland. This was a small mill built by Elkanah Hoyle about 1788 on a good central site in the valley and then enlarged about 1792. Elkanah Hoyle Junior later took over the mill but it was demolished when Ripponden Commercial Mills were built on the site in 1855.

Smallees Mill in Soyland was originally a fulling mill but was rebuilt as a cotton mill about 1801. The mill was to let in 1803 and was described as 'partly finished and partly in full work'. The mill had four storeys plus a garret and measured eighteen yards by thirteen yards one foot inside the walls while the unfinished part was to be the same size. Power was provided by a twenty-one foot by seven foot six inch water wheel which was said to generate 50 hp, which seems rather optimistic.

The tenant of the mill could also buy the spinning machinery which included:

10 mules × 300 spindles

1 throstle × 120 spindles

The construction of the mill had been undertaken by Elkanah Hoyle and Joshua Bates who had borrowed money from Swaine Brother's bank in Halifax. Hoyle was a school master and Bates an engineer. However, after £1,500 had been borrowed with the mill still not finished, Swaines foreclosed the mortgage and the mill had to be sold. Smallees Mill was not sold for two years but during that time Knight & Bamber, cotton spinners from Manchester, rented the mill. They gave up the mill in October 1805 when their machinery was for sale and the mill was to let. John Knight's machinery included:

6 mules × 300 spindles

2 mules × 298 spindles

3 throstles × 120 spindles

1 throstle × 130 spindles

Smallees Mill was then taken by John Holroyd of Ryburn House. He was a fustian manufacturer and he occupied this mill until his death in 1837.

Ripponden Mill in Barkisland was an old fulling mill which was for sale in 1782 and possibly converted for cotton spinning in 1793. The mill was then possibly run by a partnership between Ralph Holt from Heywood in the Parish of Bury and John Hoyle from Nook End in Barkisland. This partnership was dissolved in January 1801 and the cotton machinery was taken out although the mill was owned by Hoyle. The four storey mill at that time had two sixteen foot by six foot water wheels and measured eighty-three feet by thirty-three and a half feet. A mill in Barkisland, possibly this one, was converted to be used for cotton spinning and twisting in 1830. In 1833 it had an 8 hp wheel and employed fifty-two people.

Ripponden Wood Mill, later Ryburn House. Soyland. This mill was built by John Learoyd and sold to John Holroyd in 1792 for £700. It was driven by a twenty-two foot by two foot water wheel and the machinery included eleven spinning frames. The mill already incorporated a house when it was insured by John Holroyd in 1792:

	£
Cotton mill	50
Utensils and machinery	160
Dwelling house	100
Furniture	40
Stock in house	300
Stable and hay loft	25
Utensils and stock	25
	700

John Holroyd was a cotton manufacturer and therefore kept a stock of cotton at his house which was part of the mill. Holroyd was also spinning cotton at Smallees and Kebroyd Mills and eventually cotton spinning was stopped at this small mill which then became a private residence called Ryburn House.

Tom Hole Mill in Soyland was an old woollen mill but a new mill was built alongside for cotton spinning by 1803 when it was for sale. This new mill became known as Greavehead Mill and measured twenty-two and a half feet by eighteen and a half feet inside and was driven by a twenty-seven foot by two foot water wheel. The mill was owned by a Mr Ramsden from Chapeltown and the tenant in 1803 was Thomas Atkinson. Joseph Hartley occupied the mill for cotton spinning from at least 1822 until about 1830 but this mill was later used for woollen manufacturing.

Clough Mill in Soyland was originally a fulling mill and was possibly bought by Richard Learoyd in 1792 and used for cotton spinning. His son Joah Learoyd took over the mill in 1806 and in 1811 Crompton noted that he had six mules with two hundred and fifty-two spindles each.

Joah Learoyd and his brother James, carried on cotton spinning at Clough Mill for many years. The mill however, was not large for in 1833 it was only powered by a 4 hp water wheel and employed thirty-seven people. In 1841 the spinning machinery consisted of two mules and six jennies.

Severhills Mill. Soyland. This mill was built over a number of years from about 1799 by John Broadhead, a corn miller, and John Holroyd, a cotton manufacturer and merchant. The initial stage of the mill was insured for £175 in 1799 and the mill was completed in 1802 but Broadbent died in May 1804.

The original mill was four storeys high with a garret and was water powered. It does not seem that John Holroyd continued cotton spinning at Severhills for long after his partner's death and it was then let to Thomas Hadwen. According to Crompton in 1811, Thomas Hadwen was using it for mule spinning with twelve mules, each with two hundred and fifty-two spindles. By 1818 both Thomas Hadwen's lease and his machinery were for sale when he had thirteen mules and nine jennies.

The mill had several tenants after 1818. They included James and George Mitchell by 1828 and William Pogson from Booth Wood Mill by 1833. However, the mill was still being used for cotton spinning in 1835 at the end of this survey.

Soyland Mills. Soyland. A number of small corn mills, some of which were very old, made up Soyland Mills. There were therefore, several buildings which were converted or added to and used for cotton spinning at various dates. These mills were advertised for sale at various times in the 1780s.

By 1799 John Broadhead, a corn miller, had converted part of his corn mill for cotton spinning and insured his property for the following sums:

	£
Corn and cotton mill	75
Machinery	100
Stock	25
	200

Another person who was spinning cotton in one of the buildings was Samuel Riley for in 1802 he had several mules and throstles for sale. By 1804 he was bankrupt and all his machinery was for sale. This included two mules with two hundred and sixteen spindles and one with one hundred and eighty. Although he was bankrupt Riley was still at the mill in May 1805 when the mill was to let with, or without, the machinery which then included five mules and a throstle.

Joshua Bates, a cotton manufacturer, may also have leased one of the mills about 1800.

In 1811 Crompton noted the details of the machinery at Soyland Mill but not the firms. The details were:

6 mules × 204 spindles

2 throstles × 220 spindles

Lower Soyland Mill, which had been one of the corn mills, was for sale in 1818. It was still equipped with corn grinding stones but was tenanted by George Binns who was using the mill for cotton spinning. By 1822 the firm of Binns & Wrigley had changed the mill to silk and wool spinning.

Lumb Mills. Sowerby. There were two mills on this site and Upper Lumb Mill had originally been a fulling mill in the seventeenth century but, by 1805, was occupied by Abraham Whitehead for cotton spinning. By 1811 he was running twelve mules with two hundred and forty spindles each, according to Crompton. Abraham Whitehead Junior was bankrupt the following year and the mill was subsequently taken by John Whitehead. In 1833 Lumb Mill was occupied by John Whitehead, Senior and Junior, and still used for mule spinning. The mill had a 10 hp water wheel with water from Lumb Clough, and fifty-eight people were employed there.

Damside Mill. Soyland. Thomas Crompton had been spinning cotton at this small mill at Damside, Soyland until 1805. His machinery was then for sale and included four throstles with one hundred and twenty spindles each. Twelve years of the existing lease on the mill could also be purchased. Crompton assigned his

estate to Thomas Pollitt of Broadgates, Skircoat, cotton merchant, and John Oldroyd, Ripponden Wood, cotton spinner. This mill may not have been used for cotton spinning again until 1846.

Kebroyd Mills in Soyland consisted originally of two fulling mills and a corn mill. Their conversion to cotton spinning over a period of years and a multiplicity of owners and tenants makes a chronological description of their use difficult.

Kebroyd Upper Mill. This was a fulling mill which may have been partially used for cotton spinning from about 1790. It was a four storey mill in 1803, measuring fifty-one feet by thirty-one feet and driven by two water wheels, each twenty-seven feet in diameter. One section of Upper Mill was used for fulling and the two other sections were used for cotton spinning by Robert & William Graham and previously, George Wainwright. Both of these firms were tenants as the owner of all the mills was John Denton. Robert and William Graham of Making Place, Halifax and James Graham of Aldermanbury, London, were cotton manufacturers and merchants up to the end of 1803 when they were bankrupt. The machinery they had been running at Kebroyd Upper Mill included:

2 mules × 216 spindles

2 mules × 204 spindles

1 jenny × 80 spindles

Some time later, John Hadwen & Son, who had been using the Middle Mill for cotton and silk spinning, took the Upper Mill as well. By 1828 the mill belonged to Alice Denton, who may have been the widow of John Denton, and the mill was insured for £500. Hadwens insured their property in the mill for:

Mill work   £150

Machinery   1,650
Stock   200
                          ─────
                          2,000

Kebroyd Middle Mill. This five storey mill measured seventy-five feet by thirty-one feet in 1803 and was driven by a thirty-three foot water wheel. It had been built for cotton spinning by Thomas and John Hadwen, who had previously used Severhills Mill for cotton spinning. It was said to be 'newly erected' in 1803 and was described as 'Hadwen's Factory'. The Hadwens had originally come from Bolton in Lancashire where they had also been involved in cotton spinning.

A new warehouse was added to the mill in 1804 by Thomas and John Hadwen. The brothers dissolved their partnership in 1805. Thomas left the firm, which continued as John Hadwen & Sons but now included Thomas, John, Sidney and George Hadwen. Thomas Hadwen later moved to Kebroyd Lower Mill. In 1811 Crompton noted that Hadwen & Wilson had the following machinery:

20 mules × 240 spindles
8 throstles × 120 spindles

Some time in the 1820s silk spinning was added to cotton and, in the 1830s, a Pollit steam engine was installed because of the irregular water supply.

Kebroyd Low Mill. This was formerly a corn mill but by 1793 Samuel Bailey, a cotton manufacturer was using part of it for cotton spinning and insured his machinery and other utensils there for £150. By about 1800, the owner of the mill, John Denton, may have been using

Low Mill for cotton spinning. The mill in 1803 was five storeys high and measured sixty feet by thirty-six feet and had a twenty foot by six foot water wheel. Up to 1806 it had been tenanted by John Wright who was a cotton manufacturer who employed hand-loom weavers and used land near the mill for bleaching. When Wright was bankrupt in 1806 his machinery included:

2 mules × 252 spindles
2 throstles × 120 spindles
5 jennies × 106 spindles

Between 1810 and 1823 the mill was occupied by Holroyd & Denton. In 1811 Crompton noted that they had the following spinning machinery:

12 mules × 240 spindles
4 throstles × 120 spindles

Some time after 1823 Thomas Hadwen of Kebroyd Middle Mill took over the Lower Mill.

Watson Mill in Norland was originally a corn mill and later reverted to corn milling after the early boom in cotton spinning had passed. It was to let as a corn mill in 1799 but might have been converted shortly afterwards for the firm of Whitworth, Ashforth & Ashforth occupied the mill before 1805 as fustian manufacturers. The partnership was dissolved in 1805 and the following year their cotton machinery was for sale. It included:

4 mules × 252 spindles
1 mule × 140 spindles
2 jennies

William Whitworth continued cotton spinning at this mill until about 1816.

CHAPTER 13

The Keighley Area
including Haworth and Morton

THE first and only mill to be built in Yorkshire under the protection of Arkwright patents was Low Mill in Keighley in 1780. In that year there were twenty mills operating Arkwright's water frames which were owned or licensed by him. Little is known about the licences which were sold to allow other people to build mills, but it was felt that Arkwright's charges were too high. The success of Low Mill, despite the licence payments, obviously encouraged others in Keighley to build cotton mills, particularly after Arkwright's patents had been set aside.

The Worth Valley with North Beck then became one of the busiest centres for the construction of cotton mills. By 1790 eight mills had been built and in the next ten years over twenty more were added. Those mills were so close together in the town of Keighley that water from the river Worth and North Beck was fed into the mill dam of a downstream mill as soon as it had been returned from the tail goit of an upstream mill. The raising and lowering of the water levels by the various mill owners was a constant source of complaint. Despite these difficulties water power remained important as it constituted a considerable investment and even by 1835 only one of the four mills still used for cotton spinning had a steam engine.[1]

The water power available in the area was one of the reasons for the enthusiastic adoption of the industry. The others were the firm foundation of the domestic worsted industry and the relative nearness of Lancashire. A further suggested factor influencing the initial development of cotton spinning in the area might have been the ownership of some of the land by the Duke of Devonshire who also owned estates in Derbyshire where some of the earliest mills were built by Arkwright and others.

By 1792 it could be said that: 'the manufacturing of cotton is carried on in that neighbourhood (Haworth) and about Keighley in a very extensive manner.'[2]

Another valley in the Keighley area where water power was used to maximum effect was the Morton Valley where the short Morton Beck powered eight mills for various purposes before joining the river Aire two miles to the east of Keighley. Three of the mills were built for cotton spinning and one of them, Upper Mill, was still being used for that purpose in 1835. Upper Mill was said to have been built in 1779 which would have made it the first cotton mill to be built in the county, but that date cannot be verified.[3]

A small number of the pioneers of cotton spinning in Keighley came from Lancashire but they were outnumbered by local men and women. The partners in many of the firms came from the ranks of lawyers, worsted manufacturers, drapers and land-owners in or near the town of Keighley. With the benefit of a firm base in Keighley and the acquisition of some skill and capital several of the early Keighley firms extended their interests into other districts. Claytons & Walshman built Langcliffe Mill near Settle in 1784. Joseph Driver was a partner in Askrigg Cotton Mill by 1785. Thomas Parker was running Arncliffe Mill by 1792 and John Ellison ran Sutton Mill by 1810. John Greenwood, the most successful of the Keighley cotton spinners, was a partner in mills at Hampsthwaite, Burley-in-Wharfedale, Bingley, Airton and Cullingworth.

When worsted spinning by power became firmly established and trading conditions for the worsted industry improved after 1815 the gradual move back from cotton to worsted in the Keighley area accelerated. As far as the worsted industry was concerned the change to factory production came in stages with spinning being the first. It was relatively easy to start worsted spinning in the cotton mills and so over twenty Keighley mills were converted within a few years. There was a considerable transfer of assets which benefited the worsted industry although the early cotton mills rapidly became inadequate in terms of size. New and larger worsted mills were built which has meant that some of the early cotton mill buildings have remained.

The early concentration of cotton spinning mills in the Keighley area brought a need for machine makers and mechanics who could construct and maintain the preparing and spinning machines. Keighley soon became the home of a number of textile machine manufacturers which in turn stimulated the growth of the machine tool industry in the town.

The Mills

Mills on the River Worth and tributaries

Royd House Mill was built by Robert Heaton and his father-in-law John Murgatroyd in 1791, the two men having previously been worsted manufacturers in the area. The mill was burnt down in 1808 but, despite not being insured, was rebuilt and used for worsted spinning from 1810.

Bridgehouse Mill. This mill was possibly built by John Greenwood in 1785 or before on the site of a corn mill. Brooks Priestley, the tenant, married Greenwood's daughter Elizabeth, and had as his partner Charles Woodiwis. They traded as cotton spinners until July 1785, probably at this mill. In that year Brooks Priestly insured his utensils and stock with the Sun Fire Office for £300. In 1788 Joseph Blakey and a man called Lomax were spinning cotton at the mill which was still owned by James Greenwood but in 1789 William Greenwood, James Greenwood and Henry Rishworth were also spinning worsted there, possibly in another part of the mill. James Greenwood, who was John Greenwood's son, then insured the mill as a cotton mill in 1793 when it may still have been run by Blakey & Lomax:

	£
Cotton mill	200
Utensils and machinery	400
	600

In 1803 William Ellis employed twenty-five cotton workers at the mill but was the tenant as the mill was still owned by James Greenwood. The mill was changed over completely to worsted spinning about 1810 as were the majority of cotton mills in the Worth Valley.

Mytholm Mill. The firm of Newsholm, Sugden & Co. occupied this mill for cotton spinning for several years from 1791 when they were buying iron from Kirkstall Forge. In 1803 they employed forty-five people at the mill. The partners were John Wright, Lupton Wright, Henry Wright, William Newsholme and Robert Sugden. Sugden was in dispute with the other partners in 1808 and left the partnership.

In 1811 Crompton noted that the mill held sixteen frames with fifty-two spindles each. By 1813 the partners were also wool staplers but had to assign their assets to their creditors. However, they appeared to overcome their financial problems and William Newsholme & Sons continued to use part of Mytholm Mill for cotton spinning until 1831 when they were bankrupt. The three storey mill then measured sixty-six feet by thirty-four feet and the water wheel was said to generate about 16 hp. The machinery included four mules with two hundred and ten spindles each as well

as twelve throstles with ninety-six spindles each. The Newsholmes appeared to be in debt to two Manchester cotton dealers, W. H. Harrison and Joseph Thompson. After the sale the mill was converted to worsted spinning.

Ponden Mill was built on a corn mill site by Robert Heaton senior in 1791. Heaton had been a worsted manufacturer prior to this new venture. Robert Heaton died in 1793 and his mill burnt down in 1795. However, his two sons, Robert junior and William, rebuilt the mill with the help of insurance money received from the Royal Exchange Fire Office and recommenced cotton spinning. The new mill and contents were insured for the following amounts in 1795:

	£
Ponden Mill	550
Mill work	100
Machinery	420

Ponden Mill, which was built for cotton spinning in 1791–2 and later used for worsted spinning.

Stock   430
                        1,500

In 1803 the firm had twenty-five employees and in 1811 Crompton's survey found that there were sixteen water-frames with forty-eight spindles each at the mill. By 1813 Robert Heaton was losing money at the mill and eventually gave up cotton spinning. Ponden Mill was then taken by John Holt and later by John Lonsdale. In 1832 the three storey mill was for sale and there was a water wheel which measured twenty-seven feet by seven feet. Lonsdale's machinery was also for sale and included fourteen throstles with one hundred spindles each although the mill was still owned by Robert Heaton. The mill was changed over to worsted spinning some time after 1835.

Griffe Mill, Stanbury. The firm of Hollings & Ross were spinning cotton at this mill in 1793. In 1811 Crompton noted that the mill held ten water frames with fifty-two spindles each so it was not a large building.

From 1811 or before, until at least 1835 part of this mill was tenanted by Thomas Lister for cotton spinning while the other part was retained by James Ross until 1820. In that year Ross wished to sell his share. The mill then measured twelve yards by nine yards and Ross had five throstles with ninety-six spindles each. Griffe Mill was then taken for worsted spinning.

Lumb Foot Mill, Stanbury. This mill was owned by John Wignal of Keighley in 1797 when he insured the building for £200. His tenants were Jonas Turner and John Rushworth and they insured the mill work for £20, machinery for £230 and stock for £200. By 1805 this mill had been changed over to worsted spinning by Jonas Sugden.

Spring Head Mill may have been built in 1786 when John Heaton bought iron rods from Kirkstall Forge. John Heaton was a cotton manufacturer and besides the mill he built a house and cottages. He employed a large number of hand loom weavers in the area and regularly visited Manchester to sell his cotton pieces and buy raw cotton. In 1801 he insured his mill and contents for:

	£
Cotton mill	200
Mill work	100
Machinery	300
	600

In 1803 John Heaton had forty employees but died in January 1804 and left a widow and young sons, Michael, John and William. The firm continued as Dinah Heaton & Sons, cotton twist spinners, dealers and chapmen but they were rendered bankrupt in 1808 when they possibly owed money to Lister Ellis of Castlefield Mill near Bingley.

Crompton noted that 'John Heaton' had sixteen frames with fifty-two spindles each in 1811 but no further references have been found to cotton spinning at this mill.

Higher Providence Mill. This mill was built by a man called Leach from Halifax in 1801 or before. It was then leased by William and John Haggas who were cotton spinners and manufacturers. By 1811 they were running mules at this mill which was unusual for the Keighley area. They had six mules with two hundred and sixty-four spindles each and these would supply them with weft for their cotton pieces. In 1812 they also started manufacturing worsted cloth and changed the mill over to worsted spinning about 1814.

Vale Mill. This mill appears to have been built and then run by the Greenwood family for over fifty years from 1792 until 1844 when it was converted to spin worsted. In 1803, John Greenwood, who also ran Cabbage Mill in Keighley, employed one hundred and ten people at Vale Mill. Crompton noted that there were thirty-two water frames with fifty-two spindles each in 1811. Other details of what was an important mill are hard to find as the mill stayed in the same hands for many years. However, there is a small book entitled 'Account of the Time Worked at Vale Mill. With the time gained or lost by the Water Clock with any other thing remarkable.' This book has brief accounts of activity at the mill for a few years from 1820. Entries are concerned with the installation of cards and frames as well as notes on the decline in wages for the cotton weavers. From this we know that the number of spindles rose to three thousand two hundred and four in 1820. In 1833 the mill was used for spinning cotton twist of qualities from No 20 to No 36, there were two powerful water wheels and seventy-three people worked there.

Damens Mill was built in 1789 by either William or John Roper who were farmers and worsted manufacturers. In 1802 they were employing pauper children at the mill. The Roper family ran the mill for many years and in 1811 Crompton recorded that they had the following spinning machinery:

4 mules × 216 spindles
6 frames × 52 spindles

Few of the Keighley cotton spinners had mules but Ropers continued mule spinning until their bankruptcy in 1820 when the mill, machinery, cottages and land were for sale. Damens Mill was then bought by the Greenwoods and taken by William Sugden for worsted spinning in 1824.

Wire Mill was built by 1801 and run by John Walker. It may not have been used for cotton spinning for more than a few years.

Grove Mill was built by Ann Illingworth in 1795 or before for her two sons David and William. As they were possibly only young men at the time she had as a partner William Marriner from Greengate Mill. Illingworths & Marriner had been cotton spinners at Castle Mill but moved to Grove Mill when it was completed. Ann Illingworth and William Marriner insured Grove Mill and its contents for the following sums in 1795:

	£
Mill	500
Mill work	125
Machinery	125
Stock	250
	1,000

Crompton noted that Illingworths had twelve frames with fifty-two spindles each in 1811 and cotton spinning appears to have continued until 1820. Illingworths tried to sell the mill in 1818 at which time it was four storeys high and measured fifty-seven feet by thirty-two feet with a fall of seventeen feet eight inches to the water wheel. The mill was heated by steam and was said to have a large dam. Robert and John Clough bought the mill which they converted to worsted spinning.

Ingrow Corn Mill was taken for cotton spinning by Lodge Calvert in 1801. As he was a joiner by trade it is likely that he made his own spinning frames as he bought spindles, fliers and rollers for his frames from Richard Hattersley, the Keighley machine maker.

By 1811 Lodge Calvert had changed to mule spinning and Crompton noted that he had four mules with two hundred and sixteen spindles each in that year. Calvert

Greengate mill, Keighley, with Greengate House on the left
and the warehouse on the right.

later took up worsted spinning, as did most of the Keighley cotton firms, and gave up cotton spinning about 1815.

Hope Mill was one of the first steam powered mills to be built in Keighley. Thomas Corlass built the mill in 1800 and insured the steam engine and machinery for £200 while the mill was being completed. In 1801 the full insurance cover was:

	£
Hope Mill	140
Mill work	50
Machinery	100
Stock	100
Steam engine	50
Four tenements	60
	500

In 1802 Thomas Corlass was employing pauper children at Hope Mill. Thomas Corlass lost money as a cotton spinner and the mill was taken by John Mitchell for worsted spinning in 1812. By 1837 Hope Mill was occupied by Grimshaw & Bracewell who were cotton spinners and manufacturers.

West Greengate Mill was built in 1784–85 by a partnership of Abraham Smith, Rowland Watson, John Blakey and James Greenwood and in 1785 they bought iron rod from Kirkstall Forge either for the new mill or for the new spinning machinery. In 1786 the mill and contents were insured for:

	£
Cotton water mill	500
Utensils, stock and machinery	1,000
	1,500

There were several changes in the partnership until 1818 when Benjamin and William Marriner bought out the remaining partners and converted the mill to spin worsted yarn. Greengate Mill was a large five storey mill with 1,500 square yards of floor space. In 1811 Crompton noted that there were thirty-six frames with fifty-two spindles each in the mill. The Globe Fire Office insured the mill and contents for the following sums in 1812:

	£
Cotton Mill	1,000
Water wheel and mill work	400
Machinery	1,000
Warehouse	200
Stock	800
	3,400

Greengate Mill was changed to worsted spinning in 1818.

East Greengate Mill was built for cotton spinning about 1795 by either John Craven or the firm of Craven, Brigg & Shackleton.

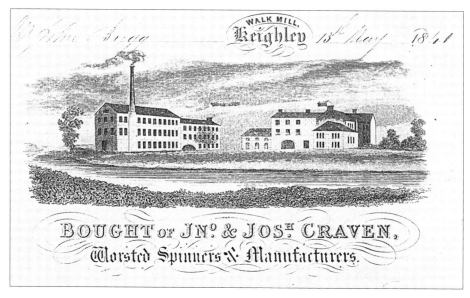

Walk Mill, Keighley. Shown on a mid nineteenth-century letterhead.

Cotton spinning continued until 1810 by which time John Craven had died and the mill was taken by A & J Hey for worsted spinning.

Walk Mill had been a fulling mill and then an early silk tape weaving mill until 1783 when part was converted into a cotton mill. The partnership which took the mill was made up of John Craven, gentleman, Thomas Brigg, piece maker and Abraham Shackleton who was also a piece maker. Many of the Keighley worsted manufacturers started cotton spinning as an additional venture at that time. In 1791 they agreed to build a new mill on part of the land near the old mill which belonged to John Craven. In 1795 they insured the mill and contents for the following sums:

	£
Walk mill	250
Mill work	50
Machinery	450
Stock	150
	900

The partners also ran Goose Eye Mill. In 1811 Walk Mill held twelve frames with fifty-two spindles each. Cotton spinning was stopped in 1812 and the mill was changed over to worsted spinning.

Low Bridge Mill was built as a steam powered cotton mill about 1800 by James Fox. After a few years the firm of John Ellison & Co, calico manufacturers and cotton twist spinners, took the mill but the partnership between John Ellison and William Wilkinson was dissolved in 1809 and re-established as Ellison & Wilkinson when William Wilkinson Junior joined the two older men. In 1811 Crompton noted that John Ellison had six mules with two hundred and twenty-eight spindles on each. After a fire at the mill in 1821 it was converted to spin worsted by Hartley Merrall.

Low Mill was the first cotton mill to be built in Yorkshire and established Keighley as a major centre for cotton spinning at the end of the eighteenth century. According

All that remains of Low Mill, Keighley.
This building probably dates from after 1780.

to a local writer the mill was started by Thomas Ramsden from Halifax and completed by Clayton & Walshman. The firm which first occupied the mill in 1780 was a partnership of John Clayton, George Clayton, William Clayton and Thomas Walshman. The Claytons were cotton manufacturers and calico printers with extensive works at Preston in Lancashire while Walshman had been a partner with Arkwright and others at Birkacre Mill near Chorley until it was destroyed by rioters in 1779. Low Mill was very much an Arkwright mill with the machinery made under his direction and some of the young workers and children sent to Cromford to learn how to use the new spinning frames. They, in turn, had to teach the rest of the carders and spinners who were taken on by Claytons & Walshman. Over a period of seven years £4,200 was paid to Arkwright for the use of his patent for the spinning frames and the instruction of the workforce. One of the few pieces of evidence relating to the royalties paid under licence to build and use the Arkwright machines comes from the partnership which ran Low Mill. In July 1786 Thomas Walshman wrote to Jedediah Strutt, who was still a partner with Arkwright, to try to have some of the royalties reduced. The payment of these must have seemed very irksome to Claytons & Walshman when growing numbers of rival spinners in nearby mills in Keighley paid nothing. The letter explains the situation:

Preston 4th July 1786
Sirs,
    Mr Claytons & myself have been expecting Mr Arkwright in Manches-

ter ever since last Feby in order that we might finally settle the paymts due on account of the priveledge granted to us under the Patents. On the 25 ulto we heard of his being there and on ye 27th waited upon him – we mentioned to him our original agreement being for 4200£ and Interest which was to be paid by installments, and that those payments have been punctualy observed by us, except the last, and the reason that was not paid at the time, was we wished to see him in person & hoped some part of the payments left behind wou'd be given up, as from the loss of the Roving Patent in 1781 all the country became Spinners as soon as ever they cou'd get machinery, to the great disadvantage of those who had purchased under you, On which account we cou'd not think the Patent of the same value as if it had then been established &. there fore hoped some concession wou'd be made – After talking awhile on these things, we paid Mr. Arkwright the money that was due in February last 660£ which makes the sums in all paid 4800£ which you will find is the full amount of Principal and all Interest – Now in this sum there was 600£ charged to us &. paid by us for overspinning, which we then thought exceeding hard & did not expect wou'd have been demanded as Mr Arkwright after the loss of his Patent, recommended to us, to do as much as we cou'd – that he wou'd be easy as to the payment for it – and that we might settle at ye last – We have performed every part of our engagement in every instance & I am sorry it has not been in your powers to perform yours on which account the only request we make is yt that 600£

for overspinning be allowed us, Mr Arkwright has kindly promised to give up His part of it, if you give up Yours &. we most cordaly hope from the equity of Your heart, that you will not have any objection –

I shall be very glad to receive your answer as soon as convenient &. am with Compts to all Your family Sir.

Your most humble Servant
T. Walshman

In 1788 Low Mill and contents were insured for the following sums:

	£
Mill	1,500
Utensils and machinery	2,000
Cotton warehouse	200
Stock	1,400
Joiners and smiths shop and steam or fire engine	400
	5,500

The steam or fire engine was a steam pumping engine which was used to pump water from the tail goit back into the mill dam. It is not clear if this pumping engine was installed when the mill was built in 1780 but it looks likely that it was installed a few years later when Claytons & Walshman were negotiating with Boulton & Watt for one of their engines in 1785 as the following letter explains.

Messrs Boulton & Watt
Gentlemen

Upon our further enquiry after fire engines we find your calculations over rated so far in the consumption of coles by the common engine, that the extra expenses attending the erection of yours with the yearly payment of your premium will not be saved, as we are situated. Therefore are about to have a common one put down. we are much obliged by your civility, & if you make any demand for the great trouble we gave you we beg that you will let us know, that we may order some of our friends to call and pay it – we are your Obednt Servants

Claytons & Walshman
Keighley, Yorkshire
July 7th 1785

As it was decided to instal a 'common engine' in 1785 this was probably the engine which was insured in 1788. Again it is not clear why it was necessary to have this extra source of power when all the other cotton mills built about the same time in Keighley relied solely on water power.

Claytons & Walshman also built Langcliffe Mill near Settle and continued running both mills for cotton spinning until after 1835. In 1803 two hundred people were employed at Low Mill which, in terms of employment, made it the largest mill in Keighley. In 1811 there were thirty-six frames with fifty-two spindles each in the mill. By 1835 the workforce was reduced to eighty and later the mill was changed over to worsted spinning. One of the early mill buildings is still standing but possibly not the 1780 mill.

Dalton or Strong Close Mill. The site for this mill was for sale in 1790. There was a fall of water of about eight feet and it was said to be in 'a neighbourhood where hands are plentiful and the spinning of cotton and worsted well understood'. The Leach family of West Riddlesden Hall owned the land but instead of selling the land, Miss Rachael Leach built Dalton Mill and took two partners to spin cotton there. One was Peter Atkinson and the other was Matthew Dalton who acted as mill manager and who gave his name to the mill. The partnership was dissolved in June 1793 and

the partnership between Atkinson & Leach was dissolved in August.

The mill was offered for auction in 1805 when it was four storeys high. The preparing and spinning machinery included twelve frames. There was also land, a house and eight cottages. John and David Cowling then bought or leased the mill and continued to run it for cotton spinning for several more years. In 1811 Crompton noted that 'Miss Leach' had twelve frames with forty-eight spindles each.

By 1818 William Clayton & Son of Low Mill had taken this mill which they then insured for the following sums:

	£
Strongclose Mill	440
Mill work	60
Machinery	1,400
Stock	100
	2,000

Claytons continued cotton spinning at this mill until after the end of this survey.

Aireworth Mill. Stubbing House Mill, as it was first known, was built in 1787 by Samuel Blakey who was a solicitor in Keighley. The first tenants were James Cawood, Joseph Wright and Thomas Binns who bought iron rods from Kirkstall Forge in 1787. They formed a partnership to manufacture screws and spin cotton at the mill. Cawood was an engineer and had been buying iron rod and nails from Kirkstall Forge for some years to make into screws. Thomas Binns had experience of cotton spinning in Lancashire. However, the mill was sometimes called Screw Mill because it was the only mill in the area where screws were made although the partnership was dissolved in July 1789. Cawood left and the other two partners were joined by Rowland Watson, another Keighley solicitor who was also a partner at Greengate Mill. In 1792 the new

Stubbin House Mill, Keighley, which later became Aireworth Mill.

Aireworth Mill, Kieghley. The central, thirteen-bay section was the earlier Stubbin House Mill, rebuilt in 1808.

partnership insured the mill and contents for the following sums:

	£
Stubbing House Mill	130
Utensils and machinery	820
Barn and stable adjoining	30
Stock in barn	20
	1,000

After the expiry of the original twenty year lease the mill was rebuilt by Samuel Blakey in 1808. Thomas Binns died in 1810 but Crompton's survey of 1811 still noted that 'Binns' was using eight spinning frames with forty-eight spindles each. Part of the mill was then let to John Greenwood.

The mill was then possibly taken by Joseph Hey until 1818 when the firm of Calvert & Clapham took the mill for worsted spinning.

Now a house, this small mill at Higher Newsholme was built for cotton spinning about 1793.

Lower Newsholme Mill was first used for cotton spinning but later became a bobbin mill.

Brow End Mill in the small hamlet of Goose Eye dates from 1791.

North Beck

Higher Newsholme Mill. This was one of two mills built about 1793 by Robert Hall who was a gentleman landowner in the area.

By 1801 the mills were owned by Joshua Field. He leased one of the mills to John Midgley & Co. for cotton spinning. The Higher Mill may not have been used for cotton spinning for many years before being used for other purposes. The building is now a house but the dam and wheel race can still be seen.

Lower Newsholme Mill. Built about 1793 by Robert Hall and used for cotton spinning. Some time later it came into the possession of John Greenwood. When the cotton machinery was taken out the mill was used for making bobbins. The building is now a house but the dam and

remains of the water wheel are still in place.

Brow End Mill. The firm of Craven, Brigg & Shackleton of Walk Mill in Keighley built Brow End Mill in the hamlet of Goose Eye in 1791 on land belonging to Thomas Brigg. The mill and contents were insured for the following amounts in 1795 when it might have been run by Thomas Broughton:

	£
Mill	200
Mill work	50
Machinery	150
Stock	200
	600

The mill was converted to worsted spinning in 1805 and used by William Rouse from Bradford. The mill building can still be seen but it is now a row of cottages.

Goose Eye Mill was built by three

gentlemen farmers in 1797. They were John Bottomley, Richard Shackleton and Thomas Shackleton. The partnership was dissolved in 1798 when John Bottomley left 'declaring that if he remained in it he should lose all the money he had'. That would probably have been the case for Richard and Thomas Shackleton were later bankrupt and the mill was demolished to make way for a large paper mill.

Wood Mill was built for cotton spinning by John Shackleton about 1795. Shackleton was another gentleman farmer who had land on North Beck where there was a good waterfall and therefore a suitable site for this mill. In 1802 pauper children were being employed in the mill. John Shackleton appears to have been successful with his cotton spinning venture and became very wealthy. No further references to cotton spinning at this mill have been found.

Holme House Mill. This mill was built on his own land by a gentleman farmer called John Horsfall sometime in the 1790s. The mill was run by Richard Horsfall and was for sale in 1804 together with the machinery. At that time it measured fourteen yards by nine yards with a fall of water of six to seven yards. The mill was bought by Thomas Teal who converted it to spin worsted yarn.

Castle Mill was built in 1783 by Joseph Smith for cotton spinning. After a few years part of the mill was taken by a partnership between his son-in-law, David Illingworth, who was a draper in Keighley, his brother William Illingworth and William Marriner. This partnership bought iron drum shafts from Kirkstall Forge in 1792. The partners moved their business to Grove Mill in 1795 when Joseph Driver, who occupied the other part of the mill from at least 1788 took the whole

of Castle Mill. Driver was also a partner in Askrigg Cotton Mill in Wensleydale.

In 1801 Driver insured this mill and contents for the following sums:

	£
Castle Mill	600
Mill work	200
Machinery	350
Stock	200
	1,350

Driver was bankrupt in 1805 and the lease of the mill was eventually put up for sale in 1807 together with the lease of his house. The mill was three or four storeys high and said to be on a powerful stream of water which was Keighley North Beck. Driver's machinery was also for sale and included:

2 spinning frames × 96 spindles

7 spinning frames × 48 spindles

The assignee was Ebenezer Thompson of Manchester and enquiries could be made to Berry Smith who was a textile machine maker in Keighley.

Castle Mill was then taken by William Wilkinson who was running twelve frames with fifty- two spindles each in 1811. The mill was changed over to worsted spinning before 1822.

Beckstones Mill was a three storey mill measured twenty-four yards by twelve yards and was still unfinished when it was offered for sale in 1810. It was possibly purchased by John and William Roper who insured the mill and contents in 1816 for the following sums:

	£
Water cotton mill	500
Machinery	150
Stock	150
	800

J & W Roper also ran Damens Mill in Keighley. Beckstones Mill may have been converted to worsted spinning about 1820. In 1835 part of the mill was taken by the firm of Fox & Bland to make power looms.

Damside Mill. The first cotton mill on the site was built in the 1790s but replaced with a larger mill in 1802. Both mills were built by Betty Hudson and the 1802 mill was steam powered. Mrs Hudson's daughter married a man called Thomas Parker who became a cotton spinner at Arncliffe Mill in Littondale and one of his sons, also called Thomas, was the manager of his grandmother's mill at Damside. The Parker family were very much involved in the Yorkshire cotton industry. One brother ran Hebden Cotton Mill, another ran a cotton mill in Gargrave while Thomas Parker Junior married a daughter of Thomas Holdforth, a cotton spinner in Leeds.

Thomas Parker Junior was bankrupt in January 1807 and the bankruptcy of his brothers followed within a few years. Damside Mill was then put up for auction and the sale particulars mention that it was newly erected, three storeys high and measured fifteen yards by ten yards. The 14 hp steam engine had a twenty inch cylinder. Amongst the machinery, which was also for sale, were twelve new water twist frames, each with sixty- four spindles, and two throstles each with ninety-six spindles. The assignees were George Richardson who was a brass founder in Keighley and James Pilkington, a Blackburn cotton merchant from whom Parker had probably been buying his cotton. The hearing was to be in Blackburn.

Damside Mill was then bought by John Greenwood and Lister Ellis who were cotton spinners, manufacturers and merchants with interests in many cotton mills in Yorkshire but based in Keighley. Parker had owed them £500 when he was bankrupt. Greenwood & Ellis paid £1,915 for the mill and steam engine and £500 for

the machinery and chattels which is roughly in line with the insurance valuation the following year. They insured the mill and contents in 1808.

	£
Cotton mill	500
Mill work	150
Machinery	800
Stock	300
Steam engine	250
	2,000

In 1811 Crompton noted that Greenwood & Ellis had sixteen throstles with ninety-six spindles each so the older water frames had been replaced with fourteen new throstles. Lister Ellis left the partnership after 1813 and in 1820 Greenwoods conveyed Damside Mill to William Sugden who then offered the mill and machinery to let. The mill still had a 16 hp steam engine. There was a great deal of cotton preparing machinery and the sixteen spinning machines were frames or throstles with a total of 1,440 spindles.

It may not have been possible to let the mill for the steam engine was taken out and the mill was later converted into cottages.

Northbrook Mill was built for cotton spinning by John Greenwood about 1782. Greenwood tried to evade Arkwright's patent by building his own machinery which did not work very well to begin with. However, he obviously succeeded and founded one of the most extensive early cotton spinning concerns in Yorkshire. In 1786 this small mill was insured for £200 and the 'utensils' which would have included the machinery for £700. This mill was taken over by George Hattersley about 1805 and expanded to produce textile machinery.

Cabbage Mill. John Greenwood, one of the most successful cotton spinners in

Yorkshire, built this mill about 1793. The machinery for the mill was brought from Stubbing House Mill and North Brook Mill where Greenwood had previously been spinning cotton. John Greenwood and his sons also ran Damside Mill in Keighley and references are confusing. However, one local writer indicates that Cabbage Mill was used for cotton spinning by the Greenwoods for over forty years. In 1811 Crompton noted that Greenwoods had sixteen throstles with ninety-six spindles each. By 1833 the mill had a 30 hp steam engine but also retained the water wheel. The mill was used for spinning 20s to 36s quality cotton yarn and employed fifty-eight people. Greenwoods sold the mill in 1860.

Sandywood Mill. This was a small water powered mill which was part of a house. Water from two small streams was brought across the Skipton Road to a wheel which was in the basement of the building. The spinning machinery was on the top floor of the house so the power had to be transmitted from the wheel by an upright shaft or long belt. The mill and house were occupied by John Oldridge from about 1800 to 1819 when cotton spinning was discontinued

Woodhead Mill, Riddlesden. Woodhead Mill was probably built as a cotton mill in 1805. It was for sale in 1808 when three years of the lease had expired. It was said that the machinery in the mill would spin two to three packs of cotton per week and the mill had the advantage of being near to the Leeds and Liverpool Canal. Woodhead Mill had a steam engine by 1808, which was early for mills in this valley but coal was available nearby from the Riddlesden pits. The tenant had been John Coates Phillips, who was bankrupt in 1809, and eventually paid seven shillings

in the pound. The mill was then probably turned over to worsted spinning.

Morton Beck

Upper Mill or Ousel Hole Mill. According to the replies made to the Factory Enquiry Commissioners in 1833 the first part of this mill was built in 1779 and the mill had always been used for cotton spinning. However, there is no substantiating evidence for that early date so it must be treated with caution. A more likely date would be around 1789. The original partners, Timothy Lister from Morton, Jonathan Barker from Windhill and Hill Barker from Morton dissolved their partnership in January 1800 when it was said that the partnership had existed for some time. Lister left the partnership in 1800 and the Barkers continued spinning cotton at this mill until their bankruptcy in 1808.

In 1803 sixty people worked at the mill, none of whom were apprentices, and in the same year the mill and contents were insured for the following sums:

	£
Mill	400
Mill work	150
Machinery and utensils	400
Stock	750
	1,700

In 1804 the total insurance cover was increased to £2,000 with increases for the mill, mill work and machinery which may have come about because of an enlargement of the mill.

The motive power arrangements for the mill were unusual in that the water from a large reservoir on the hillside by the mill was used twice to power one wheel above another. The fall was sixty feet and therefore two thirty foot wheels could be driven. The mill and machinery were advertised for sale in 1808 and also in 1809. Jonathan Barker also owned a tan-yard at Windhill and a one-sixth share in

Hewenden Mill. The final advertisement in May 1809 gave details of the spinning machinery at Upper Mill:

7 water frames × 48 spindles
7 water frames × 64 spindles
1 throstle × 96 spindles
1 throstle × 108 spindles
4 jennies × 100 spindles

Some time after 1809, John and Benjamin Knight, who also ran Great Horton Cotton Mill in Bradford, were tenants at this mill. They were bankrupt in 1827 and the mill and machinery were for sale again. The mill was then four storeys high plus an attic and measured fifty-four yards by twelve yards. The earlier frames and throstles had been replaced with :

28 mules × 216 spindles
4 mules × 240 spindles

Besides the mill, a house, barn and stable, warehouse, gas house and twenty cottages were for sale. Between 1827 and 1830 this mill was used partially for worsted spinning. However, about 1830 it was taken by Thomas and Lazarus Threlfall for cotton spinning and renamed Ousel Hole Mill. By 1833 a steam engine of 20 hp had been added while the firm employed 101 people.

Morton Bridge Mill. This mill was started in 1791 and finished the following year when William Green, Thomas Stead, Paul Lister and John Stead bought iron rods and shafts from Kirkstall Forge and insured the mill and contents:

	£
Cotton mill and machinery	600
Utensils and stock	400
	1,000

In the following two years it appears that the mill was enlarged and more machinery installed as the insurance valuation rose:

	£
Mill, counting house and smith's shop	800
Utensils, stock, goods and machinery	1,200
	2,000

By 1801 the partnership was between Thomas Stead of New Inn, Blackstone Edge, Brian Holmes of Otley and Paul Lister of Hebden Bridge. The death of one of the partners possibly brought about the necessity to sell the mill in 1801. It was described as ' new erected' with 'roving, carding and all other engines, implements and appendages complete in the cotton twist spinning business'. The mill was 'within a quarter of a mile of the Leeds and Liverpool Canal and in a very good neighbourhood for a supply of hands'. It was claimed that there was a fifty foot fall of water to drive two wheels of thirty foot and fifteen foot diameter which were capable of turning about 2,000 cotton spindles.

In 1801 the partnership between Stead, Holmes and Lister was dissolved. Paul Lister, who was a partner at Mytholm Mill near Hebden Bridge, was responsible for the remaining debts of the partnership. Unfortunately, Paul Lister, William Lister of Morton and John Longbottom of Steeton, who took over the mill, were bankrupt in 1808. They had traded as cotton spinners, dealers and chapmen and had possibly got into debt with Greenwood and Ellis, the large Keighley firm of cotton spinners and merchants.

When Morton Bridge Cotton Mill was for sale in 1808 it was four storeys high and measured sixty feet by twenty-seven feet within the walls, with each room ten feet high. It was said that the fall of water was now forty-two feet and that the mill was half a mile from the Leeds and Liverpool Canal. Despite these apparent contradictions the mill was 'very eligible, being close to the village of Morton, which affects an abundance of hands at unusually low wages, and adjoins the turn-pike road from Keighley to Otley'.

According to Crompton the mill was occupied by the firm of Haworth & Co. in 1811 when they had six mules with two hundred and four spindles each. This mill was possibly converted to another use shortly afterwards.

Dimples Mill. According to the details of the lease given when this mill was for sale in 1818, it was built in 1793 but the partners bought the land in 1791. The partners in 1796 were Robert Smithson, Thomas Leach and William Leach, who were cotton manufacturers. The partners insured the mill and stock for these amounts in 1796:

	£
Cotton mill	500
Mill work	250
Machinery	250
Stock	500
	1,500

Thomas Leach, who lived at West Riddlesden Hall near Keighley, owned considerable estates at Morton and Halifax. He was a partner in a large coal mine in Morton and in Bradford's first bank which was formed in 1777. His sister, Rachael, built Dalton mill in Keighley.

In 1811 Crompton noted that William Leach, who was managing the mill, was running sixteen frames with fifty-six spindles each.

By 1818 Thomas Leach was still running Dimples Mill although it was for sale. At that time the five storey mill measured eighty-seven feet by thirty-two feet within the walls and the machinery was driven by a twenty-seven foot by five foot water wheel. The mill was still for sale in 1821 when it was said that the purchase price could be paid in four instalments over three years. No further references to cotton spinning at this mill have been found.

Steeton Cotton Mill. Jonathan Asquith of Sutton and John Thompson of Keighley, who were cotton manufacturers, occupied this mill from 1788 or before. In 1795 they insured the mill and contents for these amounts:

	£
Cotton mill	200
Mill work	100
Machinery	250
Stock	150
	700

John Thompson was bankrupt in 1796 and the mill was taken over by the trustees, William Marriner and William Sugden from Keighley. William Davey then occupied the mill for cotton spinning until 1802 when John Bairstow, a corn miller from Steeton, Thomas Pearson, a piece maker from Steeton and Abraham England, a corn dealer from Broughton took the mill on a lease for four years. A few years afterwards the Pearson family changed this mill over to spin worsted yarn.

Sutton Mill. In 1792 Sutton Corn Mill with its two water wheels was for sale. It was said that 'One water wheel may be constructed at an easy expence for the spinning of cotton'. A cotton mill adjoining the corn mill was built but possibly not until about 1800 for it was said to be newly erected when it was for sale in 1809. The three storey mill measured thirty-seven feet by thirty-four feet and was powered by a sixteen foot six inch by five foot water wheel. The corn mill with its two water wheels was also for sale.

John Ellison from Low Bridge Mill in Keighley took the mill and insured the building and contents in 1810:

	£
Mill	300
Mill work	100
Machinery	300

Stock   100
                                   ─────
                                     800

Sutton mill was converted to spin worsted by 1820.

Cowling Cotton Mill. Joseph Halstead was a cotton manufacturer in Cowling in 1822 and by 1837 had also started spinning cotton at a mill in Cowling.

Ickornshaw Cotton Mill. This mill was built in 1791 on land bought by John Dehane and the water wheel was installed by a Mr Dawson from Clitheroe. This mill became known as Binn's Mill and was used for mule spinning with eight or ten mules about 1800. Thomas and Abraham Binns insured their small cotton mill and contents for the following sums in 1799:

	£
Cotton mill	150
Mill work	50
Machinery	150
Stock	50
	400

Thomas Binns was Abraham's nephew and he was also spinning cotton at Stubbing House Mill in Keighley. The yarn was woven locally and the cloth sold in the Manchester market by a salesman called James Gibson. It was possibly changed over to worsted spinning after the death of Abraham Binns in 1812.

CHAPTER 14

The Saddleworth Area

SADDLEWORTH PARISH was a parochial chapelry of the Parish of Rochdale but for civil administration lay within the West Riding of Yorkshire and therefore has been included in this survey. As a region it looked westwards towards Lancashire and firms adopted the 'room and power' system more widely than in the rest of Yorkshire. This system was helpful in allowing firms with little capital to become established but it makes accounting for individual firms and mills extremely difficult as there was multiple occupancy and a number of firms only traded as cotton spinners for a few years.

The two dominant textile industries in the area during the period under survey were wool and cotton. Changes from one fibre to another in the mills were very common and were even more frequent amongst the domestic weavers. The reason for this was explained by John Buckley, a prominent clothier, to the Parliamentary committee investigating the state of the woollen industry in 1806.[1] He explained how the cotton business had expanded so much in Saddleworth that there was a scarcity of woollen weavers. It was normal for weavers to change from one yarn to another 'if one trade be more brisk than the other they changed frequently'. The reason why so many weavers had changed to cotton weaving at that time was that it was more profitable. Many of the early wool scribbling mills were changed over to cotton spinning in the years around 1800 but reverted to wool, only to be changed back to cotton again in the 1830s

or later. Until the second part of the nineteenth century it was rare for a mill to be built specifically for cotton spinning in the Saddleworth area. Most of the mills listed below had formerly been fulling and scribbling mills.

John & James Buckley & Co. had the first cotton spinning mill in the area about 1782 at which they employed parish apprentices.[2] However, the cotton industry did not develop here to any great extent until after 1800. During the 1780s there were very few cotton workers in Saddleworth Parish and even by 1820 there were only about ten cotton mills in this area but this late development meant that the mills were likely to have steam power. As Oldham was only a few miles away, the Saddleworth district became part of the Lancashire industry in every way and had few, if any, links with the rest of Yorkshire. The close proximity to Lancashire was responsible for the continued growth of the industry after the end of this survey.

The development of the cotton industry in Saddleworth brought a new pattern of organisation to textile production in the area. The woollen industry, which was already well established, was comprised of small family clothiers who rarely employed anyone from outside their own family circle.[3] The new cotton industry, on the other hand, was organised into larger units with power driven spinning mills and large numbers of domestic weavers dependant on a few manufacturers for their work. It may be that the extensive use of the 'room and power' system which developed was a means whereby small family businesses could venture into cotton spinning. Certainly, the contemporary trade directories list numbers of cotton spinners who have not been traced through other sources. This would indicate that many of them were small concerns, possibly running a few machines on one floor of a cotton or fulling mill.

The Mills

Calf Hey Mill was built about 1790 by Benjamin Gartside and about 1800 part of the mill may have been occupied by John Ogden who was then followed by John & Joseph Grafton. About 1816 this mill was turned over to woollen manufacturing.

Horest Mill. William Hegginbottom & Brothers occupied this mill for cotton spinning from 1821 and soon added power loom weaving. In 1833 thirty-eight people were employed and the mill had a 6 hp steam engine. The mill continued to be used for cotton spinning until the late 1870s.

Old Tame Mill was a woollen mill which was used briefly for cotton spinning about 1814 by James Schofield.

Pingle Mill. This was a fulling mill run by Joseph Lawton but after about 1805 his son James, together with John Gartside started mule spinning with four mules in 1812. After about 1815 the mill was changed back to fulling and scribbling.

Woodhouse Mill. This cotton mill on the river Tame was for sale in 1805, having probably been occupied since 1802 by James Wilde who was then bankrupt. The mill had four storeys plus an attic and measured nineteen yards by eleven yards four inches within the walls. The water wheel was twenty-seven feet by five feet

six inches and was housed completely within the building. The rent for the mill was £135 pa plus £4 pa for the covered goit. Application had to be made to John Buckley of Upper Mill who was one of the assignees.

This mill was used as a dyeworks from 1818.

Hull Mill. This was built as a cotton mill in 1787 by John Scholefield and occupied by George, Anthony and Henry Salvan. They were succeeded in 1792 by James and Edmund Buckley. The Buckleys had left by 1810 and the mill changed to wool processing by 1814.

Shore Mill was a woollen mill on the river Tame which was first used for cotton spinning in 1788 by James and John Buckley. Further tenants were James Lawton and Robert Mellor but from 1815 it was occupied by Edmund Buckley until about 1830 when it was changed to woollen. This may have been the mill in Delph where Crompton noted that there were eighteen mules with two hundred and twenty-eight spindles each in 1811. The mill building can still be seen.

Lumb Mill. This mill may have been the one owned by James Lees of Delph in 1804 and was used for wool scribbling and cotton spinning. One section was leased by Samuel Collier from Dob Cross for carding and roving cotton at an annual rent of £105. Collier insured his machinery and stock with the Royal Exchange Assurance Company for £500 and £50 but on 14th September 1804 attempted to defraud the company by burning down the three storey water powered mill. Another tenant was William Hartley who lost goods to the value of about £500. Collier was sentenced to death but was later reprieved.

John Lees and James Schofield may then

Shore Mill, Saddleworth. Originally a woollen mill, but changed to cotton spinning in 1788.

have taken the mill and insured their cotton machinery and stock for the following amounts in 1804:

	£
Mill work and machinery	300
Stock	50
	350

The mill was changed back to wool processing about 1810.

Gatehead Mill was the first cotton mill to be built in the Saddleworth area and one of the first in Yorkshire as it was built in 1781. The mill was run by John & James Buckley & Co. The other partners were possibly John Wrigley a clothier and John Horsefield a cotton weaver. The mill was turned over to fulling by 1789.

Knarr Mill was used for cotton spinning from about 1824 to 1840 by Ralph Thornley although it had previously been a woollen mill.

Yew Tree Mill in Thorns Clough was a woollen mill but part was used briefly for cotton spinning by John Wood about 1818.

Old Mill had been a corn mill and then a scribbling mill before being used for cotton spinning by John Dransfield in 1817. It was later used by John Binns.

Andrew Mill. In 1813 Andrew Mill was used for cotton spinning and fulling. Tenants included Messrs Buckley, Roberts, Kenworthy and others. About 1822 John Neild took over the cotton section which in 1832 was four storeys high and where he had eight mules with two hundred and seventy-two spindles each and employed forty-seven people. Power was derived from a twelve foot eight inch diameter water wheel and an 8 hp steam engine.

Bentfield Mill. This mill was described as 'new erected' when it was for sale in

1807 so it may have been built about 1800. It was occupied by Robert Hadfield or his assignees which suggests that Hadfield was bankrupt. The mill was three storeys high and measured ten yards by seven yards within the walls. The water wheel, which was inside the mill, was twenty-four feet in diameter and fed from a large reservoir. Houses and land were also for sale but the mill was changed over to wool scribbling.

Charlotte or High Grove Mill. The mill was converted to spin cotton about 1828. At that time the mill was four storeys high and had a 20 hp steam engine as well as a 30ft water wheel. There was still some use of water power in 1833 and George Bramall was running the mill and employed fifty-six people.

Wright Mill. This was an old fulling mill on the river Tame which was turned over to cotton spinning about 1810. Binns & Buckley were the first occupants for this purpose and they were succeeded, possibly after an interval, by James Seel. He employed forty-four people in 1833 and the mill was driven by a 20 hp water wheel.

Woodend Mill was being used for cotton spinning by David Thackeray in 1792 but burnt down in 1811. After the mill was rebuilt John Booth took the mill and continued there into the 1830s.

Carr Hill Mill. This was probably a new cotton mill built by Thomas Cresswell alongside an older scribbling mill in 1805. It was then leased by Andrew Binns. In 1808 the mill was three storeys high and measured forty-five feet by thirty-one feet with a thirty-six foot by two foot water wheel. In 1813 it was for sale together with Andrew Mill. It was then occupied by Andrew Binns and Nathaniel Buckley. Nathaniel Buckley & Sons continued to

use this mill for cotton spinning until after the end of this survey.

Hopkin Mill. Three men, Samuel Andrew, Thomas Harrop and Robert Dransfield were all listed as cotton spinners at Hopkin Mill in 1816 but no further references have been found.

Strines Mill. John, James and William Winterbottom insured their cotton and woollen mill, together with the contents, for the following sums in 1802:

	£
Cotton and woollen mill	300
Mill work	150
Machinery	350
Stock	100
	900

Winterbottoms were still spinning cotton at part of this mill in 1822 but part of the mill had been taken by John and Peter Lees for mule spinning by 1805. In 1805 they had 7 mules all with two hundred spindles or more. Part of the mill may have been taken by Joshua Seville who had twelve mules with two hundred and twenty-eight spindles each in a mill near Lees in 1811. In 1821 the mill was three storeys high and measured seventeen yards by ten and a half yards. Power was provided by a twenty-four foot by thirteen foot water wheel and a small steam engine.

Lowbrook Mill had been a wool scribbling mill on the river Medlock which was changed to cotton spinning by 1786 when 'Savall & Loes' insured their cotton mill for £100 and utensils and machinery for £200. In 1811 Abraham Seville had ten mules with two hundred and sixteen spindles each and later the mill was taken by Peter Seville. When the mill was advertised for sale in 1821 it was three storeys high and measured seventeen yards by ten and a half yards. There was a twenty-four by thirteen foot water wheel and a small steam engine. The use of the mill after 1821 is uncertain.

Waterhead or Mill Bottom Mill. From 1816 or earlier a number of firms were using Waterhead Mill for cotton spinning. These included James Mayall in 1816, Burton & Co. and Thomas Schofield together with other firms by 1822. In 1830 Joseph Lees was the sole occupant and this firm later became Lees & Mills.

At Austerlands. This steam powered mill was built in 1819 and occupied by Benjamin and William Beaumont for cotton spinning. In 1834 the mill had a 16 hp steam engine and William Beaumont & Co. employed sixty-six people there although more would have been employed at a section of the mill which was leased to another company for cotton spinning.

At Scouthead. A mill on this site may have been used by Joseph Wrigley who, according to Crompton, was running seven mules with two hundred and sixteen spindles each in 1811. The new mill was built for cotton spinning and weaving in 1825. There was a 24 hp steam engine with two thirds of the power being used by Samuel Wrigley & Brothers for cotton spinning and one third of the power used by John & Richard Allen for power loom weaving. One hundred and forty-seven people were employed at the mill in 1833.

Newhouses Mill. This was a small mill occupied from about 1820 by Thomas Buckley for cotton spinning.

Pastures Mill. This mill was built in 1808 by George Robinson and used for cotton spinning by John Waring & Brothers and later by John Waring & Son. In 1833 the mill had a 12 hp steam engine and forty-six people worked there.

At Woodbrook. This mill was built in 1811 and occupied at first by John Robinson and later by John Robinson & Sons. In 1833 the mill had a 20 hp steam engine and ninety-four people worked there.

Woodbrook Old Mill was used from about 1814 by Joseph Lees and later by J. & S. Robinson for cotton spinning. It was still cotton in 1835.

Radcliffe Mill on Wood Brook was built as a woollen mill by John Radcliffe in 1806 but in 1811 Crompton noted that Edward Radcliffe was running four mules with two hundred and sixteen spindles each which may have been a short term venture into the cotton trade.

Shelderslow Mill. A mill in this area was run by George Buckley in 1811 when he had four mules with two hundred and fifty two spindles each. This later mill was built by John Buckley in 1823 who then let part of the mill to a variety of tenants for a number of years and occupied part himself. In 1833 Shelderslow Mill was being used for mule spinning. Eighty-six people were employed at the mill and it was powered by a 12 hp steam engine.

Spring Mill was occupied by various members of the Wrigley family for cotton spinning from 1811 or earlier. In that year Crompton noted that there were five mules with two hundred and sixteen spindles each at the mill. J. Wrigley was running the mill during the 1820s and 1830s.

County End Mills. Two mills on this site were used for cotton spinning. By 1789 the first mill was occupied by Daniel Thackray but rebuilt about 1818 by John Booth of Woodend, John Andrew of Lees, Samuel Andrew of Knowles Lane and James Lawton of County End with John Booth the main partner. The second mill was built about 1823 by John & Samuel Andrew. Both mills continued to be used for cotton spinning after 1835.

Lydgate Mill. This mill was built in the 1790s for cotton spinning by Joseph Wrigley with steam installed by 1799. In 1811 Crompton noted that the firm was running ten mules with two hundred and sixteen spindles each. One of the Wrigleys continued at this mill but then built a new mill nearby which Hilton & Lees occupied for cotton spinning from about 1824.

At Quick Edge. This woollen mill was used by John Whithead for cotton spinning for a few years before 1814 and from about 1825 by John Buckley. A larger cotton mill was built about 1840.

Quick Mill. One mill was built in 1824 and used for cotton spinning from 1825 by William Kenworthy & Sons. In 1833 they were using the mill to card, prepare, spin and reel cotton wool into yarn for the export trade and card, slub and spin wool. One hundred and three people worked at the mill which had a 20 hp steam engine. The mill was five storeys high and measured thirty yards by fifteen yards. Another cotton mill was built alongside some time in the early 1830s.

Valley Mill was in operation by 1823 and run by Charles Kershaw for cotton spinning until after the end of this survey.

Brookbottom Mills. William Buckley was running two mules with two hundred and sixteen spindles each at a small mill in this area in 1811. The site may have been used for two adjoining steam powered mills which were built in 1825 and 1829. The steam engine was rated at 20 hp, while Giles and Mark Andrew, who had occupied one of the mills, employed one hundred and eleven people in 1833. The older mill was occupied by James Buckley, also for cotton spinning.

CHAPTER 15

West Craven

SOME of the traditional area of Craven has been covered in the chapter on the Yorkshire Dales but enough remains for it to be treated separately and it is a useful name under which to classify all the mills which were built in the area to the west of Skipton. In 1900 a writer describing Craven mentioned that: 'towards the end of the last century and the beginning of this, almost every village in Craven had its cotton mill where now hardly a stone is left to tell the story of a vanished industry.'[1]

On investigation this has been found to be completely true. There were very few Craven villages which did not have a water powered cotton mill for some period. Small villages and hamlets such as Holden, Howgill and Booth Bridge had mills as did such towns as Barnoldswick and Earby. About fifteen cotton mills had been built or converted from corn mills by 1800 and a few were added later.

During the boom in cotton mill building in the 1780s and 1790s a number of land owners offered sites specifically for that purpose, usually stressing the advantages of the particular location. One was on the river Ribble, at an unspecified place but 'contiguous to several villages'. The land owner added that 'Those who propose to build upon a small scale are requested not to give themselves the trouble of applying'.[2] Another was at the small hamlet of Paythorne, a few miles north of Gisburn, where a corn mill on the river Ribble was 'an excellent position for a cotton factory' particularly as at that time the price of labour was 'very reasonable'.[3] Land at Grindleton was also available for the construction of a cotton mill in 1796.[4]

One significant feature for the Craven area was the completion of the Leeds & Liverpool Canal through to the Lancashire cotton areas. It was opened through the tunnel at Colne to Burnley on the 3rd May 1796 and within a few weeks carriers were advertising that goods would be taken on by road from Burnley to Manchester and Rochdale. The Craven area through to Skipton was thus helped to integrate with the Lancashire cotton industry and some of the marginal villages retained their cotton mills for many years because of the advantage of having access to the canal.[5]

The early rush to build cotton mills in Craven had mixed results. Some firms collapsed after a few years while others survived into the twentieth century. There was a rash of marginal concerns with little capital and mills in unsuitable locations. The mills at Holden and Marton, for example, had ceased to function by 1820 or earlier, as had some of the small mills in other villages. However, hand-loom weaving became an important industry before the introduction of power-looms and the mule spinning firms flourished well into the nineteenth century. If the Skipton area is correctly placed within Craven, cotton was an important industry until recent times.

The Mills

Lothersdale Mill. This cotton mill was built on the site of an earlier corn mill in 1792. The firm which build and ran the mill, Chippendale, Parker & Co. had the following partners:

Thomas Parker

Thomas Chippendale, Skipton

Dr Wigglesworth, Cononley

Edmund Spencer, Cononley

Richard Croasdill, Marton Scar

They insured their property for these amounts in 1793:

	£
Cotton mill including water wheel	700
Machinery	500
Stock	300
	1,500

In 1795 the total insured value was reduced to £1,000. The partnership was dissolved in June 1798 and Thomas Parker was able to buy some of his partner's shares. Parker married Dr Wigglesworth's niece and their daughter married a John Wilson from Scotland who took control of the mill and later, in 1835, changed it over to worsted spinning. Part of the mill is still used for textile purposes and, although added to over the years, retains many of the features of an early mill complex.

Mitchell's Mill, Barnoldswick. Hartley, Bracewell & Co. occupied a small cotton mill in Barnoldswick in 1800 which they insured for the following amounts:

	£
Cotton mill	200
Mill work	10
Machinery	150
Stock	40
	400

This may have been the 'Old Cotton Mill' owned by William Mitchell in 1812 which he insured with the Sun and Norwich Union.

	Sun £	Norwich Union £
Cotton mill	100	400
Machinery	500	
Drying house for cotton warps	50	
Stock	50	200
	700	600

The mill was originally four storeys with about fifteen hundred square feet of floor area but Mitchell extended the mill and by 1827 had added steam power. In 1831 the insurance value was:

	£
Cotton mill, engine house, sizing house etc	700
Mill work	100
Machinery	1,500
Stock	100
	2,400

Mitchell appears to have been a successful cotton manufacturer who also owned Parock Mill in Barnoldswick.

Lower Parrock House Mill, Barnoldswick. In 1808 this mill was owned by Henry Lambert, a cotton spinner and manufacturer from Barnoldswick. He rented the mill to William Hall for cotton spinning and insured the mill for £250 and the mill work for £50. William Hall, in turn, insured his machinery for £230 and stock for £120. Lambert was running the mill himself in 1810 and following his bankruptcy the mill was for sale in 1813. It was then three storeys high and measured eight yards by three yards while adjoining it was a two storey sizing house which measured six yards by three and a half yards. Lambert also owned Gillians Mill and a spinning shop and warehouse.

By 1831 this mill was owned by William Mitchell and tenanted by John Smith.

Gillians Mill, Barnoldswick. William

The complex of buildings at Gillians Mill which included handloom weaving shops and a spinning factory.

and Henry Lambert were running this mill in 1790 when they insured the mill, property and contents for the following sums:

	£
Water cotton mill	50
Going gears and machinery	20
Utensils and goods	50
House and warehouse	20
Utensils and goods	60
	200

By 1795 there had been some expansion and a 'spinning factory' added. Henry Lambert was bankrupt in 1813 and Gillians Mill was for sale. The water powered mill was then three storeys high and measured twelve yards by six yards. There was also a two storey warehouse which measured nine yards by five yards. Gillians House was also for sale and next to it was the 'spinning factory' which was four storeys high and measured eleven yards by eight and a half yards.

By 1831 this small mill was being run by John Smith. He insured his machinery in the mill for £100 and his stock for £50. The spinning factory near to the mill was also used by Smith and was described in the following way:

On a building consisting of three cottages and a small place for taking in the work from the weavers on the ground floor, second floor is a spinning room, no other process carried on therein, third floor and attic are hand loom weaving shops.

	£
Building	100
Machinery in spinning room	30
Looms and gearing in upper rooms	70
	200

The ruins of Booth Bridge cotton mill, which was rebuilt in 1814 following a fire.

At Barnoldswick. In 1821 a building was advertised for sale in Barnoldswick which had formerly been used for spinning cotton weft and occupied by James Cook. There was also a worsted mill and out-buildings on the site which adjoined the turnpike road from Colne to Settle.

At Elslack. In 1789 thomas Dewhirst, a worsted manufacturer from Marton, which is a village to the west of Skipton, bought a building at Elslack which he converted into a water powered cotton mill. This mill was used for spinning cotton twist which was sold in the Blackburn weaving area and in Manchester. In 1801 Wilson & Dewhirsts insured their property for the following sums:

	£
Cotton mill at Elslack	80
Mill work	20
Machinery	190
Stock	190

Building used as a spinning shop	20
Machinery	10
Stock	10
	520

Dewhirsts probably left this mill in 1813 when they rented two larger mills at Embsay and no further references have been found.

At Marton-in Craven. John Bond occupied a small cotton mill and spinning shop here in 1801 when he had this insurance cover:

	£
Cotton mill at Marton-in-Craven	40
Mill work	30
Machinery	100
Stock	50
Spinning shop near	30
Machinery	20
Stock	30
	300

Bond probably used the mill for spinning warp yarn on water frames and the spinning shop to produce weft on jennies. He was also a publican in Marton and was engaged in cotton spinning until at least 1810.

Booth Bridge Mill near Earby. John Broughton of Thornton-in-Craven insured this small cotton mill and contents in 1798:

	£
Cotton mill	50
Machinery	150
Stock	50
	250

Although the mill was small over twenty people were employed there in 1803.

The mill burnt down in 1813 but was rebuilt the following year and equipped with a new thirty foot water wheel. The mill was said to be only one mile from the Leeds and Liverpool Canal and 'near the populous town of Earby, where hands may be had in plenty and at a cheap rate'. Application to lease the mill had to be made to John Broughton.

By 1822 the mill had been taken by Richard Green and was still being used for cotton spinning at the end of this survey.

At Earby. Joseph Cowgill and William Harrison were cotton spinners and manufacturers until 1806 when their partnership was dissolved. They had a small mill at Earby which William Harrison was trying to let in 1810. This may have been a corn mill which was only used for a few years for cotton spinning.

Kelbrook Cotton Mill, near Earby. This mill was used for cotton spinning on mules by John Wormwell prior to 1818 when his machinery was for sale and the mill was to let. The mill was later taken by Henry Jackson and continued to be used for cotton spinning.

Howgill Cotton Mill was established about 1790 by Thomas Hague Newcome and Samuel Westerman. Their partnership was dissolved in 1795 and Westerman then insured the mill and contents:

	£
Mill	200
Mill work	100
Machinery	200
Stock and goods	300
Store and wash house	100
	900

In 1803 more than twenty people were employed at the mill. Westerman was still running the mill in 1822 and the mill continued to be used for cotton spinning after 1835.

Feazor Mill, Waddington. Thomas Taylor bought a disused fulling mill and built a three storey cotton mill on the site in 1792. The mill was then leased to John Shepherd and later Shepherd & Hartley. Feazor Mill was water powered and in 1817 was used for preparing and spinning cotton with six water frames. Shepherd was bankrupt in 1825 but the mill had been bought the previous year by Joseph Fenton of Bamford Hall who was a Rochdale banker. After Shepherd gave up the mill it was taken by Corbridge & Brown. They insured their stock in the mill for £100 in 1828 while Joseph Fenton insured the mill, mill work and machinery for £160, £40 and £200.

By 1831 Corbridge & Brown were also running Grindleton Mill and during the 1830s a four storey extension was built and mule spinning added. Feazor Mill was later run by a number of firms for cotton waste spinning and suffered two fires until it was finally abandoned in the 1880s.

Holden Cotton Mill. This mill was built by Miss Jane Dixon of Lancaster for her nephew, Robert Tipping of London, in 1796–97. John Parker of Chancery Lane, a former partner at Low Moor Mill, Clitheroe, also had an interest in the mill. It was a three storey mill with two wings of two storeys each. The main rooms measured fifty-seven feet by thirty feet and the wings twenty-two feet by fifteen feet. Most of the early trade was with Blackburn but the mill was used for mule spinning from an early date. It may have been that money was owed to John Horrocks of Preston because for some reason the mill was for sale in 1799. Shortly afterwards the buildings were adapted to calico printing by Robert Tipping and George Flemming. Tipping was bankrupt about 1813 which caused his aunt's bankruptcy and the mill was sold. The buildings were then used as Holden Workhouse.

Low Moor Mill, Clitheroe. This was the first cotton spinning mill built in Clitheroe in 1782 by Livesey, Hargreave & Co. who were calico printers at Mosney, near Walton-le-Dale. The firm, which rivalled Peel, Yates & Co, had many bleaching works in Lancashire but built this mill just over the Yorkshire boundary. Livesey, Hargreave & Co. failed in 1788 and the mill was given up for some years. In 1832 a cotton spinning mill was for sale at Low Moor. It was occupied by Messrs Garnett & Horsfall and was a small part of the Low Moor complex which included three mills, gas and engine houses and one hundred and ninety- eight cottages.

Grindleton Mill was originally a corn mill but in 1792 it was leased to Robinson Shuttleworth, a banker and cotton spinner from Preston. He added a carding mill which was run in conjunction with Twiston Mill and a jenny workshop at Chatburn. Shuttleworth was bankrupt in 1796 and carding stopped. It appears that cotton machinery was not used at this mill again until 1831 when Isaac Corbridge and William Brown from Feazor Mill insured the mill and contents for these sums:

	£
Water cotton mill at Grindleton	270
Mill work	20
Machinery	100
Stock	10
	400

By 1837 Grindleton Mill was being used for corn milling again.

Bottom Factory, Holden Clough near Holden. This mill was possibly used as a water powered carding and jenny mill in the 1790s. By the 1840s the buildings were being used by hand-loom weavers and eventually the building went out of use.

Sawley Corn Mill. This mill was taken by Peel, Yates & Co. for block printing, dyeing, bleaching and finishing some time about 1795. Thomas Peel lived nearby but in July 1811 the lease was for sale as the company started to reduce the number of works they operated.

CHAPTER 16

East Yorkshire

ALTHOUGH the area of Yorkshire to the east of the Pennine valleys is not usually associated with the production of textiles there were a number of cotton, flax and worsted mills. In the 1790s a number of corn and other mills were advertised for sale where it was suggested that they could be converted to cotton spinning. This happened at Easingwold and Pocklington in 1791 and 1792. A few years previously the following advertisement had appeared in the *Leeds Intelligencer*:

[Water Mills]
Wanted a mill with a constant and regular supply of water near a town or village where a number of children may be had on easy terms. A situation in Derbyshire or the East Riding of Yorkshire would be preferred. Apply by letter post paid to Mr John Gore, Liverpool.

30th January 1787

Besides the eventual cotton mills there were also a number of worsted and flax mills. Bell Mills near Driffield were used for worsted spinning and Balk Mill near Thirsk was a flax mill. Certainly linen weaving was important in the seventeenth and eighteenth centuries, especially in the villages along the edge of the Hambledon Hills. However, six cotton mills have been traced and there may have been more. Most of these were short lived and most were small apart from the mills at Ripon and Thirsk. They all came into existence in the 1790s at a time when so many

cotton spinning concerns were started throughout Yorkshire. All were being used for other purposes by about 1820. The largest cotton mills to be built in East Yorkshire, at Hull, were not started until 1836 and therefore lie outside the scope of this survey.

The Mills

At Great Ayton. James Davison, who was a cotton manufacturer in Great Ayton, insured his machinery and stock in a cotton mill in the village in 1795 for the following sums:

	£
Machinery	600
Stock	200
	800

Davison had the same insurance cover in 1797 but the mill was possible soon changed over to flax spinning.

At Wansford near Driffield. Work started on this mill in 1789 when iron was bought from Kirkstall Forge by Bainton, Boyes & Co. It was finished in 1790 and run by Thomas Bainton, Christopher Bainton and John Boyes. The mill was on the side of the Driffield Canal and measured eighty feet by thirty feet. At that time the company were cotton and worsted manufacturers and insured their mill and contents for the following amounts:

	£
Mill	2,500
Weaving Shop	3,000
Stock in warehouse	2,000
	7,500

John Boyes Junior was still spinning cotton

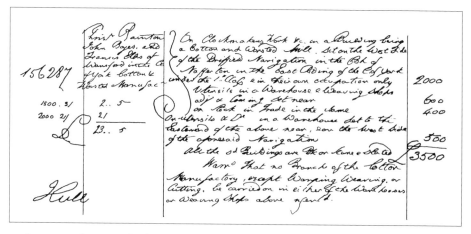

Insurance for Wansford Mill, Driffield, with Royal Exchange Fire Office. 1797

at Wansford in 1822 but his main business was in making carpets. The mill was said to be 'the only establishment of the kind in this part of the country'. The business was given up about 1823

At York. In 1793 cotton machinery was for sale in York as well as the lease of a 'cotton manufactory'. The machinery was 'fixed up in two large rooms, well lighted, in the City of York, each room containing 17 yards by 7 ½ wide, which the purchaser may be accommodated with at a reasonable rent for the term of five years by virtue of a lease unexpired therein.

For particulars enquire of Mr J Dickens, machine maker, Leeds, or Mr Sanderson, Attorney in York.

Besides the carding, drawing and roving machinery there were the following spinning machines:

1 mule × 144 spindles
1 mule × 96 spindles
2 jennies × 106 spindles
1 jenny × 84 spindles

There was no mention of any water or steam power so this could have been a horse mill although the range of machinery makes that seem unlikely.

At Millgate in Thirsk. The corn mill in the centre of Thirsk was being used for cotton spinning by 1797 when it belonged to Robert Bell. Cotton spinning continued until at least August 1809 when the mill was to let. At that time there were 5,536 spindles running in the mill. Later that year the machinery was for sale and it then included:

16 mules × 216 spindles
10 mules × 228 spindles

The total spindles were 5,736 which does not tally with the previous advertisement but such discrepancies are not unusual. Potential purchasers had to contact Mr Wasse who may have been the mortgager for the enterprise.

It was said that the mill could be used for linen or paper making but it may have been used for making tobacco when cotton spinning stopped.

At Pocklington. A mill alongside the corn mill in Pocklington was being used for cotton spinning in 1792 when it was said that the corn mill could also be used for cotton spinning. However, it would seem that the cotton mill was soon after-wards converted to worsted spinning and run by O'Connor, Weddell & Hart.

At Easingwold. A cotton mill was built or converted from a corn mill at Easing-wold some time before 1800 but cotton spinning did not continue after 1823.

CHAPTER 17

South Yorkshire

THIS REGION includes the parishes of Conisborough, Doncaster, Ecclesfield, Penistone, Royston and Sheffield where several cotton mills were built. The eagerness to start in the business of cotton spinning, which started first in West Yorkshire, spread to South Yorkshire in the 1790s but fewer than ten mills were built. Cotton spinning did not last long in Conisborough, Ecclesfield and Stocksbridge although the two mills in Ecclesfield represented a considerable investment. Ecclesfield Cotton Mill had a steam engine by 1803 and the mill and all its equipment was insured for £2,500 that year.[1] The firm which occupied the mill, Hobson, Norwood & Blinkhorn, were also cotton manufacturers.[2]

The largest mill in the area was Sheffield Cotton Mill which, from 1793, quickly became a large complex of buildings with workshops, warehouses, additional spinning mills and workers' dormitories. Steam power was used from the outset and a large number of children were housed at the mill where they were employed as carders and spinners. In 1793 the buildings, equipment and stock were insured for £8,200.[3] This mill was built by a firm of London merchants who also had extensive cotton spinning interests in Stockport.[4] A new mill was added in 1799 and the old mill was demolished to make way for a further six storey building in 1812 which was lit by gas. By 1815 the mill held fifty-six throstles which, together with six mules, gave a total spindle capacity of 8,016 thus making this mill one of the largest in Yorkshire. Unfortunately the firm running the mill was not successful and despite attempts to start worsted and silk spinning went out of business about 1815. As there was little textile manufacturing industry in the area the mill buildings were put to other uses, one of them being used as a poor house.

One town in South Yorkshire which did have a flourishing textile trade was Barnsley. The linen trade started there in 1744 and became firmly established by the end of the century.[5] By 1789 there were seven firms of linen manufacturers in the area, one of which was run by George Scales. Before 1812, a possible relative, William Scales was running a mill at Burton Smithies for cotton spinning and corn milling. No further references have been found to cotton spinning in the area, but there was probably some cotton weaving as four of the six bleachers in Barnsley in 1822 bleached cotton as well as linen cloth.[6]

The other town which became well known for cotton weaving was Doncaster but then only because the Rev Edmund Cartwright set up a weaving factory there about 1787. The enterprise failed because of inherent defects in the power loom which Cartwright had patented and the cotton industry never became established.

The Mills

At Balby. All the cotton preparing and spinning machinery of a mill at Balby near Doncaster was for sale at the end of 1792 which indicates that the mill was operating prior to that date. The spinning machinery included:

3 mules × 120 spindles
1 throstle × 72 spindles
2 jennies × 106 spindles
2 jennies × 100 spindles
7 jennies × 84 spindles
2 jennies × 72 spindles

There was also a lathe and other tools for repairing the machines. For further particulars prospective purchasers had to apply to George Rhodes on the premises or to John Rhodes at Sunny Bank, near Leeds.

This appears to have been a local attempt to start cotton spinning in an area well away from the main textile centres but possibly to supply local weavers. Unfortunately no further references have been found.

Sheffield Cotton Mill. This was a large mill which was established by 1788. It was built by the firm of Wells, Heathfield & Co. Joseph Wells and Thomas Heathfield were London merchants who also owned extensive cotton spinning works in Stockport. In 1793 they insured this mill and contents for the following sums:

	£
Warehouse	200
Utensils and stock	1,500
Moating and picking rooms	
adjoining	200
Utensils and stock	100
Smithy and joiner's shop	100
Utensils and stock	100
Cotton mill	1,000
Machinery	3,000
Stock	1,500
Steam engine	500
	8,200

By 1796 another mill had been built with an additional steam engine. Neither of these engines was supplied by Boulton & Watt so they were possibly of local manufacture. The large, or older mill, was five storeys high and contained 1,436 square yards of floor area. The new mill was four storeys high and contained 768 square yards of floor area. Both mills also had attic and basement rooms. The firm probably employed parish apprentices from other towns for in 1804 they insured a building used as a 'dormitory' together with the household goods, wearing apparel, books etc on the premises. This dormitory would have been for the one hundred and twenty-nine apprentices they employed in 1803. At that time from three to four thousand spindles were running in the mill. By 1803 the partnership had changed. Joseph Wells had left and Thomas Heathfield had as new partners, Jane Martin and John Middleton

Heathfield, Middleton & Co. added worsted combing and spinning by hand in 1813 and later silk manufacturing but this diversification did not stop them going bankrupt in 1815 and cotton spinning ceased.

The mill premises had been considerably expanded in 1812. A new brick built six storey fire proof mill had been built on the site of the old mill. This measured one hundred and forty-two feet by twenty-eight feet inside the walls and it had been prepared for gas lighting. It was said that there was room for 8,000 water twist spindles and 3,000 mule spindles. To power the new mill two new Boulton & Watt engines had been bought:

1 × 36 hp 30¾" × 6ft cylinder
1 × 20 hp 23¾" × 5ft cylinder

Sometime after 1815 part of the mill was used as a workhouse.

At Sheffield. The only other reference to cotton spinning in Sheffield comes from Crompton who noted that a Jonathon Taylor was running 1,600 throstle spindles somewhere in the area but no further details have been found.

At Conisborough. This cotton mill was advertised for sale in 1792 when it had been occupied by Thomas Beckett and again in 1795. It was a four storey mill measuring thirty-nine feet by thirty-four feet with a twenty-seven foot overshot water wheel and held six spinning frames with a total of four hundred and four spindles. There were also six jennies, two mules and four looms. The ground floor of the mill was taken up by a 'compleat smith's shop' with all utensils.

The firm occupying the mill may have been Messrs James Robinson & Sons for they would provide particulars of the mill. The cotton machinery could be bought separately from the mill and the purchaser could also have two houses, a large 'factory' (possibly a cotton weaving shop) and sixteen acres of land.

The mill was for sale again in 1802 with the same complement of machinery but with the addition of a 'fire-engine' with a fourteen inch cylinder. It was also stressed that the mill was on the turnpike road from Doncaster to Rotherham. No further references have been found.

At Burton Smithies, near Barnsley. The mills around Barnsley were usually used for flax spinning but this one at Burton Smithies had been used for cotton spinning up to 1812. The firm which occupied the mill, William Scales and Co were cotton spinners and corn millers until their bankruptcy in 1812. Their cotton spinning machinery consisted of:

  4 throstles × 108 spindles
  1 throstle × 120 spindles
  1 throstle × 132 spindles

The mill may then have been used for flax spinning.

At Tickhill. In January 1804, Joseph Steel, a check manufacturer from Stockport in Cheshire, registered a cotton factory with the West Riding magistrates. This factory employed over thirty people but was probably a small spinning mill which only had a limited life.

Ecclesfield Cotton Mill. It is not clear when this mill started running but the mill and contents were insured by Hobson, Norwood & Blinkhorn for the following sums in 1796:

	£
Cotton mill	800
Mill work	300
Machinery	600
Stock	200
	1,900

By 1803 the partnership had changed and the firm of Mellor & Hobson occupied this mill. The insurance valuation then increased as a steam engine and more machinery was added. The mill was changed over to flax spinning by 1815.

Stocksbridge Cotton Mill. This mill had been occupied prior to 1804 by Jonathon Denton. The five storey mill, which measured eighteen yards by thirteen yards, was driven by a water wheel twenty-four feet in diameter from a twenty-one foot fall. Denton was bankrupt and interned in Rothwell Gaol so the mill and his machinery for spinning cotton twist were for sale as well as a house, stable, and eight acres of land. Denton's assignees were still trying to sell the mill and machinery in 1807 although the mill, according to the advertisement, was then only nine yards wide. As an inducement to purchasers it was said that the mill was 'in a good neighbourhood for the supply of hands. Coals may be had near to the mill on reasonable terms'. The mill was near the Rotherham to Manchester road but does not appear to have been used for cotton spinning after this time.

Hartcliffe Mill, Denby. A partnership

between Nathanial Shirt, John Wood Senior, John Wood Junior and James Hough traded in cotton warps and weft in Penistone and Silkstone up to 1808. By 1810 Hough had left the partnership and the remaining partners split up to continue as cotton spinners and dealers in cotton warps and weft at Hartcliffe mill separately.

CHAPTER 18

The Yorkshire Dales

I T IS EASY to understand how the well established Lancashire cotton industry spilled over the county boundary to Calderdale and Saddleworth in the 1780s and 1790s but more difficult to understand why cotton mills should be built in Littondale, Nidderdale or Wensleydale. The reasons however, are similar. Although Lancashire merchants and manufacturers played a part in the development of the industry in Calderdale and Saddleworth they were in a small minority. In the rest of Yorkshire they had even less influence. In terms of initiative and capital it was very much a locally developed industry and throughout the three Ridings the individuals and partnerships who built and operated the early mills were usually drawn from the ranks of local bankers, doctors, wool and worsted manufacturers and landowners.

This was very true in the Dales where large numbers of corn mills came onto the market following the changes in agriculture at the end of the eighteenth century. The vacant mill sites, together with the new carding and spinning technology and markets for cotton yarn led to the mobilisation of capital and the building of mills on a large scale. For many of the corn mills which dated back to medieval times, the change to cotton spinning was a final chapter in what had been a long story. Today, on a visit to the Yorkshire Dales armed with an Ordnance Survey Map, it is possible to find several of the early mills which were built on the corn mill sites.

A feature of the development of the cotton industry in certain Dales locations was the way in which the building of one mill on a prime site was followed by the building of others. Langcliffe Mill was built outside Settle in 1784 and was followed by four more cotton mills in and near the town as well as other mills at Rathmell, Stainforth and Long Preston. Skipton and the surrounding villages became a centre for the cotton industry from the earliest days. High Mill was built in 1782 and by 1800 there were a number of small horse mills, mule factories and weaving shops in the town where calico and muslin were produced. Belle Vue Mills were rebuilt for cotton spinning in 1831 and further large cotton mills were built between 1850 and 1880. A number of villages outside Skipton also had cotton mills before 1800 and many of those were still in production by 1835. Embsay and Eastby for example, had four mills before 1800. With the development of mechanised spinning and hand-loom weaving in Skipton came a need for and establishment of cotton merchants in the town. Other towns which formed separate centres were Addingham, Grassington and Ingleton.

The opportunities for employment in the new cotton mills for young people were most welcome in many Dales villages. Besides the reduction in employment in agriculture there was less access to common land and wages were decreasing in the local lead and coal mining industries. Help with the family income from children employed as carders and spinners was gladly received.

The north-west dales had a number of well established cotton mills by 1800. Sedbergh had two mills and by 1822 it was said that 'The principal manufacture . . . is cotton, and there are two mills at which a considerable number of persons are employed'.[1]

Two more cotton mills which were still in use in 1835 were in the parish of Thornton-in-Lonsdale. The mill at Burton was built for silk and cotton spinning about 1797 and the mill at Westhouse some time before 1802. Both mills were used for mule spinning and were larger than average. Other villages to the south which also had cotton mills were Ingleton, Clapham, Austwick and Bentham.

The river Aire and its tributaries down to Gargrave provided sites for nine cotton mills. Malham Mill was built in 1785 between the village and Malham Cove. Scalegill Mill was built in 1794 and Airton Corn Mill was first used for cotton spinning about 1786. Both these mills were still operating in 1835 but were run by large firms of Keighley and Skipton cotton spinners and manufacturers. Small mills were built at Otterburn in 1793 and near Eshton Bridge in 1797 but were not used for many years. Bell Busk Mill was built in 1794 and was used for cotton spinning until about 1862. It was owned by the Skipton firm of Garforths and later used by them for mule spinning. Three mills had been built in Gargrave by 1800 and they, together with cotton weaving, provided so much employment that in 1822 it was said that 'The principal business of the place is the cotton manufacture'.[2]

The early success of the cotton industry in the Western Dales depended on a number of factors. Vacant mill sites within reasonable distance of Lancashire together with a ready supply of labour and an existing wool textile industry brought cotton to the villages. The three, four and five storey mills which were built spun yarn for the Blackburn weaving area to begin with but, as we have seen earlier, quickly became centres for an extensive local hand-loom weaving industry. The cotton workers were concentrated in the townships but the whole area was part of the West Riding domestic worsted industry where yarn had been given out as far as the Lancashire border. Thus the new factory based cotton industry was imposed on an existing domestic worsted industry.

The Wharfe Valley provided sites for some of the earliest cotton mills and also some which were very successful. Five mills were built before 1790 in Otley, Ilkley, Addingham and Hartlington. Those were followed by about fifteen more by the turn of the century. They provided a stimulus to the development of the towns and villages when they were first used for cotton spinning and later when many were changed to worsted spinning. The mills built varied from small horse mills in the centre of Otley to one of the largest cotton mills to be built in Yorkshire – Greenholme Mill at Burley-in-Wharfedale which continued to be used for cotton spinning and later weaving until about 1850.

Addingham owed much of its growth to the establishment of both cotton and worsted mills from 1787 onwards. In that year John Cockshott, a cotton manufacturer, and John Cunliffe built a mill which was intended to be a cotton mill but was used for worsted spinning until 1824.[3] After building Low Mill Cockshott and a partner called Henry Lister started spinning cotton at High Mill in Addingham about 1790. Robert Hargreaves had been spinning cotton in the same mill from about 1787. Cockshott's son-in-law, Abraham Dean, built Townhead Mill for cotton

One of the remaining buildings on the site of Millthorpe Mill built in 1790–1.

spinning in 1799 and another Addingham man, Anthony Fentiman, who was a weaver, built a new mill for himself in 1803. These four mills were at various times used for both cotton and worsted spinning and contributed to the growth of Addingham as an industrial village and led to the addition of further textile mills later in the nineteenth century.

Further up the valley more cotton mills were built at Hartlington, Hebden, Skyreholme, Linton, Grassington, Kettlewell, Arncliffe and Halton Gill. Both Arncliffe and Kettlewell Mills managed to survive for many years after other early cotton mills had gone out of use or had been converted to spin worsted yarn. Both villages had a flourishing cotton weaving industry which provided a local market for the mill spun yarn. Kettlewell Mill closed down about 1856 and Arncliffe Mill about 1870.

The Mills

Lonsdale

Millthorpe Mill was built by a partnership of experienced cotton spinners in 1790 when they were advertising for men to help build the new machinery. The partnership consisted of Peter Garforth Senior, Peter Garforth Junior and William Sidgwick who also owned High Mill in Skipton. In 1792 they insured this mill and contents for the following sums:

	£
Cotton Mill	800
Utensils, stock, goods, machinery	2,000
	2,800

In the following year they insured utensils and goods in a warehouse near the mill for a further £1,500 and in 1804 more details of the mill were given for insurance purposes:

Birks Mill, just outside Sedbergh.

	£
4 storey cotton mill, 1,336 sq yds	800
Mill work	200
Machinery	1,000
Stock	800
	2,800

Millthorpe mill was not used for a few years after 1808, However, by 1813 John & James Upton of nearby Birks Mill had bought the mill from William Sidgwick and J. B. Garforth on mortgage. Uptons increased the amount of preparing and spinning machinery in this mill and appeared to be very successful. The mortgage was transferred to the Skipton bankers Chippendale, Netherwood & Carr in 1817 and paid off by 1821. The insurance cover by then had increased to £6,000 and this large mill continued to be used for cotton spinning until after 1835.

Birks Mill was occupied by John Upton for cotton spinning from at least 1809 but it may have been built before then. In 1810 the three storey mill and contents were insured for these sums:

	£
Mill	700
Mill work	300
Machinery	400
Stock	600
	2,000

John Upton and later James Upton continued to use Birks Mill for cotton spinning until at least 1835. They also ran Goit Stock Mill near Bingley from about 1834.

Burton-in-Lonsdale Cotton Mill was running by 1797 and was also used for silk processing. The firm of Simpson & Dodson insured their property and goods in that year for the following sums:

	Royal Exchange £	Phoenix £
Cotton and silk manufactury	600	
Mill work	400	

A farm building which was part of Westhouse Mill.

Machinery	1,000
Stock	100
Counting house with picking room and store rooms over	50
Joiners shop, smithy and picking and mixing room for cotton and silk	50
Utensils in	25
House	50
Barn and stable	25
	2,300

2,300

Simpson & Dodson offered the mill for sale in 1800 when it was described only as a cotton mill. It was then five storeys high, seventy-eight feet by thirty-five feet, with a sixteen and a half foot by ten foot water wheel. As an inducement to a purchaser it was said that 'payment of the purchase money (if required) may be made by instalments and at periods to suit the purchaser'. Also it was said that the mill was well supplied with water, near to a populous village and within a mile of a coal supply.

The mill was then taken by John Green who continued to use it for cotton spinning until 1808 when he was bankrupt. His spinning machinery then included:

14 twist frames × 48 spindles
6 twist frames × 48 spindles (unfinished)
10 mules × 216 spindles
8 mules × 216 spindles (unfinished)

Green was followed by Henry Smithies & Sons who were also cotton spinners and manufacturers. They occupied the mill until at least 1835.

Westhouse Mill at Thornton-in-Lonsdale was in operation by 1802 and possibly before. Robert Burrow was described as the master at the mill until he died in 1812. In 1814 the machinery in the part of the mill run by Mr G Burrow was for sale. The machinery included ten mules with 3,336 spindles as well as all the preparing machinery. However, the machinery may not have been sold for the Burrow family continued to spin cotton at this mill until at least 1835.

Ingleton Mill was built for cotton spinning in 1791 and was run by the partnership of George Armitstead, a cotton spinner from Clapham, Ephriam Ellis, a joiner from Ripon, William Petty, a bridal-bit maker and Thomas Wigglesworth a flax dresser. The partners paid £120 for the barn at the old corn mill and built this mill on the site near to the church buying iron for the mill from Kirkstall Forge in Leeds. The partners traded as George Armitstead & Co. and insured the mill and contents for the following amounts in 1797:

	£
Cotton mill ± smithy and joiners shop	600
Mill work	400
Machinery	400
	1,400

The mill and the machinery for spinning cotton twist and weft were for sale in 1807 when the mill might have been owned by Thomas Lister Parker but leased by John Armitstead. The mill was then used by John Coates & Son to spin flax and tow but they converted it back for cotton spinning in 1837.

Clapham Mill was originally a corn mill but was insured in 1786 for cotton spinning for £200 and the utensils for £700. The partners were George Armitstead, a yeoman, Thomas Wigglesworth of Padside Hall, Parish of Hampsthwaite, flaxdresser, William Petty of Darley, Parish of Hampsthwaite, bridle-bit maker and Ephriam Ellis of Dacre, Parish of Ripon, joiner. It was being used for cotton spinning by 1786 by George Armitstead who later took Ingleton Mill together with his partners. By 1807 it was for sale together

with Ingleton Mill. Clapham Mill was small and sometime afterwards was converted into a saw mill.

Clapham Wood Cotton Mill was for sale in 1798 when it was said to be newly built. It may have been run by Thornbers of Settle but no further references have been found.

Austwick Cotton Mill was in the village of Wharfe and was built near an existing corn mill by Jeremiah Taylor and Robert Parkinson sometime about 1792 as they bought the corn mill in that year and also the corn mill at Lawkland. Taylor & Parkinson, who came from Dutton near Ribchester, had traded as cotton manufacturers, dealers and chapmen, but were bankrupt by 1795 and their mill was for sale. At that time the four storey mill was described as 'new-erected and well built'. It measured sixty feet by thirty-six feet and was powered by a thirty foot by four foot water wheel. There was a reservoir and about half an acre of land with three cottages. The machinery consisted of ten water spinning frames, two more which were unfinished and the necessary carding engines, drawing and roving frames. There were also four flax spinning frames. The corn mill at Lawkland in the same parish was also for sale. Obviously plans for expansion there never materialised.

The mill was then taken by the Burrow family from Westhouse Mill and Robert's son, John, became the manager of Austwick Mill. In 1797 the mill was insured for the following sums:

	£
Mill	150
Mill work	150
Machinery	450
Stock	250
	1,000

This family continued to run the mill until at least 1835

Bentham Mill was built sometime in the 1790s by William, John and Joseph Birkbeck of Settle. In 1799 their tenants were Thomas Danson & Co. who employed 52 people at the mill in 1804. Within a few years the mill was converted to spin flax.

Swaledale

Richmond Cotton Mill was built in 1793 by Francis Michael Trapps of Saint Trinions near Richmond and Thomas Hauxworth of Richmond, a merchant. The mill was built alongside an existing corn mill at Whitcliffe Mill Close and in the lease from Richmond Corporation the name of Francis Trapps of Nidd Hall was substituted for that of Francis Michael Trapps. The Trapps of Nidd Hall were also involved in cotton spinning at Scotton Mill in Nidderdale. This mill in Richmond was only used for cotton spinning for four years and then it was changed over to flax. The mill was three storeys high and measured eighteen yards by twelve yards and had a fifteen foot fall of water from the river Swale.

Wensleydale

Gayle Mill was built about 1784 in the village of Gayle about half a mile from Hawes. The mill was built by Oswald and Thomas Routh who were local hosiers and land owners. It was turned over to flax spinning by 1813 when John Readman

Gayle Mill, just outside Hawes, was built for cotton spinning but later changed to flax.

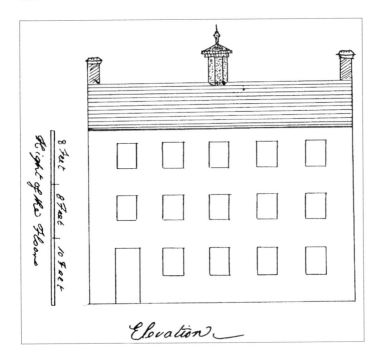

8 Feet | 8 Feet | 10 Feet

Height of the Floor

Elevation

Askrigg Mill,
from a sale
notice, 1814.

from Gayle Mill was bankrupt. The mill can still be seen at the side of Gayle Beck.

Askrigg Mill was built by John Driver, a lawer, his brother Joseph from Keighley and John Dinsdale of Nappa Hall. The year was about 1784 and Askrigg Cotton Mill was built near Paddock Beck in Askrigg just below the corn mill. The three men provided capital of over £1,000 each but the business lost money and there were disputes between the partners. John Driver died, John Dinsdale was rendered bankrupt and eventually, in 1789, the mill was offered for sale. It was said that, 'The premises are well adapted for being converted to any of the branches of machine spinning.' However, John Driver's wife, Agnes took control of the mill while Joseph Driver acted as manager. He was also spinning cotton at Castle Mill in Keighley where initially he rented part of the mill and then all the mill. In 1805 he also became bankrupt so was able to devote himself to running Askrigg Mill.

The mill was for sale or to let in 1814. At that time it was three storeys high, fifty-six feet by thirty feet and with a 'capital' water wheel. Eventually there was a change of ownership and the mill was turned over to flax spinning about 1820.

Yore Mill on the falls at Aysgarth started running in 1785 although the construction of the mill commenced the previous year. The land for the mill was bought by a large partnership made up of John Pratt Esq of Askrigg, William Winstanley of Walton in Lancashire, corn merchant, John Harrison of Hawes, hosier, Abraham Sutcliffe of Settle, doctor, William Birkbeck of Settle, merchant, John Birkbeck of Settle, merchant, Christopher Picard of Cowan Bridge, Gentleman and Robert Dickinson of Lancaster, engineer. The leading partners were the Birkbecks who probably provided most of the finance. In 1787 the mill was insured for the following sums:

Askrigg Mill today, with a two-bay extension on the right.

	£
Yore Cotton Mills	400
Stock	1,000
Utensils and machinery	500
A building	100
Utensils and stock	500
	2,500

The first three items were also insured for £1,000 with the Liverpool Fire Office.

It was Yore Mill which, as an example of the new industrialisation of the Yorkshire Dales, drew the dislike of the Hon John Byng, later Lord Torrington:

What has completed the destruction of every rural thought has been the erection of a cotton-mill on one side, whereby prospect and quiet are destroyed . . . here now is a great flaring mill, whose backstream has drawn off half the water of the falls above the bridge. With the bell ringing, and the clamour of the mill, all the vale is disturbed.

The mill was for sale in 1811 when it was four storeys high and measured fifty seven feet by twenty seven feet within the walls. The machinery included fourteen spinning frames with 864 spindles as well as cards and drawing frames but the advertisement explained that the mill would hold double that amount of machinery. A similar advertisement appeared again in 1814. The under usage of the mill suggests problems and within a few years production stopped at Yore Mill. The original mill burnt down in 1853 and was rebuilt as a worsted mill and it is this later mill which can be seen today

At Masham. A cotton and worsted mill was operating about 1800.

Bishopton Mill, Ripon. occupied a site next to a corn mill between the river Laver and the river Skell just outside Ripon. The water supply was first taken from the Skell but later a dam was built on the Laver about half a mile upstream. The mill was built in 1793, and at one time there was a bell with the inscription 'Bishopton Cotton Mill 1793' hung at the top of the mill and rung to bring the workers to the mill. It was possibly first occupied by the firm of Coates, Batty, Martin & Pearson with Samuel Coates as the leading partner. In 1797 the three storey mill and contents were insured for the following sums:

	£
Bishopton Cotton Mill	800
Mill work	400
Machinery	1,100
Stock	500
	2,800

The mill was changed over to flax spinning some time after 1810.

High Mill, Ripon was occupied for cotton spinning by the firm of Atkinson & Johnson from about 1793 but no further details have been found. It was possibly used for cotton spinning until about 1810 when it was said that 'Some cotton mills have lately been erected near the town, which afford employment for a large portion of its poor.'

Duck Hill Mill, Ripon, was originally a corn mill but enlarged by a man called Jones and used for cotton spinning in the 1790s. It was four storeys high and had an undershot water wheel but no further details have been found.

Nidderdale

Wath Mill was a small corn mill which was used briefly for cotton spinning. The only reference found indicates that it was converted back to corn milling before 1820

Hollin House Mill may have been built on a new site or on the site of an old corn mill. However, in 1795 Parker, Grainge & Co. were occupying the mill for spin-

ning cotton. They insured the mill and contents for:

	£
Hollin House Cotton Mill	200
Mill work	100
Machinery	490
Stock	210
	1,000

The mill was insured for the same sum in 1797 when the partners were described as cotton manufacturers. They may therefore have been engaged in cotton weaving in the locality as well as spinning yarn. Grainge had been involved with plans to build a cotton mill in Pateley Bridge in 1792 but in 1812 joined another partnership to spin flax at Glasshouses Mill nearby. Hollin House Mill was converted to flax spinning by 1818.

Scotton Mill was originally a corn mill but was being used for cotton spinning by 1795. John and William Trapp had cotton preparing machinery and four cotton frames in the mill in that year. The Trapps were also involved in cotton spinning in Richmond but by 1798 Scotton Mill was converted or rebuilt as a flax mill.

Wreaks Mill was also built alongside an existing corn mill some time before 1793 by the partnership of Benjamin Blezzard, Mary Arthington and Henry Lister Junior who were all from Leeds. The insurance cover for this large mill and its contents in 1793 was:

	£
Cotton Mill	1,800
Utensils and machinery	4,000
Warehouse, stables and offices	200
	6,000

Henry Lister left the partnership in 1795. Benjamin Blezzard and Mary Arthington offered the mill for sale in January 1799 and, later that year, dissolved their partnership. A full description of the mill was given:

[HAMPSTHWAITE]
To Be Sold

(By Private Contract)

The cotton twist mill situate upon the River Nidd, at Hampsthwaite, near Ripley, Yorkshire, being a firm, new-erected stone building, five storeys high including an attic, one hundred and twenty feet long, thirty feet wide and twelve high within the walls: with two water wheels (one for the spinning room, the other for the card room), shafts and great gears, with dams, goits etc including the advantage of all the water in the river with a soak corn mill upon the same: The spinning room contains thirty-six frames, forty-eight spindles each, in the second storey; the cards, drawing, roving and reeling frames compleat in the third; and eleven mules from one hundred to one hundred and twenty spindles each in the fourth.

Wreaks Mill was advertised for several months before it was bought by Wilmer Mackett Willett from Rushforth Hall near Bingley. Willett was a partner in the firm which ran Castlefield Mill near Bingley until he moved to Wreaks Mill. During the time that Willett owned this mill it held about 1,400 spindles and about 150 people worked there, thirty-six of them at night.

Much of Willett's trade was with Manchester for he had a warehouse or its contents there insured for £2,000 in 1803. When he was bankrupt the following year his bankruptcy hearing took place in Manchester

The mill was eventually assigned to John Greenwood and Lister Ellis who were cotton spinners and merchants in Keighley and other places. Greenwood and Ellis brought not only new capital to the concern but also expertise in cotton spinning. John Greenwood also bought Swarcliffe Hall near to Wreaks Mill and became the local squire. Ellis eventually left to join Colbeck & Wilks at Westhouse Mill where they were flax spinners.

In 1811 Crompton noted that there were three thousand throstle spindles at Wreaks Mill and by 1820 that number had increased to three thousand five hundred and eighty-four. By 1833 the mill was still water powered with one wheel of 40hp fed by the river Nidd. Eighty-one people worked there, spinning cotton of counts between 28 and 36.

Castle Mill, Knaresborough was built on the site of a paper mill by John and Charles Lomas, the owners of the paper mill, and John and Leonard Green, who were tanners. A fifth partner, Robert Thornton from Airton Cotton Mill was brought in as the expert on cotton spinning. The mill was built in 1791 but the Greens left the partnership after two years and the mill was for sale. The mill held carding, roving and drawing machinery which was used to prepare cotton for the twenty-six spinning frames which had 1,408 spindles between them.

The mill was sold to Messrs Curtis, Driffield, Oliver, Dearlove & Co. Thomas Driffield was acting as manager of the mill when returns were made to the West Riding Justices in August 1803. The mill at that time did not employ any apprentices but had about 750 spindles. Later that year the number of spindles had increased to 868 at which time the mill employed twenty-five male and thirty-five female workers. Curtis, Driffield & Co. continued cotton spinning at Castle Mill until about 1811 when it was changed over to flax

Ancliffe Mill
before the top
two storeys
were removed
in the 1920s.

spinning as the linen industry was very important at that time in Knaresborough.

Littondale

Halton Gill Corn Mill was turned into a cotton mill for a few years about 1800. This was normal for corn mills in the area but transport problems to this remote hamlet must have been immense.

Arncliffe Cotton Mill. In 1785 Arncliffe Corn Mill was for sale. It was said to be 'a very convenient situation for establishing a cotton manufactory'. A few years went by until this cotton mill was built on the site of the corn mill by Thomas Parker in 1792. He insured the mill and contents for the following sums:

	£
Cotton Mill	500
Stock and machinery	1,500
	2,000

Parker had been a publican in Keighley, firstly at the Buck Inn and then the Devonshire Arms before becoming a cotton manufacturer. The early partnership between the two Thomas Parkers, senior and junior, was dissolved in 1801. Thomas Parker junior was bankrupt in 1815 and all his possessions together with Arncliffe

Mill were for sale. The four storey mill measured twenty yards by nine yards and there were ten cottages, a house and land included.

The total possessions of a cotton manufacturer in a remote Dales valley were listed in the auction notice:

The whole of the spinning manufactory and farming stock also the household furniture, of the said Thomas Parker, comprising carding engines, with cards, drawing, fly and spinning frames and throstles; winding machine, cotton picker, warping mills, bobbins, skips, cans, straps, lathe and tools, a large quantity of iron and brass, smiths and joiners implements, upwards of two hundred calico warps, reeds, healds etc. Six valuable draught horses and gears. New Waggon. Two carts, cow, pigs, several lots of hay etc. Four post bedsteads, feather beds and bedding. Mahogany tables, chairs, desk etc. Piano-forte, sofa and clock. Carpets, glasses, china and glass, and a variety of kitchen furniture and brewing vessels.

James Cliffe took the mill after Parker and

The house in Arncliffe which was converted from the mill
when the top two storeys were removed.

Cottages in Grassington which were converted from Scawgill Mill about 1812.

Linton Low Mill. Now a house, originally a corn mill but used for cotton spinning
in the early 1800s.

at some time also ran Kettlewell Cotton Mill. In 1833 Arncliffe Mill had a 9 hp water wheel and forty-five people worked there. Richard Brennand was a later occupant and cotton spinning continued until about 1875. The mill was converted into a house after the top two floors were removed but the rows of mill windows and the dam behind can still be clearly seen.

Wharfedale

Kettlewell Cotton Mill was converted from the old corn mill by Richard and William Calvert of Kettlewell about 1805. The three storey mill had a 4 hp water wheel in 1839 and was used for cotton spinning until 1856.

Scaw Gill Mill, Grassington. This small mill was built by a partnership between

Moses Wright and William Hardacre on land owned by John Lupton about 1792. Wright & Hardacre were spinning cotton there in 1805 but their partnership was dissolved in 1809. The mill was then taken by William Gill and, after a short while, by Richard Chester but the mill was then turned into cottages. However, when the cottages were for sale in 1813 it was suggested that they could easily be converted back into a mill. The cottages are still occupied and are not far from the centre of Grassington.

Linton Low Mill was a corn mill near the river Wharfe but driven by a small tributary. The first tenant to use the mill for cotton spinning was Samuel Gill who had, until 1804, James Parker of Gargrave as his partner. The mill was used for cotton spinning for many years but details of the

Hebden Mill from an old photograph.

firms which occupied the mill have not been found. The mill may have been owned by a Mr Wildman of Threshfield for he would show prospective purchasers round the mill when it was for sale in 1804 and 1813. Low Mill was changed over to spin worsted by 1830 when it was owned by Birkbeck & Co. Today it is a house.

Hebden Mill. A number of firms occupied Hebden Mill for cotton spinning in the first few years of its existence and it is difficult to give a date when it was built. A trade directory noted that J Holmes & Co. were spinning cotton at this mill about 1793. There was a corn mill nearby but the cotton mill was considerably larger and was insured by Waddilove, Lister & Co. in 1795 for:

	£
Cotton Factory	1,200
Mill work	200
Machinery	600
Stock	500
	2,500

At some time a man called Grainge from Hollin House Mill in Nidderdale may have been a partner at Hebden Mill. The mill and machinery were for sale in 1800 and at that time there were ten spinning frames, ten cards together with the preparing machinery and also two mules. It was said that the mill would hold eighteen frames on each floor. John Holmes of Burnsall or Joseph Constantine of Hebden would show the premises to any prospective purchaser. William Parker then took the mill but he was bankrupt in 1809, two years after his brother at Damside Mill in Keighley. Hebden Mill was then taken by a third brother, James Parker of Gargrave and run by the firm of Hepworth & Parker. James Parker was bankrupt in 1811 so Hebden Mill was once again for sale. In 1812 the three storey mill measured ninety-seven feet by twenty-nine feet within the walls and was powered by a twenty-four foot by six foot water wheel. The machinery consisted of:

Carding Room
    12 × 18in carding engines
    2 drawing frames
    4 roving frames
    1 devil or cotton picker
    1 batting frame
Spinning Room
    16 frames × 48 spindles
GroundFloor
    Mechanics tools etc

In 1821 the mill was for sale or could be leased. Application had to be made to Richard Waddilove or Abraham Chamberlain from Rilston, Joseph Constantine from Thorpe or John Holmes from Burnsall who would show people the mill and machinery. The mill then held sixteen throstles with one thousand four hundred and forty spindles and it was said that 'as there is no other Manufactory in the Neighbourhood, hands may be obtained on very moderate terms'.

Sometime later the mill passed into the hands of Walter Bramley who bought the mill with the help of a mortgage from J B Sidgwick & Co. of Skipton. Bramley was spinning cotton at Hebden Mill in the 1830s. He insured his property for the following amounts in 1837:

	£
Water cotton mill lit by candles	400
Water wheel	200
Mill work	100
Machinery	1,000

Hartlington Mill, which has been completely rebuilt,
but retains its original features and water wheel.

Stock   200
Machinery unfinished and to repair 100
                                        2,000

The mill was demolished in recent years but part of an outer wall can still be seen.

Hartlington Mill was built alongside an old corn mill sometime in the 1780s and both buildings can still be seen today as they have been rebuilt in the original style. Both mills were for sale in 1789, when there was also a warehouse, and had been occupied by William Myers who was a cotton spinner and manufacturer. The mill was then probably bought or leased by James Brown & Co. of Skipton who insured the property and contents for the following sums in 1800:

	£
Cotton mill	150
Mill work	100
Machinery	400

Stock   25
Store room, hanking room and
  counting house adjoining   20
Utensils in above   180
                                         875

The partners in 1804 were James Brown, Edward Moorhouse who was a grazier from Skipton and Richard Atkinson, a stonemason also from Skipton.

When James Brown's lease expired in 1812 Hartlington Mill was advertised to let. The three storey mill had three rooms, a low room forty-three feet six inches by sixteen feet and two upper rooms forty-three feet six inches by twenty-seven feet. There was also a warehouse and a counting house. Any prospective tenant could also have the cotton preparing and spinning machinery which included nine carding engines and eight cotton twist spinning frames with 48 spindles. It is not clear which company ran Hartlington Mill after

Skyreholme Old Mill after it was coverted to paper making.
Taken from an old photograph.

High Mill, Addingham, after conversion to flats.

1812 but the mill was changed over to worsted spinning in 1835.

Skyreholme Mill was established by 1805 but very few references to the mill or the firms which occupied it have been found. However, in 1822 the mill was to let and could be taken with or without the cotton machinery which consisted of eight mules with two hundred and four spindles each as well as the preparing machinery. The mill building was then twenty yards by nine and a half yards and three storeys high. Henry and Thomas Bramley bought the mill in 1826 and built a further mill in 1831. This second mill commenced working in 1832 but was not completely filled with machinery in 1833. The mills were powered by two water wheels, one thirty-nine feet by five foot six inches and the other twenty-four feet by seven feet. Probably the larger diameter wheel ran the

newer mill which was fed by a goit and launder across the road. In 1833 Bramleys were running mules and throstles and employed eighty-one people. Cotton spinning stopped about 1850 and the mill was later used for making paper. Only the foundations of the older mill by the river can be seen.

High Mill, Addingham was originally a corn mill but was enlarged in 1787 to spin cotton and possibly worsted. Richard Hargreaves & Co, who were cotton manufacturers, insured the mill and contents in 1787 for the following sums:

	£
Water mill house for spinning cotton	300
Machinery	250
Utensils and stock	250
	800

Shortly afterwards Richard Hargreaves

moved to Linton Mill to spin worsted and John Cockshott and Henry Lister, who were cotton manufacturers in Addingham in 1786, started cotton spinning at High Mill. In 1798 they insured their premises and machinery for the following sums:

	£
Water cotton mill and water wheel	300
Utensils and stock	330
Storehouse adjoining	50
Utensils and goods therein	20
	700

John Cockshott was also a partner with John Cunliffe in the first worsted mill to be built in Yorkshire in 1787 which was also in Addingham. Cockshott was bankrupt in 1794 and his part of High Mill was then taken by William Bland. Prior to 1797 part of the mill had been taken by Robert Pearson. He was a grocer, draper and cotton spinner at Addingham and Draughton. Unfortunately, Pearson was bankrupt in 1797 and had to assign his estate to a cotton merchant in Blackburn. Besides his stock of drapery, hats and hosiery in his shop in Addingham, his machinery from this mill was for sale and included:

    3 carding engines 30″ new covered

    4 roving billies 42–52 spindles each

    19 spinning jennies 84–106 spindles each

    1 cotton picker

It would seem from the list of machinery that this part of the mill only used power for picking, carding and possibly roving. High Mill was then used for cotton, flax and also some worsted spinning. The three storey section of the mill used for cotton spinning was to let in 1822 and it was taken by William Bland who used it for spinning cotton and worsted until 1869. The mill is now flats and the line of the impressive weir across the river can still be seen.

Townhead Mill, Addingham. Ambrose Dean built this mill about 1799 and insured the mill and contents in that year.

	£
Water cotton mill and warehouse	150
Mill work	10
Machinery	120
Stock	120
	400

In 1802 Dean bought land at Townhead for £900 which gave him more control over the beck. However, the stream must not have been powerful enough for his purposes so he negotiated for a 6 hp Boulton & Watt steam engine in 1803 through the firm of Halliday & Cockshott of Baildon. Halliday was a worsted spinner at Tong Park Mill and was joined there by John Cockshott from Low Mill in Addingham on his bankruptcy. Cockshott was Dean's father-in-law. The engine was to be taken from Birmingham to Leeds by canal and then to the canal warehouse at Silsden. From there it would have had to be brought over the moor to Addingham.

In 1811 Crompton noted that Ambrose Dean was using ten throstles with 126 spindles each in his mill. Besides spinning yarn Dean was also a calico manufacturer and sold his cloth in Manchester. The mill was for sale in 1820 together with its water wheel and Boulton & Watt engine. The mill was then taken by John Cockroft until after 1835.

Fentiman's Mill was built on Back Beck, the other small stream which runs through Addingham. Anthony Fentiman had been involved in cotton spinning since at least 1793 in Skipton and may also have changed Beamsley Corn Mill over to cotton spinning. In the late 1790s he started work preparing the dam and water courses for this mill which bears the date and initials, 'AFM 1802'. In 1799 three mules with one hundred and six spindles each

This corn mill at Beamsley was used for cotton spinning about 1810.

were left for Fentiman in Skipton and may have been intended for this mill. Fentiman was still running the mill in 1818 but by 1838 it was owned by one of the Cunliffe family and run by John Booth. The mill has been run as a saw mill since about 1860 and is still a busy saw mill today with the original building obscured with later extensions and piles of timber.

Beamsley Mill was converted to cotton spinning several years before 1818, possibly by Anthony Fentiman. When it was for sale in 1818 it was said that of late it had been used as a cotton mill but that formerly it had been a corn mill and was 'well calculated for either purpose'. William Winterburn from Bolton Bridge would show prospective purchasers round the mill and other particulars were available from a Mr Preston in Skipton. The mill at that time was twenty-seven feet square and had a twenty-six foot wheel. It was for sale again in 1821 when it was said that it could be used for corn, cotton or worsted.

Low Mill, Addingham was the first worsted mill to be built in Yorkshire in 1787 and the second in the country. It was originally planned as a cotton mill but it appears that the original partners, John Cockshott and John Cunliffe, installed water twist frames and altered them to spin worsted yarn instead of cotton. The original mill was offered to let in 1822 when it was four storeys high and had two water wheels. Worsted spinning may have continued until 1824 when the mill was taken by Jeremiah Horsfall who built an additional mill and re-equipped for cotton spinning. The new mill had its own water wheel, the new spinning machinery consisted of mules and the mill was lit with gas. In addition Horsfall started power-loom weaving by 1826. This new mill attracted the attention of rioters from Lancashire and other places who attacked the

Part of Low Mill, Addingham, showing the new mill built in 1824 for cotton spinning by Jeremiah Horsfall (demolished 1972).

mill in April 1826. A local newspaper reported that all power looms had been broken within ten miles of Blackburn and that looms had been smashed nearby at Mr Mason's Mill at Gargrave. Although the rioters spent two days attacking Low Mill the preparations made by the owner and the military stopped any serious damage.

In 1827 the mill was insured for the following sums:

	£
Cotton mill lit by gas	1,300
Mill work	450
Machinery	2,300
Stock	350
	4,400

In 1833 the mills were still being used for mule spinning and power-loom weaving. The two water wheels produced about 40 hp and shortly a steam engine was added.

Horsfall continued cotton spinning and weaving at Low Mill until 1841.

Ilkley Cotton Mill. It is not clear when this mill was built but it was possibly 1787. The two partners who built the mill were Thomas Hawksworth and Jonathon Curtis. Hawksworth had been in partnership with Thomas Hill from Baildon until January 1788 but in June he was bankrupt and described as a calico maker, stuffmaker, dealer and chapman. Jonathon and Samuel Curtis were worsted spinners in Bingley up to 1793 when they had to assign their assets to their creditors. However, Ilkley Cotton Mill was insured by Hawksworth & Curtis in 1788 for the following sums:

	£
Water cotton mill, warehouse, counting house and offices	400

Utensils, stock and going gear   600
                                     1,000

The mill appeared to be used for both cotton and worsted spinning, perhaps from the start. The section of the mill used for worsted spinning by Jonathon Curtis was for sale in 1795 and again in 1797 so the impact of his previous financial difficulties was slow in catching up with him. The mill may then have changed over completely to cotton spinning as it was used for that purpose by Robert Atkinson in 1838. The mill was later turned into cottages and, when Wells House Hydro was built in 1856, these were demolished and new buildings put on the site

Spinning Mill, Burley Woodhead.
This mill never seems to have had a name and today, although occupied as two houses is still called the 'Spinning Mill'. It was running in 1795 when it was insured by Henry Marshall and William Lister.

		£
Cotton factory		250
Utensils		90
Stock		200
		540

This was probably a jenny mill with only the carding machinery run by power with water supplied from the dam behind the mill. The mill was to let in 1800 when it was said to be newly erected. The machinery consisted of two carding engines, two billies, twelve jennies and other machinery necessary for spinning cotton weft. Further particulars could be had from Benjamin Langwith in Burley or Henry Lister, the owner, in Addingham. This was probably the same Henry Lister who was spinning cotton at High Mill in Addingham. It is not clear what happened to this small mill after 1800.

Hazel Grove Mill, Burley Woodhead.
Henry Marshall and William Lister also insured this mill in 1795:

A small jenny mill at Burley Woodhead built about 1795.

	£
Cotton factory	230
Mill work	55
Machinery	75
Drying stove	15
Utensils and stock	5
	380

By 1809 this mill was owned by William Popplewell of Cowhouse near Bingley and tenanted by Joseph Hudson who was a cotton spinner. In 1817 the mill was changed over to worsted spinning.

Greenholme Mill, Burley-in-Wharfedale. This was one of the most successful mills in terms of the length of time that cotton spinning was carried out on the premises. A second mill was built after a few years and one of the regions leading cotton firms occupied the mill until half way through the nineteenth century when it was changed over to worsted spinning.

The first mill was built in 1792 by a partnership of:

Thomas Davison, Stockton-on-Tees
George Merryweather, Burley-in-Wharfedale
Jonas Whitaker, Leeds

They insured the mill and contents in 1792 for the following sums:

	£
Burley Cotton Mill and blacksmith's shop	2,650
Stock etc	2,550
	5,200

The six storey mill was run by a labour force of children brought from London and by 1797 a special lodging house had been built for them and separately insured: 'On a building occupied as a house and school room with chamber over as lodgings for the children employed in the factory – £360.'

The total insurance cover for the firm's

Hazel Grove Mill at Burley Woodhead which was built about 1795 and changed to worsted in 1817.

This building may be part of the cotton spinning mill built at Oaks Farm near Otley about 1788.

assets, which included the partner's houses, as well as work people's cottages, came to £12,000 in 1796. However, by 1805, when only the mill and machinery were included, the sum was £10,000.

	Sun Fire	Imperial Office
	£	£
Mill and workshops	625	625
Mill work	900	900
Machinery	2,575	2,575
Stock	600	600
Unfinished machinery	300	300
	5,000	5,000

Davison left the firm soon after the mill was built and the partnership between Whitaker and Merryweather did not appear to be amicable. The firm traded as Jonas Whitaker & Co. and in February 1805 Whitaker appeared to take over the management of the firm. In 1810 George Merryweather took his share of the workforce and marched one hundred and

ninety apprentice boys and girls to Manchester to set up in business there. After Merryweather left, John Greenwood and Lister Ellis, two very experienced and wealthy cotton mill owners from Keighley, joined Whitaker and formed a new partnership. They brought additional capital and a large new mill was built in 1811 and gas lighting was installed. Another sign of increased working capital was the increase in the insurance value of the stock of cotton and yarn from £750 in 1807 to £6,500 in 1813. Lister Ellis wanted to retire from the partnership in 1812 and his one third share in the mill and property was bought by William Ellis for £42,000.

Although the notes are difficult to decipher, Crompton's survey of spindle capacity in Yorkshire mills in 1811 showed that 18,000 spindles were running at Greenholme Mill which would have made it the largest mill, in terms of spindles, on his list. All the spindles were on throstles which were the spinning machines Greenwood & Ellis ran in their other mills.

A weaving shop was added by 1819 but that was superseded by 108 power looms added in 1825 and worked in the small mill which was also used for mule and throstle spinning. In 1833 there were six mules, each with 400 spindles, which were used for spinning weft yarn. In 1833 power was still provided by two wheels fed by the river Wharfe which generated 108 hp.

A sale notice for 1848 gave the size of the mills as:

Old Mill
   5 storeys – 105ft × 28ft 6ins
   1 wheel 16ft × 14ft
New Mill
   5 storeys – 211ft × 34ft
   2 wheels 22ft × 18ft, 30ft × 18ft
Weaving Shed: 96ft × 64ft

In 1835 three hundred and eighty people worked at Greenholme Mill.

At Ellar Gill, Otley. Gervais Marshall of Leeds and William Mounsey of Otley were hosiers with shops in Leeds and Otley. In 1787 they had goods and machinery in a room in Silver Mill in Otley where they were possibly involved in jenny spinning or rented room and power from Yeadon & Walker. However, by 1788 they had moved to this small cotton mill outside Otley on land near Oaks Farm. The land had been bought in 1783 by William Mounsey, a stocking manufacturer from Otley, Samuel Mounsey, a stocking weaver from Otley and John Popplewell, a schoolmaster from Bramhope. The mill was on the side of Hell Hole Gill, or as it is called today, Ellar Gill. In 1791 they insured goods in Leeds and Otley but also:

	£
Cotton mill at Oaks Farm	100
Machinery, goods and stock	300
	400

From 1791 and certainly by 1795, Marshall & Mounsey were involved with cotton weaving in a room over stables and a barn attached to a house in Otley occupied by William Mounsey. By 1795 also, more details of the insurance were given:

	£
Cotton water mill at Oaks Farm	100
Mill work	40
Machinery	220
Stock	40
	400

By 1800 this mill had been converted for wool scribbling and was occupied by Joseph Baldwin. Two years later Marshall & Mounsey were selling four cotton twist spinning frames, five cards and other cotton preparing machinery so it looks as though they gave up cotton spinning at that time. The mill is now owned by an engineering firm but can still be seen from the road.

Otley Mills. According to replies made to the Factory Enquiry Commissioners in 1833 this mill was built for cotton spinning in 1793 but it may have been earlier for Garforth & Sidgwick bought spindles from Kirkstall Forge in 1787. A number of firms were involved with cotton spinning in Otley until about 1810 but it is not clear which one occupied Otley Mills. There were several buildings on this site and so there may have been several tenants at the same time. However, Peter Garforth and Samuel Sidgwick insured their utensils, stock, goods and machinery in this mill and warehouses for £1,500 in 1792 and so they probably occupied the mill prior to that. The partners also had in interest in cotton mills in Skipton, Bell Busk and Sedbergh. Their partnership at Otley was dissolved in April 1793. Garforth & Sidgwick had been tenants at the mill which was owned by Mary Fairbank, a widow from Otley, and Joseph Chippendale from

Part of Silver Mill at Otley, which was used for both cotton spinning and silver plating.

Wakefield. By 1795 the mill had been let to the firm of Walker, Maude & Co. who continued cotton spinning there until 1797 or later.

The mill was advertised to let in 1809 when the main building was four storeys high and measured sixty-nine feet by twenty-six feet within the walls. There were two rooms at the east end of the mill which had been used for spinning cotton weft and measured sixteen feet by twenty-seven feet. The main water wheel was eighteen foot by twelve foot but there was another wheel to drive the paper glazing mill and leather dressing mill alongside. Marshall & Mounsey may have bought part of Otley Mills about 1800 and possibly bought part or all of the mill some time later. It is possible that George Foster and Jonathan Cawood rented parts of the mill for Crompton noted that Foster had

twenty mules and Cawood had ten mules somewhere in Otley. However, the mill was let to James & Thomas Craven from 29th April 1810 and used for worsted spinning.

Silver Mill, Otley was occupied by Yeadon & Walker for cotton spinning and silver plating from about 1784, when they were buying iron rods from Kirkstall Forge, until the mid-1790s. John Walker was a partner at Low Mill near Fewston from 1791 and also had a tan yard in Otley. John Yeadon may have been an engineer as he bought iron rods from Kirkstall in 1782. The partnership at Fewston then built High Mill so John Walker let Silver Mill to John Ritchie who had been a mercer in Otley and who ran the mill until his death in 1799. John Ritchie's manager tried to bring in a partner so that

Low Mill, West End, *c.* 1900.

he could take over the mill but without success. Yeadon, Walker & Co. then sold the mill to Thomas Butler who advertised for a tenant in 1800. The mill was three storeys high, twenty yards long and seven yards wide. A new reservoir had been built and a new thirty-six foot water wheel installed. Shafting and gears suitable to the needs of the new tenant could be added. An unknown tenant was found who continued cotton spinning but by 1822 the mill was used solely for worsted spinning. Today part of the mill still remains together with the adjoining cottages.

Tadcaster Cotton Mill was built by Hartleys, Atkinson & Isles on the river Wharfe near to the bridge in 1792. The three storey mill was built in brick and measured seventy-nine feet by thirty-six feet. When the mill was for sale in 1815 it was powered by a 20 hp steam engine which was said to be on the Boulton & Watt principle. Although this mill was built as a cotton mill it was changed over to corn milling after a few years and to flax spinning by 1825.

Washburn Valley

Low Mill, West End. In 1791 a group of Otley tradesmen bought Thruscross corn mill for its water rights. Alongside the corn mill they built this cotton mill on the river Washburn. the corn mill was bought subject to a mortgage of £379 but by 1793 the partners had been able to repay the loan. In 1795 the partnership consisted of:

John Walker, Otley, silver plater

William Maude, Otley, silver plater

Joseph Hardcastle, Otley, grocer

Robert Thompson, Thruscross, joiner

Richard Holdsworth, Keighley
(late of Otley), maltster.

Hardcastle was Walker's brother-in-law and Thompson helped prepare the machinery and install it in the mill. By 1796 the value of the business was estimated to be £4,550. In 1797 the partners insured their mill and machinery for the following amounts:

	Royal Exchange £	Phoenix £
Cotton mill	500	500
Machinery	500	1,200
	1,000	1,700

By about 1800 the partners were so successful that they decided to build a second mill just above West End church which they called High Mill. The partners built thirteen cottages at Low Mill and by 1803 employed about ninety people. At that time the mill was running spinning machinery with 1,540 spindles but as some of these were worked by hand the machinery was probably a mixture of water frames and jennies.

By 1805 John Walker and his wife had bought the shares of the other partners and still owned the mill when their tenant Christopher Smith changed it over to flax spinning before 1814.

High Mill, West End. This mill was built by the partners from Low Mill about 1800 as their cotton spinning activities increased. By 1803 thirty people were working at this mill and it held spinning machinery with 516 spindles. Cotton spinning continued at this mill until 1812 when it was turned over to flax spinning.

Little Mill, West End. This mill was run by Daniel Garforth for cotton spinning from 1807 or before. In 1811 Crompton noted that 'D Garfoot' of West End had twenty throstles with seventy-two spindles each giving a total of 1,440 spindles. This mill was turned over to flax spinning soon afterwards.

Ribblesdale

Stainforth Cotton Mill. According to a local writer the corn mill at Stainforth was

Little Mill, West End, c. 1900.

Langcliffe Mill, Settle. The original 1784 mill and later additions.

converted to cotton spinning in 1793. The firm which occupied the mill was Redmayne & Armitstead who were also spinning cotton at Giggleswick at about the same time. However, no further references gave been found so it is likely that this venture did not continue for many years.

Langcliffe Mill was built on the site of a corn mill by George and William Clayton and Thomas Walshman in 1784. These men came from Lancashire where they had extensive cotton manufacturing and printing interests. They had built the first cotton mill in Yorkshire at Keighley in 1780 and the two mills appeared to become their main concern.

The demand for cotton twist was such that Claytons & Walshman started spinning before the mill was completed. Some of the machinery was made at the mill but some was brought from Keighley where they had operated spinning frames under Arkwright's patent. The first payments to the spinners were made in October 1784 but at the same time payments were still being made to the carters, labourers and craftsmen who were completing the mill. When repairs had to be made to the water wheel the spinners wages went down but then rose when overtime had to be worked to make up for the lost production. Some workers were brought from Keighley and payments to local men for forging spindles, turning rollers and making the flyers continued into 1789.

In 1787 Claytons & Walshman built cottages near the mill to attract more families to this rather isolated site over a mile from Settle.

Notice is hereby given that Messrs Clayton & Walshman, cotton manufacturers, in order to accommodate work people are now erecting a number of convenient cottages at Langcliffe Place, which will be ready to enter at Mayday next.

Any people with large families that are desirous to have them employed, and can come well recommended, may be assured of meeting with every reasonable encouragement, by applying to Messrs Clayton & Walshman, at Langcliffe aforesaid, or at their cotton works at Keighley.

April 10th 1787

The following year the cottages and tenements for the workers were included in the insurance cover provided by the Royal Exchange Fire Office:

	£
Langcliffe Mill	1,500
Stock and machinery	1,500
Warehouse and cottages	200
Stock	500
Tenements near	250
	3,950

In 1803 Claytons & Walshman employed 150 people at Langcliffe Mill which, with the 200 employed at Keighley Low Mill, made this one of the largest cotton spinning concerns in Yorkshire.

In 1818 machinery in what was described as a new mill was insured for £5,200. This new mill was lit by gas so considerable additions and alterations had taken place.

By 1833 cotton weaving had been added and a 30 hp steam engine installed to supplement the 40 hp water wheel, although it was stressed that the steam engine was only used in dry weather. The firm then employed 203 people at the mill. Langcliffe Mill continued to be used for cotton spinning and weaving by Claytons until at least 1849.

Settle Bridge Mill was possibly converted from a water driven forge early in 1785 by a partnership of Thomas and William Buck of Settle, whitesmiths, Thomas Wilkinson of Leeds, pocket book maker, David Jay of Leeds, apothecary and Thomas Ritchie of Leeds, book-keeper. Later that year the partnership of Wilkinson, Bucks, Jay & Co. was dissolved. Any debts payable to the firm were to be paid to Thomas Wilkinson of Leeds who may have been the same person who was

Settle Bridge Mill built about 1784.

involved with cotton spinning there at Hillhouse Bank.

By 1800, when the mill was for sale, it was being used for both cotton and worsted spinning. There was also an early combing machine in use for combing wool ready for worsted spinning. The cotton machinery was used for spinning cotton twist and was said to be 'constructed upon an excellent plan'. The owner of the mill at that time was William Buck who had also possibly been involved in machine making at the mill with his brother Thomas.

The mill was then taken by Edmund Armitstead who was a cotton merchant. The mill appears to have been used for cotton spinning until at least 1834 when it was for sale again. The machinery at that time consisted of six and a half pairs of hand mules with 4,160 spindles. Although described as hand mules some of the motions would have been power driven. The mill is now holiday flats and can easily be seen from the road.

King's Mill, Settle. This mill had been used as a corn and snuff mill prior to its adaptation for cotton spinning. It was possibly described as Settle Mill when it was for sale in 1793. It was said that this mill and another corn mill in Giggleswick were 'well situated for erecting extensive works for spinning cotton, wool or flax'. John Thornborow (Thornber ?) from Colne insured the mill and contents for the following amounts in 1795:

	£
Cotton Mill	200
Mill work	31
Machinery	266
Stock and goods	165
	662

The mill was leased to John Procter & Son who continued cotton spinning there for many years. In July 1830 there was a serious fire which necessitated the

King's Mill, Settle, rebuilt in 1830 after a fire.

Runley Bridge Mill, Settle, built in the early 1780s.

rebuilding of the mill. Water power was still used however, and in 1833 there was a 30 hp wheel deriving power from the river Ribble. In the same year one hundred and twenty-five people were employed at the mill.

Runley Bridge Mill, Settle. James Brennand was spinning cotton at this mill in 1788 when he insured the mill and contents with the Sun Fire Office for the following sums:

	£
Twist mill	400
Utensils and stock	600
	1,000

By 1795 the mill had been taken by John Thornborow (Thornber ?), who was also running High Mill in Settle. He insured a mill called Lower Mill and two cottages which was probably this mill.

	£
Mill and two cottages	200
Millwrights work	40

Machinery	98
Stock and goods	300
	638

In 1803 John, James and Thomas Thornber, who were cotton merchants, owned the mill but about that time let the mill to Messrs Procter who were also spinning cotton at High Mill. In 1810 the mill was advertised for sale following the death of James Thornber. One building was three storeys high, eighteen yards by nine and a half yards with a twenty-one foot by four foot wheel. Another three storey building, which was used for spinning weft, was eleven yards by seven and a half yards and a third building which was also three storeys was ten yards square with a nearly new fifteen foot by four foot wheel. Four cottages were included with the mill. Runley Bridge Mill was still being used for cotton spinning in 1835 although it had been partially rebuilt after a fire in 1825. The fire had been caused by a defect in a cotton preparing machine called a

Fleet's Mill, Long Preston, built about 1790. It is now used as a barn, but the outlines of the original windows can still be seen.

devil. The mill buildings can still be seen as they were converted to agricultural use after 1850.

At Giggleswick. A small mill was built on the site of some cottages near Catteral Hall. This mill may have been run by Taylor & Son until the 1820s.

Giggleswick Mill. This mill was built on the site of a corn mill which was advertised for sale in the 1780s and 1790s as a suitable site on which to build a cotton mill. It was possibly converted to cotton spinning about 1793 and run by Redmayne & Thornber who were also calico manufacturers. William Redmayne had been a cotton manufacturer in Blackburn before moving to Settle where he died in 1805. The mill was then run by his two sons Giles and Robert Redmayne who had a

mercer's shop in Settle. The Redmayne brothers were bankrupt by 1816 and the mill which was by the church was then demolished.

At Rathmell near Settle. Armitstead, Brown & Co. were running one of the small cotton spinning mills in Rathmell in 1793. This mill was just above Lumb Brig at Capelside. By 1797 the mill was for sale and was said to have been 'lately occupied by Messrs Brown & Clark'. The mill was then four storeys high and measured twenty-seven feet by thirty-six feet. The machinery included six jennies so perhaps just the preparing machinery was water powered.

At Rathmell near Settle. Thomas Holden & Co. were running the other mill at the same time. It was built near to the corn mill near a house belonging to a Mr Mansegh. The remains of one of the mills can still be seen.

Lower Mill, Long Preston. This cotton spinning mill was built about 1790 and occupied by the firm of Serjeantson & Tatham from 1793 or earlier until 1812 when it was for sale. This mill was possibly built on the site of an old corn mill and was four or five storeys high. It was equipped with a 5 hp steam engine and the machinery in 1814 included mules and throstles as well as all the cotton preparing pickers, engines and frames. This mill and the corn mill were demolished in 1881 by a great flood. No further references have been found to cotton spinning at either mill after 1814.

Fleets Mill, Long Preston. This mill was also run by Serjeantson & Tatham over the same period as Lower Mill although a contemporary trade directory listed one of the mills as being run by Robert Serjeantson. This mill was three storeys high and was water powered but was very unusual in that it also had a wind mill to raise water back to the large dam. In 1814 it was said that the two mills would 'turn off about 10cwt of calico weekly between them' and that hands may be obtained at 'reasonable wages'. Prospective purchasers had to apply to John Wood at the Auction Mart in Manchester or Mr Heelis in Long Preston. Fleets Mill is now a barn but the long rows of windows, typical of cotton mills of the time, can still be seen on one side and the dams and water courses can still be traced.

Airedale

Malham Mill, which was very near Malham Cove, had been converted from a corn mill in 1785 by a partnership consisting of:

Richard Brayshaw, excise officer, Liverpool
Robert Hartley, draper, Colne
Robert Moon, wool stapler, Colne
William Hartley, shalloon manufacturer, Colne

Brayshaw had recently bought the corn mill and owned other land and cottages in the area. The partners agreed to pay £50 each towards the cost of building the new mill. In 1786 the four partners insured the mill for £300 and their utensils for £200. Unfortunately the partners had a disagreement so it was decided to partition the mill in 1796 with part to be taken by Richard and John Brayshaw who had:

6 spinning frames × 60 spindles = 360 spindles
2 spinning frames × 48 spindles = 96 spindles

and the other part to be taken by Robert Hartley who had:

2 spinning frames × 60 spindles = 120 spindles
2 spinning frames × 48 spindles = 96 spindles

Robert Hartley had wanted to split the mill and Peter Garforth of Skipton and William Marriott of Marsden in Lancashire had been called in to decide how the mill should be divided. Internal walls were to be built and agreement had to be reached on the speed of the front rollers on the spinning frames relative to the speed of the water wheel. The rollers were one inch in diameter and would have to run at a speed of between fifty and seventy revolutions per minute. The former partners also had to agree not to interfere with the shafting in the mill or change the quantity or speed of the machines.

Hartley was bankrupt in 1800 and his third part of Malham Mill was for sale together with his machinery. The complete mill was four storeys high and measured twenty-seven feet by twenty-one feet with an eighteen foot fall of water to the wheel. The newspaper advertisement mentioned that the mill was only five miles from the Leeds and Liverpool canal and that 'from the village of Malham hands may be engaged on reasonable terms'. The auction was to take place at the Angel Inn, Colne.

Richard Brayshaw leased his section of the mill to William Cockshott about 1797 and in 1815 it was leased again to the Cockshott brothers with part being taken by John & Joseph Lister from Haworth who were cotton twist spinners, possibly at Griffe Mill near Stanbury. In 1815 the mill and contents were insured for £2,400 and cotton spinning continued until the 1840s. By 1850 the mill was in ruins and was said 'by no means adds to the beauty of the scene'. The mill was then pulled down and the stone used to build a barn.

Scalegill Mill, Kirkby Malham. This was probably a corn mill which was used for cotton spinning by Roger Hartley in 1792 when he insured the mill for £200 and the utensils for £600. The mill was then rebuilt in 1794/5 and came on to the market in March 1795 when it was described as:

> All that new erected cotton mill, called Scalegill Mill situate about four miles from Airton aforesaid, with a very convenient dwelling house, outhouses, and other appurtenances adjoining to the mill.
>
> The above mill was intended for a cotton mill and is a very desirable situation for any person wishing to go into trade; it has a constant and powerful supply of water, and in a situation where wages are low.

Richard Shackleton of Airton would show the premises to prospective purchasers and other particulars could be obtained from Mr Sidgwick or Mr Netherwood in Skipton or Messrs Hartley & Swale in Settle.

Scalegill Mill near Kirkby Malham.

Christopher Netherwood was the leading partner when the mill was built. He was a merchant and banker and in 1804 he insured the mill and contents for the following sums:

	£
Scalegill Mill	400
Mill work	200
Machinery	200
Stock	50
	850

There was also a warehouse nearby which contained more stock and which was insured for £50 with the stock valued at £200. By 1809 the mill appears to have been enlarged again for the insurance valuation had increased considerably. In addition the mill was now four storeys high and there was a house adjoining.

	£
Mill	800
Mill work	300
Machinery	1,000
	2,100

For some reason the mill was not in use in 1809 but the following year it was leased to Joseph Mason. When Christopher Netherwood's sons came of age they ran the mill themselves in 1818 but by 1820 the mill had been sold to John Dewhirst & Co. of Skipton. According to the insurance valuations Dewhirsts spent a considerable amount of money on new machinery and continued spinning cotton at Scalegill Mill until after the end of this survey. The mill and dam can still be seen.

Airton Mill. The old corn mill was for sale in 1785 and was bought by William Alcock who was a banker in Skipton. Alcock sold an interest in the mill to Margaret Williams, a widow of Kirkby Malham, and several others. The corn mill was then used by John Brown for cotton spinning about 1786. The partners then built a new mill alongside especially for cotton spinning. This was built about 1789 and taken by Robert Thornton. Thornton

Airton Mill.

joined a partnership at Castle Mill in Knaresborough that year but continued cotton spinning at Airton Mill until 1795 when he was bankrupt. The cotton spinning machinery used by Thornton was then for sale. It included six carding engines, two drawing frames, two roving frames and six spinning frames with forty-eight spindles each.

In 1797 the mill was owned by John Williams of Dartford in Kent, who was Margaret Williams' son-in-law, and John Hartley of Airton while the firm leasing the mill was Hartley, Maher & Co. The mill and machinery were insured for £1,000 while Hartley & Maher insured their stock for £500. Williams leased the cotton and corn mills to William Ellis from 1st May 1803 for a term of thirty-one years at a rent of £113 pa. William Ellis transferred the lease to John Greenwood and Lister Ellis, the Keighley cotton spinners and merchants, in December 1807. As they only rented the mill they only insured the mill work, machinery and stock for the following sums in 1808:

	£
Millwork	200
Machinery	900
Stock	400
	1,500

In 1817 there was a crisis in the Greenwood & Ellis family businesses and all their properties and leases were for sale but none were sold. Airton Mill was then seventy-two feet long by thirty-one feet wide and was equipped with a twelve foot by three foot water wheel. Greenwood & Ellis had also installed a sixteen foot by six foot wheel as they had increased the fall of water to the mill. Their preparing and spinning machinery was also for sale and included seventeen frames with one thousand six hundred and thirty-two spindles. John Greenwood & Co. continued cotton spinning at Airton Mill until 1825 when

the lease and machinery were taken by John and Isaac Dewhirst from Skipton. By 1837 James Garforth had taken the mill. In 1836 a new mill was added with steam power and gas lighting. The mill has now been turned into flats.

Bell Busk Mill. Peter Garforth Junior and Thomas Hallowell had been using Carleton Old Hall as a spinning shop before they built Bell Busk Mill in 1794. In that year the mill was insured for £1,500 and the will work for £200. By 1797 machinery and stock to the value of £2,000 and £500 had been added as were a warehouse and joiners shop. The main mill was five storeys high with 1,965 square yards of floor space.

By 1805 Garforth had withdrawn somewhat from the partnership as he had other cotton mills at Sedbergh, Skipton and Bingley. Bell Busk Mill was therefore tenanted by the firm of Thomas Hallowell & Co.

James Braithwaite Garforth took over the running of Bell Busk Mill after his father's death. By 1816 a small steam engine had been added and the insurance cover was as follows:

	£
Mill	1,000
Mill work	400
Steam engine and pipes	200
Machinery	3,200
Stock	200
	5,000

By 1833 this mill was being used to spin 40s quality cotton yarn on mules. Power was supplied by a 16 hp water wheel and the small 6 hp steam engine. Altogether one hundred and seven people were employed at the mill which was turned over to silk spinning about 1862.

Eshton Bridge Cotton Mill. This mill appears to have had a short life. The

Bell Busk Mill was built in 1794 and used for cotton spinning until about 1862 when it was changed over to silk.

partnership of John Orrell, Cornelius Lister and John Blackburn rented the mill and land from Matthew Wilson of Eshton Hall in 1797. Unfortunately they were bankrupt in 1798 and the mill and their machinery were for sale. The partners had the usual cotton preparing machinery as well as jennies and frames. Although there was a dam and mill race to the water wheel at the mill and it was said to be to the north side of Eshton bridge no trace of this mill has been found.

At Otterburn. A man called Dale Preston was spinning cotton at a mill in Otterburn in 1793 but no trace of the mill has been found nor have any other references.

High Mill, Gargrave. According to a contemporary trade directory Joseph Mason was spinning cotton at this mill in 1793. In 1818 he insured this mill and contents for:

	£
Cotton mill	250
Mill work	150
Machinery	800
Stock	200
	1,400

In 1819 this mill was mortgaged to Chippendale, Netherwood & Carr who were bankers in Skipton. Cotton spinning continued until after the end of this survey.

Low or Goffa Mill, Gargrave. Betty Hudson, who built Damside Mill in Keighley, insured this mill and its contents for the following sums in 1800:

	£
Cotton mill	500
Mill work	100
Machinery	500
Stock	400
	1,500

Mrs Hudson's daughter married a man called Thomas Parker and it was Parker

High Mill, Gargrave.

Low Mill, Gargrave, with later extension.

and his son, also called Thomas, who ran mills in Keighley, Arncliffe and Gargrave although Mrs Hudson probably supplied the capital. The firm running this mill initially was a partnership between Parker's brother, James and Samuel Gill, a cotton spinner from Grassington. That partnership was dissolved in 1804 whereupon James Parker & Co. ran the mill.

All the members of the Parker family in the cotton trade were eventually rendered bankrupt with James Parker the last in 1811. This mill was then for sale. The three storey building on the river Aire measured seventy-four feet by twenty-nine feet within the walls and had a 'most powerful water wheel'. The spinning machinery consisted of:

14 water twist frames with 48 spindles each

1 throstle with 148 spindles

There was also a house suitable for the mill owner or manager and twelve cottages. It is not clear if cotton spinning continued at this mill after 1811.

Airebank Mill, Gargrave. This mill on the river Aire was built in 1791, possibly by Thomas Mason Senior. Another Thomas Mason, who may have been his son, continued cotton spinning at this mill until well after the end of this survey. In 1823 he insured the mill and contents for the following sums:

	£
Cotton mill at Gargrave	400
Mill work	250
Machinery	1,250
Stock	150
	2,000 (?)

The high insurance value of the machinery in 1823 might reflect the cost of the power looms which were introduced about that time. Certainly the mill was attacked in April 1826 by rioters who objected to the looms and many of them were smashed.

In 1833 the mill had a 20 hp water wheel and employed thirty-three people.

Threaplands Mill. Benjamin Shiers, a grazier and cotton spinner, insured this small water powered cotton mill and contents in 1800:

	£
Threaplands Cotton Mill	100
Mill work	50
Machinery	300
Stock	50
	500

Shiers was still running the mill in 1812 but the mill was then leased by a man called Whitaker. In 1818 the four storey mill was for sale and no further references have been found.

Rilston and Hetton Mill was an old established corn mill which had been built half way between the two villages. A firm trading as Robert Bradley & Co. made enquiries about buying or leasing the mill and a local landowner called Richard Waddilove may have been associated with them in some way. Bradley does not appear to have occupied the mill himself but leased it to John Heaton who bought iron rod from Kirkstall Forge in 1794.

In 1798 there were proposals from Robert Bradley & Co. to build a new mill on the site of the old mill and it was hoped that the Duke would either pay for the construction of the new mill or take some land in exchange. The mill was to be four storeys high and measure seventy-five feet by thirty feet. It is not clear if the new mill was built but it probably was as the cotton machinery in Rilston Mill was advertised for sale at the end of December 1803 and would have taken up a building of the size which was proposed. The machinery for sale included:

3 cotton carding engines 30 inches wide

2 drawing frames

3 mules with 204 spindles each
1 throstle with 108 spindles
1 throstle with 72 spindles
1 stretching frame with 96 spindles
1 cotton picker
11 pairs of looms for weaving velveteens

For particulars people had to apply to Robert Wallace at Sedbergh, James Wallace at Rilston Mill, who may have been the tenant, or George Dixon at Halifax. By 1806 the mill was out of production and in a dilapidated state. It was pulled down in 1826.

Hetton Mill was offered for sale by auction in March 1795. It was said to be a new building which was upwards of eight yards long and used for spinning cotton weft. The machinery consisted of:
6 jennies with 100 spindles each
1 carding engine
1 roving billy with 42 spindles
1 beater of cotton which was driven by the water wheel
There was also a stove, pipes, a press and weights.

It was said that the mill was well supplied with water and could be entered immediately and there was also a house for sale. Hetton was described as six miles from Skipton and three miles from the Leeds and Liverpool Canal. Any application had to be made to Joseph Ellis at Hetton.

The mill and house were offered for sale again in April 1798 when it was said that Joseph Ellis was the owner and that the cotton twist mill was in Manheads Close. No further details have been found.

High Mill, Skipton. According to replies given to the Factory Enquiry Commissioners in 1833 High Mill was built in 1782. The original partners were Peter Garforth, John Blackburn and John Sidgwick. Blackburn soon dropped out of the partnership which was continued as Garforth & Sidgwick who were brothers-in-law. They were cotton spinners and manufacturers on quite an extensive scale with, for example, £2,000 worth of stock insured in their warehouse in Skipton in 1791. In 1793 they insured their cotton mill and contents for:

	£
Cotton mill, warehouses and counting houses under one roof at the Spring, Skipton	800
Utensils, stock and goods	2,000
	2,800

The four storey mill had a floor area of 1,344 square yards in 1804 and was equipped with water frames. These had been made by the engineers and joiners employed by Garforth & Sidgwick. At this time Peter Garforth Senior was also spinning cotton at mills in Sedbergh, Bell Busk and Bingley.

In 1814 the water frames were replaced with throstles which were used to spin 27 to 32 quality twist yarn which was woven locally by their hand-loom weavers. To cover the increased cost of the new throstles the insurance for the machinery in the mill was increased to £3,000 in 1816.

William Sidgwick took over the control of the firm about 1805 and in 1825 built a new mill adjoining the old one which was then only used for storage. The new mill was steam powered with an engine of about 30 hp. The water wheel provided from 15 hp to 30 hp. By 1833 power-loom weaving had been added to throstle and mule spinning and the firm of J B Sidgwick employed two hundred and sixty-four people.

Belle Vue Mills, Skipton. John, Isaac and James Dewhirst were cotton spinners and manufacturers in Skipton in the early part of the century. They had their yarn spun at mills in Embsay and Kikby Malham until Belle Vue Mills were rebuilt for cotton spinning. Dewhirsts used warehouses in Skipton and also had premises for sizing and drying their cloth. They attended Manchester markets to sell their calicoes with premises at 14 High Street. Belle Vue Mill was built in 1828 for worsted spinning but burnt down in 1831. It was then rebuilt for cotton spinning. The new mill was steam powered and lit with gas. Dewhirsts extended the mill between 1832 and 1834 as the insurance cover rose from £2,150 to £15,560. The mills were added to considerably during the rest of the century and became the largest textile mills in Skipton.

Whitfield Syke Mill, Embsay. This mill was built about 1795 and in 1800 was insured by John Hammond and Thomas Tattersall for:

	£
Cotton mill	200
Mill work	80
Machinery	170
Stock	200
	650

Hammond & Tattersall continued cotton spinning until 1809 when the partnership was dissolved and the mill was offered for sale. In 1809 Whitfield Syke Mill was three storeys high, measured forty-six feet by twenty-eight feet and was driven by an eighteen foot by three foot water wheel. Besides the usual preparing machinery there were six mules with 1,144 spindles between them. The firm also had four more mules in a building in New Market Street, Skipton with a further 854 spindles.

A new partnership between John Hammond and William Oldridge ran the mill until 1811 when it was dissolved. John Hammond died in 1817 but prior to that the mill had been taken over by Greenwood & Whitaker of Burley-in Wharfe-

dale. They had advanced money to Hammond on Mortgage which could not be repaid, which explains why they came to own the mill. William Oldridge lived in one of the cottages near the mill and appears to have been the manager from about 1811. Oldridge later leased this mill from John Greenwood and in 1833 he was spinning weft on mules. The mill was driven by a 3 hp water wheel and employed twenty-one people. The mill burnt down in 1837 and was finally demolished in 1903 to make way for a new reservoir.

Good Intent Mill, Embsay. It is not clear when this mill was built but the first reference found was when Thomas Bramley & Sons occupied it in 1816 for cotton spinning. The mill was built over some cottages and a stable with a water wheel at the southern end. Bramleys insured their property for the following sums in 1820:

	£
Mill and cottages	280
Mill work	20
Machinery	200
Stock	600
	1,100

By 1833 the insurance cover had increased to £1,600. Thomas Bramley & Sons were calico manufacturers, as were most of the local cotton firms, and sold their cloth in Manchester. When the mill was for sale in 1844 it was three storeys high and measured eighty-one feet by twenty-seven feet. Power was supplied by a 9 hp water wheel and a 10 hp steam engine. The preparing machinery was on the bottom floor while the blower and four pairs of mules were on the other two floors. Also for sale was land, seven cottages, a house and a new building which was used as a warehouse but could be adapted for power-looms or throstles. The mill burned down about 1857.

Crown Spindle Works, Embsay. This mill was built in 1826 by Richard Shackleton and used for both cotton spinning and the making of spindles for the expanding textile industry. The outline of the old weaving shed can still be seen, while the main mill building is now cottages.

Sandbeds Mill, Embsay. Land at Sandbeds suitable for building a cotton mill was for sale in 1786. It is not clear when the mill was built but it may have been occupied by John Merryweather, a hosier from Leeds, from before 1794 when he was buying iron rod and bar from Kirkstall Forge. He was bankrupt by 1796 with the creditors' meeting in Manchester. The mill was then taken by Thomas Whitehead. He insured the mill and contents for the following sums in 1795:

	£
Cotton factory	200
Mill work	200
Machinery	300
Stock	300
	1,000

Thomas Whitehead also had property in Skipton which was used for cotton spinning, possibly on hand mules or jennies. This mill was later taken by Edward Whitehead & Co, but was to sell or let in 1805. The four storey mill then measured sixty-six feet by thirty feet and was powered by a 'large' water wheel. Besides the usual preparing machinery the firm had seventeen mules with three thousand spindles. Edward Whitehead occupied this mill until about 1815 when it was for sale again together with the machinery. If the mill was not sold it could be leased for £60 per year. John Dewhirst took up the lease and continued cotton spinning at Sandbeds Mill until after the end of this survey.

Millholme Mill, Embsay. This mill was built by William Baynes and Allen Edmondson in 1792/3. The millwright was Thomas Sheperd from Bradford and rod iron was bought from Kirkstall Forge in Leeds in 1792. The firm traded as Baynes, Barker, Spencers & Co. and promptly insured the mill and contents with the Royal Exchange Fire Office:

	£
Building at Millholme	500
Utensils and trade goods	1,000
	1,500

By 1795 they were insuring an additional building for £100, hand mules and machinery for £200 and stock for £300. The firm also manufactured cotton cloth using hand-loom weavers. William Baynes owned the mansion and estate of Embsay Kirk. The mill had been built on his land and he wished to sell the whole estate together with this mill in 1797 and also in 1810. It was stressed, however, that the mill could not be seen from the mansion. About 1813 the lease was taken by John Dewhirst from Skipton and the firm of John Dewhirst & Sons continued cotton spinning at this mill until at least 1835.

Eastby Cotton Mill was started in 1794 and finished in 1796 when the machinery commenced running. It was built on the site of a corn mill by the firm of Netherwood & Carr who were bankers and merchants in Skipton. However, by 1797 William Chamberlain had either bought or leased the mill. He insured the mill and contents for the following sums that year:

	£
Eastby Cotton Mill	100
Mill work	50
Machinery	200
Stock	50
	400

Chamberlain was a cotton manufacturer and ironmonger in Skipton where he also had extensive mule spinning shops and hand-loom weaving shops. This mill would therefore give him his own supply of yarn. In 1801 a new mill was built and the total insurance cover was increased:

	£
Mill and tenement	600
Mill work	300
Machinery	1,000
Stock	300
	2,200

The old mill was four storeys high with a floor area of 938 square yards while the new mill was five storeys high with a floor area of 1,161 square yards. Both mills also had attics and basements. Chamberlain was running a mixture of mules, jennies and water frames in 1804. By 1809 the new mill was just being used as a warehouse, the old mill was re-equipped and two water wheels were installed. Chamberlain tried to let Eastby Mill in 1813 with the machinery available at valuation but did not do so. By 1820 the mill buildings were being used as security for a mortgage. William Chamberlain was later followed by George Chamberlain who still had premises in Skipton. The mill was eventually demolished.

Draughton Cotton Mill. This mill was built about 1788 and in 1793 the unexpired part of the lease of the ground floor of the mill was available. Also included was a house, a spinning shop and the machinery, which included:

1 double carding engine 44 inches wide

2 roving billies
1 cotton picker
2 jennies with 112 spindles each
6 jennies with 106 spindles
2 jennies with 84 spindles

The owner or tenant was possibly Robert Pearson from Addingham who was a draper and cotton spinner in that village until 1797 when he was bankrupt. Draughton Mill was then bought by or assigned to William Myers who tried to sell it in 1799. The mill appeared to have power only for the picking and carding processes but it continued to be used for cotton carding and spinning until at least 1822 when George Pickup was running the mill.

Cononley Cotton Mill. It is not clear when this mill was built but in 1796 William Watson insured the mill and contents for:

	£
Cotton mill	100
Mill work	30
Machinery	140
Stock	30
	300

By 1810 the machinery at this small mill was for sale. It was said that Mr Nathan Horrocks would show prospective purchasers the workshop tools and the machinery which included three mules with one hundred and twenty spindles each. No further references have been found.

CHAPTER 19

Consolidation

THE RAPID AND DRAMATIC GROWTH of the cotton industry must have been apparent to all sections of the communities in the West Riding and also in parts of the East and North Ridings of Yorkshire. Even the remote rural districts were touched by the developments which, in many ways, had a greater impact there than in the urban areas. Within twenty years from 1780, cotton mills were built in nearly every township where there were streams from the Pennine uplands. The mills were usually new structures and often larger than any others in the neighbourhood. They employed large numbers of people and were often the first to use steam power. Many worked twenty-four hours a day and with their artificial light and rows of windows stood out everywhere on the landscape. They changed patterns of employment and scales of wages. Young people and women could achieve a degree of independence through their own earnings while traditional trades such as hand-loom weaving expanded rapidly with the availability of large local sources of cheap yarn. The drain of labour to the mills affected agricultural wages while the working conditions for cotton operatives prompted the first factory employment legislation.

There cannot have been many people who were unaware of the growth of the cotton industry in a locality. Mill buildings were followed by rows of cottages and houses for workers and overseers. At times special dormitories or apprentice houses were also built. Transport by road and canal expanded as did the firms of millwrights, machine makers, steam engine and boiler manufacturers. Commercial activities to deal with warehousing, merchanting, insurance and banking also expanded to meet the needs of the new cotton firms. There was an unprecedented growth of industry which had an impact on nearly every community.

Within Yorkshire there were four main effects from the establishment of the cotton industry. The first was to confirm the foothold of the industry on a permanent basis in the western valleys near to Lancashire. The second was to introduce relatively permanent industrial production to some rural areas. Thirdly it facilitated the subsequent development of factory based worsted spinning through the transfer of assets and labour and the fourth effect was to stimulate the growth of the coal, iron and engineering industries together with associated commercial developments.

Of the large numbers of cotton mills which were built in this early period, many were no longer in use for cotton spinning by 1835. In general terms they fell into the following categories which illustrate the overall decline in numbers:

1. Rural Mill – failed, no future industrial use, perhaps dismantled or used for housing

2. Rural Mill – changed to worsted/woollen/flax/silk

3. Rural Mill – succeeded, continued until 1900 or later

4. Urban Mill – failed, changed to other industrial use

5. Urban Mill – changed to worsted/woollen/flax/silk

6. Urban Mill – succeeded, continued until 1900 or later

At times the distinction between a rural and urban location is difficult to define as some rural mills prompted the further development of the community surrounding them. Examples against the trend can also be found where additional factors such as availability of capital and easy transport worked in favour of particular mills and locations. However, from the evidence in the previous chapters it would appear that by 1835 most mills had fallen into the categories given above and many examples can be found to illustrate this.

1. Yorkshire Dales villages and hamlets went through a process of industrialisation by the cotton industry which was unique. For example, cotton mills were thrown up with great speed in practically every village in Wharfedale, Airedale and Ribblesdale. Two of these were the mills built just outside Long Preston near the road to Slaidburn. These three and five storey mills had water wheels but also a steam engine and power from a windmill.[1] One of the mills had mules and throstles in 1814 but both appeared to go out of use at that time and there is little trace of them on the first Ordnance Survey map.

2. An example of a cotton mill in a rural situation which was changed to worsted spinning was Brow End Mill in the hamlet of Goose Eye about three miles from Keighley. It was built in 1791 and used for cotton spinning for nearly fifteen years. However, the early boom in cotton spinning did not last and, as one of the partners at this mill was also a worsted manufacturer, the cotton frames were replaced with worsted frames in 1805.[2]

3. The Upper Calder Valley provided many sites for early water powered cotton mills and the proximity of Lancashire ensured an extended life for many of them. It is difficult to select a typical rural mill but Lord Holme Mill, or Gibson Mill, as it is better known, was used for cotton spinning from about 1803 to 1900. The Gibson family occupied it for much of the

Greengate Mill, Keighley. The original cotton mill in the foreground with later additions for worsted spinning.

time. Eventually its remote situation on the highest reaches of Hebden Water, a tributary of the Calder, made cotton spinning there uneconomic.

4. Examples of town mills which were built for cotton spinning and were then not used for other textile purposes are rare. Most of the cotton mills built in Leeds, Halifax and Keighley were later taken over by worsted and woollen firms, but there was one notable exception in Sheffield. This large mill, built in 1793, was possibly an attempt to emulate the apparent success of the large steam powered mills in Leeds and Manchester. However, its isolation from the other textile centres meant that after cotton spinning was given up about 1815 it was no longer used for textile purposes and was eventually turned into a poor house.

5. Many urban cotton mills changed to worsted spinning or wool scribbling, particularly in the Keighley, Bingley and Leeds areas. There the cotton industry was superimposed on the existing worsted and woollen trades which were able to benefit by taking over failed cotton mills, particularly when mechanised worsted spinning became profitable. By 1835 only four of the forty or so earlier cotton mills were still being used for that purpose in the Keighley area.[3] Greengate Mill, for example, was built as a cotton mill in 1784/5 but, when in 1817 complete ownership came into the hands of the young Marriner brothers, Benjamin and William, they changed it to spin worsted yarn as they were also worsted manufacturers.

6. Although the Saddleworth and Calder Valley areas generally provide most examples of cotton mills used for nothing else for the whole of the nineteenth century, many of the larger mills in places like Sowerby Bridge were built after the end of this survey.[4] Many of the mills which were in existence by the 1830s were not in towns, so the example here is Belle Vue Mill in Skipton which was still used for making sewing cotton until recent years. As it was originally built as a worsted mill in 1829 but rebuilt as a cotton mill after a fire in 1831, it is an example of the second wave of cotton mill construction. The owners, John Dewhirst & Brothers had previously leased mills at Embsay near Skipton and Scalegill and Airton Mills near Kirkby Malham. They appeared to have more capital than other local firms and were able to build Belle Vue Mill when many other cotton firms were going out of business.

The first frantic wave of cotton mill building lasted until about 1810 but in that time there were many ups and downs. The awareness of the apparent ease with which Arkwright and his licensees accumulated profits no doubt inspired the growth of the Yorkshire cotton industry at a time when so many factors were in its favour. The first Arkwright licensees did not start building mills until 1777, with the only known example in Yorkshire, being built in 1780. The other mills built in the next three or four years were local attempts to copy Arkwright and a few mills were started before his patent expired. When his spinning patent was defeated in 1783 the way was open for the growth which has been described in previous chapters. However, within a few years the mills which had been built on the Arkwright pattern and at long distances from Manchester were overtaken by newer mills equipped with mules. The benefits of building mills in the towns and dales of Yorkshire proved to be transitory and the new trade did not live up to the hopes of the owners and tenants. As technology, transport and the scale of the industry changed, the advantages of sites in many parts of Yorkshire became disadvantages. The success of mechanised spinning in the cotton industry soon led to its adoption in the wool, worsted and flax industries which were already well established in most parts of Yorkshire. The worsted industry had suffered a brake on its growth during the 1790s when there was a possible local contraction.[5] However, this hiatus in its growth was only temporary and the new local cotton industry soon felt the competition for capital and productive capacity, particularly in the towns. The earlier opposition to power driven machinery in Lancashire was overcome and with the expiry of Watt's patents, large steam powered mills were built in large numbers in the cotton towns to the west of the Pennines. Thus there were both changes within the cotton industry itself which affected the marginal producers and also increasing competition for the transferable assets of individual firms which meant that overall there was a decline in the number of cotton mills in Yorkshire after 1810. This decline lasted until the 1830s when a second wave of development in the western margins of Yorkshire occurred because of the continued growth of the adjacent Lancashire industry.

As the structure of the northern cotton industry changed after about 1810, more and more Yorkshire firms and mills went out of production or changed to other fibres. One or more of the following factors was enough to tip the scales against the marginal firms and bring about the sale of their mills. The first was the cost of transport, for Manchester was always the trading centre for the Yorkshire firms. Transport costs started to rise during the French wars which meant that raw materials and finished goods from Yorkshire were more expensive. The costs of producing yarn and cloth fell as productive techniques improved which meant that transport costs became a larger part of unit costs and this discriminated against the more remote mills.

Many of the Yorkshire mills were small and the firms which ran them had inadequate financial resources to weather periods of bad trade or to re-equip to take advantage of periods of good trade. The firms in the Yorkshire Dales which did manage to struggle on until the 1850s were usually the larger ones such as Greenwood & Whitaker at Burley-in-Wharfedale or Claytons at Langcliffe Mill near Settle.

With the development of more efficient steam engines and the lowering of the cost of each unit of horse power, steam powered mills near to supplies of coal had advantages over those relying solely on water wheels. There were limitations on the amount of power which could be obtained from particular streams and from the types of water wheel in use. Larger dams could be built but they cost money and suitable land was not always available. Improved machinery needed to be driven at higher speeds which meant difficult gearing and problems with power transmission from a water wheel.

The small scale of many of the Yorkshire firms mitigated against their ability to expand as they did not have the financial resources. Although machinery did last for years, larger versions were produced with more spindles and other improvements were made. Thus firms which could afford new machinery achieved greater outputs. Examples have been found of different firms operating spinning frames at the same time which had, on the one hand, forty-eight spindles per frame and, on the other, one hundred and forty-four.[6] Often firms had to keep their old machinery as they did not have the physical space in their mills for the new, larger throstles and mules. The sixty or seventy foot by thirty foot, three of four storey, Arkwright standard mill, was widely copied but after a few years was superseded by larger structures. Some firms were able to rebuild or add new buildings and the successful ones often had a collection of buildings which had been added to over the years. New spinning mills were build while the older buildings were turned into warehouses.

One of the reasons why mills were built in certain areas was to secure a supply of child labour. The supply often proved inadequate and additional labour had to be imported to be housed and fed in accommodation provided at the mill. Eventually the treatment of apprentices prompted legislation which restricted their use. The child labour was used to attend to the early water frames and throstles. The new mills had mules which became power assisted and spun finer yarns. They depended on the skill of the operator, who initially was usually an adult male, and they also took less power. The early advantages of cheap water power and cheap child labour in the countryside became less important in the face of cheaper steam engines and the more versatile mule. As an example, Whitaker & Merryweather employed two hundred and sixty apprentices as well as adult workers at their mill in Burley-in-Wharfedale in 1802.[7] By 1833 they only employed a total of three hundred and seventy-six workers of all ages at the original old mill and a new mill which was over twice the size.[8] Eventually the impact of these changes and the growth of the east Lancashire towns drew cotton workers away from such areas as Craven to work in the new mills over the county border.[9]

In certain areas of Yorkshire there was a rapid decline in the hold of the cotton industry which can be specifically accounted for. In the Bradford-Keighley area, the economic depression of 1826 brought two events which accelerated the

already steady decline. The first was the failure of one of the leading Yorkshire firms of merchants which had serious repercussions for the local cotton spinners and manufacturers. This firm was the partnership of Jabez Butterworth, Joseph Horatio Butterworth and Sidney Aquila Butterworth who traded as Butterworth Brothers. Their business premises were at Shelf, between Bradford and Halifax, and at Lawrence Lane in London.[10] They traded in both worsted and cotton goods and many cotton manufacturers in Bradford and Keighley sold their pieces through this firm. The general commercial crisis of 1826 was so associated locally with the name of this firm that it became known as the 'Butterworth Panic'. It was said at this time that 'scores of honest tradesmen were reduced to poverty'. In the Great Horton area of Bradford, cotton and calico manufacturers lost £6,000 in the collapse of Butterworth Brothers.[11]

To add to the worsening financial problems of local cotton firms caused by the general decline in demand the water powered mills suffered from a serious drought. That lasted for at least two months in May and June of 1826 and seriously affected their ability to do any work as well as slowing traffic on the canals.

By 1835 the Yorkshire cotton industry had been in existence for over fifty years but had undergone major changes. It was very much more concentrated in the western part of the three counties with the main centres in Saddleworth, the Calder Valley and the Craven district. Most of the mills in the Dales were no longer used for cotton nor were those in Leeds. The Keighley mills had nearly all been changed over to spin worsted yarn while in other parts flax and silk spinning had taken over from cotton. The industry was to continue however, and was just receiving a new lease of life from the discovery that mixed fabrics of cotton warp and worsted weft could be successfully dyed.[12] Cotton was also successfully combined with woollen weft to make union cloth from about 1838.[13] Thus technology changed the industry again and by the end of the century the Yorkshire cotton industry was still flourishing and had even increased its share of the trade but it was a different industry from that of 1800.[14]

Notes

NOTES TO CHAPTER I

1. D. T. Jenkins, 'The Cotton Industry in Yorkshire, 1780–1900, *Textile History*, 10 (1979), pp. 75–95.

2. E. Baines, *History of the Cotton Manufacture in Great Britain* (1835), p. 418.

3. D. T. Jenkins, *The West Riding Wool Textile Industry, 1700–1835* (Edlington, 1975).

4. J. Hodgson, *Textile Manufacture and other Industries in Keighley* (Keighley, 1879).

5. J. Parker, *Illustrated Rambles from Hipperholme to Tong* (Bradford, 1904), p. 88. Parker for example quotes the failure of the Shelf firm of merchants, Butterworth Brothers as 1845 when it was 1826–see the *Bradford Courier and West Riding Advertiser* 2 Feb. 1826.

6. R. H. Hills, *Richard Arkwright and Cotton Spinning* (London, 1973), p. 13.

7. J. Travis, *Notes of Todmorden and District* (Rochdale, 1896), p. 23.

8. For a detailed account see C. Aspin, *James Hargreaves and the Spinning Jenny* (Preston, 1964).

9. E. Baines, op. cit., p. 159.

10. S. D. Chapman, *The Early Factory Masters* (Newton Abbot, 1967), p. 48.

11. E. Baines, op. cit., p. 159.

12. R. S. Fitton and A. P. Wadsworth, *The Strutts and the Arkwrights* (Manchester, 1964), p. 78.

13. S. D. Chapman, 'The Arkwright Mills', *Industrial Archaeology Review*, VI, I (Winter, 1981–2), p. 17.

14. E. Baines, op. cit., p. 160.

15. H. Catling, *The Spinning Mule* (Newton Abbot, 1970), p. 40.

16. LM 14 July 1785

17. LM 24 Jan. 1786.

18. LI 9 June 1789.

19. LI 22 March 1791.

20. LI 19 Apr. 1791.

21. LI 9 Jan. 1792.

22. LI 13 Dec. 1791.

23. LI 23 Jan. 1792.

24. LI 9 Jan. 1792.

25. LI 4 Apr. 1796.

26. LI 4 Apr. 1796.

27. Chambers MSS, MS20, WYAS, Leeds.

28. HJ 17 Dec. 1808.

29. H. J. Turner, *Haworth Past and Present* (Brighouse, 1879), p 147.

30. LI 21 Jan. 1811.

31. J. Hodgson, op. cit., p. 232.

32. LI 28 Jan. 1793.

33. G. Firth, 'The Genisis of the Industrial Revolution in Bradford 1760–1830'. Ph.D. Thesis University, Bradford University, 1974, p. 37.

34. A. Raistrick, *Old Yorkshire Dales* (Newton Abbot, 1964), p. 94.

35. LI 3 Aug. 1779.

36. LI 13 May 1788.

37. T. W. Hanson, *The Story of Old Halifax* (Halifax, 1920), p. 223.

38. W. Marshall, *The Review and Abstract of the County Reports to the Board of Agriculture* (1808, reprinted Newton Abbott, 1968), p. 353.

39. E. Baines, *A Directory of the County of Yorkshire* (1822), p. 405.

40. LI 20 July 1795.

41. T. Brayshaw and R. M. Robinson, *A History of the Ancient Parish of Giggleswick* (London, 1932), p. 207.

42. Report on Agriculture 1793, quoted in G. H. Brown, *On Foot Round Settle* (Settle, 1896).

43. C. B. Andrews (ed.), *The Torrington Diaries*, Vol. III (1792), *A Tour of the North* (reprinted 1970), p. 108.

44. LI 30 May 1796.

45. LI 2 May 1796.

46. H. B. Rodgers, 'The Lancashire Cotton industry in 1840', *Transactions of the Institute of British Geographers*, 28 (1960), p. 140.

47. E. M. Sigsworth, *Black Dyke Mills* (Liverpool, 1958), p. 15.

48. LI 12 Aug. 1788.

49. Extract from the *Leeds Mercury*, 1802,

quoted in R. G. Wilson, *Gentlemen Merchants* (Manchester, 1971), p. 92.

NOTES TO CHAPTER 2

1. LI 5 Jan. 1790.
2. LI 3 Aug. 1812.
3. See S. D. Chapman, The Arkwright Mills–Colquhoun's Census of 1788 and Archaeological Evidence, *Industrial Archaeology Review*, VI, I (Winter 1981).
4. Returns Made to the Clerk of the Peace under Statutes Geo III 87, 1802. An Act for the preservation of the health and morals of apprentices and others employed in cotton and other mills. WYAS Wakefield.
5. HJ 7 Jan. 1804.
6. Samuel Crompton's Notebooks.
7. R. L. Hills, 'Hargreaves, Arkwright and Crompton. Why Three Inventors', *Textile History*, 10 (1979), p. 120.
8. G. W. Daniels, 'Samuel Crompton's census of the Cotton Industry in 1811', *Economic History* (1930), p. 108.
9. HJ 3 Dec. 1803.
10. LI 25 Nov. 1811.
11. D. T. Jenkins, *The West Riding Wool Textile Industry 1770–1835* (Edington, 1975), p. 75.
12. LI 2 Dec. 1793.
13. J. Sutcliffe, *A Treatise on Mills and Reservoirs* (Rochdale, 1816), p. 67.
14. LI 4 March 1811.
15. HJ 5 March 1808.
16. HJ 13 Aug. 1808.
17. LM 3 Oct. 1818, LI 29 June 1795.
18. HJ 24 Sept. 1803.
19. LI 19 Sept. 1796.
20. LI 10 July 1809.
21. LI 4 July 1808.
22. LI 2 Feb. 1801.
23. G. R. Binns, 'Water Wheels in the Upper Calder Valley', *HAS*, January 1972.
24. J. H. Priestley, 'Mills of the Ryburn Valley', *HAS*, 3rd Series (1941).
25. LI 18 Feb. 1805.
26. J. Hodgson, *Textile Manufacture and other Industries in Keighley* (Keighley, 1879), p. 238–9.
27. Hodgson, op. cit., p214.
28. LI 28 Jan. 1793.
29. John Cross to Boulton & Watt, 24 Dec. 1792. B&W MS BCRL.
30. John Cross to Boulton & Watt 24 Dec. 1792. B&W MS BCRL.
31. John Cross to Boulton & Watt, 4 July 1793. B&W MS BCRL.
32. John Cross to Boulton & Watt, 4 June 1794, B&W MS BCRL.
33. John Cross to Boulton & Watt, 4 June 1794, B&W MS BCRL.
34. LI 4 July 1796.
35. W. B. Crump, *The Leeds Woollen Industry 1780–1820* (Leeds 1931).
36. D. T. Jenkins, *The West Riding Wool Textile Industry 1770–1835* (Edington, 1975), p. 82.
37. Ibid, p. 84.
38. T. W. Hanson, *The Story of Old Halifax* (Halifax, 1920), p. 239.
39. S. D. Chapman, *The Early Factory Masters* (Newton Abbott, 1967), p. 59.
40. R. A. Guest, *A Compendious History of the Cotton Manufacture* (1823; reprinted 1968), p. 31.
41. LI 29 Jan. 1798.
42. LI 30 Dec. 1793.
43. SunCR44/724793/1801.
44. LI 9 Sept. 1805.
45. LI 28 Apr. 1800.
46. LI 2 Apr. 1798.
47. LI 27 May 1799.
48. Factories Enquiry Commission, Supplementary Report 1834 (167) C2, Mill Number 126.
49. LI 24 Jan. 1814.
50. B&HC 14 Sept. 1826.
51. HJ 6 March 1802.
52. HJ 12 July 1802.
53. LI 6 Apr. 1807.
54. HJ 9 Feb. 1805.
55. Crompton's Survey, 1811.
56. Report of the Minutes of Evidence on the state of children employed in the manufacturies of the United Kingdom, 1816 (397) III.
57. Hodgson, op. cit., p. 251.
58. Minutes of evidence collected by the committee on the petition of Mr Samuel Crompton, 24 March 1812.
59. Crompton's Survey 1811, BCM.
60. S. D. Chapman, 'Sources and Methods for the Study of Capital Formation in Britain, 1750–1850', *Economic History Review*, xxiii (1970), p. 268.
61. R. L. Hills, *Power in the Industrial Revolution, Manchester* (1970), p. 131.
62. LI 10 March 1800.
63. Crompton's Survey 1811, BCM.
64. RE30/148266/1795.

65. LI 28 Oct. 1805.
66. HJ 25 Nov. 1809.
67. Jenkins, op. cit., p. 109.
68. SunOS319/489479/1785.
69. LI 6 Jan. 1789.
70. LI 28 Sept. 1790.
71. LI 2 May 1796.
72. LI 6 Oct. 1800.
73. LI 9 March 1812.
74. LI 30 May 1796.
75. LI 2 Jan. 1797.
76. LI 12 Oct. 1801 and 19 Sept. 1803.
77. SunCR64/778830/1806.
78. WYAS Leeds, MS DB 23.
79. SunCR54/746731/1804.
80. LM 21 Feb. 1818.
81. SunCR100/868473/1812.

NOTES TO CHAPTER 3

1. S. D. Chapman, 'Sources and Methods for the Study of Capital Formation in Britain 1750–1850', *Economic History Review*, xxiii.
2. E. M. Carus-Wilson, 'An Industrial Revolution in the Thirteenth Century', *Economic History Review*, xi (1941).
3. D. T. Jenkins, *The West Riding Wool Textile Industry 1770–1835* (Edington, 1975), p. 8.
4. J. H. Priestley, 'Mills of the Ryburn Valley', *HAS*, 1933.
5. RE32A/161553/1795.
6. LI 12 May 1800.
7. LI 4 June 1792.
8. LI 16 Sept. 1799.
9. LI 21 March 1814.
10. LI 9 May 1814.
11. LI 8 Aug. 1803, 10 Feb. 1812.
12. RE28/143710/1795.
13. RE32A/160109/1797.
14. LI 27 May 1799.
15. LI 3 March 1800.
16. LI 29 Sept. 1800.
17. SunCR94/847906/1811.
18. LM 8 Aug. 1818.
19. For example LI 9 Dec. 1793.
20. SunCR126/955165/1819.
21. LI 28 Jan. 1793.
22. LI 28 Jan. 1793.
23. LI 22 Apr. 1793.
24. Chambers MSS 20. WYAS Leeds.
25. LI 13 Dec. 1791.
26. LI 30 Jan. 1792.
27. LI 21 Feb. 1803.
28. LI 5 March 1804.

29. LI 10 July 1809.
30. LI 28 Oct. 1799.
31. LI 13 July 1797.
32. LI 12 Nov. 1792.
33. LI 7 Jan. 1793.
34. LI 5 May 1789.
35. HJ 23 Feb. 1810.
36. J. Hodgson, *Textile Manufacture in Keighley* (Keighley, 1879), p. 27.
37. Hodgson, op. cit. p. 220.
38. Deed of partnership. 3 Nov. 1784 Marriner MS LUL.
39. Dartmouth Estate Terrier 1805, Estate Office, Slaithwaite.
40. For a full discussion of the activities of the Leeds woollen merchants see R. G. Wilson, *Gentlemen Merchants* (Manchester, 1971).
41. SunCR7/638227/1795.
42. SunCR13/653283/1797.
43. J. H. Priestley, 'Mills of the Ryburn Valley', *HAS* (1941), pp. 1–19.
44. HJ 23 June 1804.
45. HJ 22 June 1805.
46. LM 3 Oct. 1818.
47. Extracts of Titles to Lands in Bradford, Vol. 2.71, BCL.
48. LI 17 June 1793.
49. Birkbeck MS, WYAS Leeds.
50. SunCR126/955165/1819.
51. LI 3 Nov. 1800.
52. HJ 3 Dec. 1803.
53. H. A. Cadman, *Gomersal Past and Present* (Leeds, 1930), p. 74.
54. LI 14 Sept. 1807.
55. M. M. Edwards, *The Growth of the British Cotton Trade 1780–1815* (Manchester, 1967).
56. RE32A/167265/1799.
57. SunCR60/765181/1805.
58. LI 8 Apr. 1805.
59. LI 15 June 1812.

NOTES TO CHAPTER 4

1. T. Brayshaw and R. M. Robinson, *A History of the Ancient Parish of Giggleswick* (London, 1932), p. 209.
2. LI 22 Apr. 1788.
3. J. Tann, 'The Textile Millwright in the Early Industrial Revolution', *Textile History*, 5 (1974), p. 80.
4. Temple Mill in Rishworth near Halifax was said to be capable of running over 16,000 cotton spindles but never reached that target. LI 10 March 1800.
5. R. Guest, *A Comprehensive History of the*

Cotton Manufacture (London, 1823; re-printed 1968), p. 31.

6. J. Hodgson, Textile Manufacture and other Industries in Keighley (Keighley, 1879), p. 213.

7. Deed of Partnership. 3 Nov. 1784, Marriner MSS LUL.

8. Hodgson, op. cit. p. 36–7.

9. LI 7 June 1791.

10. LI 8 Nov. 1791.

11. Report of the Minutes of Evidence on the state of children employed in the Manufactories in the United Kingdom, 1816 (397), Vol III.

12. Hodgson, op. cit., p. 213.

13. A. Sutcliffe to W. Birkbeck, 20 June 1786. Birkbeck MSS, WYAS, Leeds.

14. J. Parker, Illustrated Rambles from Hipperholme to Tong (Bradford, 1904), p. 88.

15. Report from the Select Committee appointed to consider the state of the Woollen Manufacture of England, 1806 (268), Evidence of William Illingworth.

16. Brayshaw, op. cit., p. 201.

17. LI 10 Apr. 1787.

18. LI 21 Jan. 1805.

19. LI 19 Nov. 1804.

20. Factories Inquiry Commission. First Report 1833 (450) Evidence of Thomas Brown.

21. Old West Riding, Vol. 10, 1990, p. 26.

22. Reports of the Society for Bettering the Conditions and Increasing the Comforts of the Poor, Vol IV, London 1805, Supplement No. 11.

23. Report of the Minutes of Evidence taken before the Select Committ on the State of Children employed in the Manufactories of the United Kingdom 1816 (397), Evidence of William Sidgwick.

24. HJ 12 Feb. 1803.

25. For example, HJ 5 Dec. 1807.

26. HJ 3 June 1809.

27. HJ 15 Oct. 1803.

28. LI 22 March 1791.

29. HJ 25 Apr. 1807.

30. HJ 18 Nov. 1809.

31. Reports of the Society for Bettering the Conditions and Increasing the Comforts of the Poor, Vol IV, London 1805, Supplement No. 11.

32. HJ 24 Feb. 1810.

33. LI 27 Sept. 1791.

34. LI 23 Aug. 1791.

35. LI 15 March 1791.

36. E. M. Sigsworth, 'William Greenwood and Robert Heaton', Journal of the Bradford Textile Society (1951–2), pp. 61–72.

37. LI 23 Nov. 1795.

38. John Cross to Boulton & Watt, 24 Dec. 1792. B & W MSS BCRL.

39. A. Sutcliffe to W. Birkbeck, 14 Aug. 1786. Birkbeck MSS, WYAS, Leeds.

40. LI 3 Oct. 1796.

41. LI 7 March 1803.

42. A. D. Gayer, W. W. Rostow and A. J. Schwartz, The Growth and Fluctuation of the British Economy 1790–1850 (Oxford 1933), p. 125.

43. Hodgson, op. cit., p. 228.

NOTES TO CHAPTER 5

1. G. Jackson, Hull in the Eighteenth Century (London, 1972).

2. M. M. Edwards, The Growth of the British Cotton Trade 1782–1815 (Manchester, 1967), p. 108.

3. Ibid, p. 113.

4. LI 1 March 1802, 21 June 1802, 6 Sept. 1802.

5. LI 7 Oct. 1804, HJ 8 Oct. 1803.

6. H. J. Priestley, 'Mills of the Ryburn Valley', HAS (1941).

7. LI 14 Aug. 1797.

8. LI 22 Aug. 1808.

9. LI 1.16.1809.

10. LI 13 Nov. 1809.

11. W. Winstanley to W. Birkbeck, 23 Jan. 1804, Birkbeck MSS, 14, WYAS, Leeds.

12. HJ 4 Apr. 1807.

13. For a more complete account, on which this section is based, see G. Ingle, 'A History of R. V. Marriner Ltd, Worsted Spinners, Keighley', Unpublished M.Phil Thesis, University of Leeds (1974).

14. S. Dumbell, 'Early Liverpool Cotton Imports and the Organisation of the Cotton Market in the Eighteenth Century', Economic Journal, xxxiii (1923), pp. 326–73.

15. M. M. Edwards, op. cit., p. 227.

16. Ibid, p. 10.

17. Factories Enquiry Commission, Supplementary Report 1834, C.1. Mill Number 16.

18. Edwards, op. cit., p. 73.

19. Ibid, p. 144.

20. E. M. Sigsworth, Black Dyke Mills (Liverpool, 1958), p. 11.

21. For a more complete account on which this section is based, see G. Ingle, 'A His-

tory of R. V. Marriner Ltd, Worsted Spinners, Keighley', Unpublished M.Phil Thesis. University of Leeds (1974).

22. G. A. Miller, *Blackburn, The Evolution of a Cotton Town* (Blackburn, 1951), p. 330.

23. G. Unwin, *Samuel Oldknow and the Arkwrights* (Manchester, 1924), p. 100.

24. LI 20 May 1788.

25. LI 12 Aug. 1788.

26. LI 12 Aug. 1788.

27. G. A. Miller, op. cit., p. 393.

28. Robert Heaton's Diary, Heaton MSS, BCRL.

29. Edwards, op. cit., p. 172.

30. SunCR52/741797/1803.

31. SunCR86/831408/1810.

32. SunCR95/861636/1812.

33. SunCR111/902671/1815.

34. SunCR118/928065/1817.

35. Factories Enquiry Commission, Supplementary Report 1834, Part II, C 1. Mill Number 16.

36. SunCR7/638227/1795.

37. SunCR13/653283/1797.

38. LI 23 Dec. 1805.

39. LI 9 Nov. 1807.

40. Pigot Commercial Directory for 1816. Manchester 1816.

41. Holden Triennial Directory 1809/10/11, Vol 2, p. 58.

42. J. Hodgson, *Textile Manufacture and other Industries in Keighley* (Keighley, 1879), p. 232.

43. J. Parker, *Illustrated Rambles from Hipperholme to Tong* (Bradford, 1904), p. 88.

44. HJ 16 Feb. 1805.

45. HJ 22 June 1805.

46. H. Speight, *Old Bingley* (London, 1897), p. 222.

47. LI 20 July 1795.

48. LI 27 Apr. 1807.

49. LI 9 Aug. 1813.

50. LM 9 Jan. 1819.

51. LM 14 Jan. 1804.

52. LI 28 Jan. 1799.

53. M. M. Edwards, op. cit., p. 130.

54. LI 19 Sept. 1794, 8 Feb. 1796.

NOTES TO CHAPTER 6

1. LI 13 July 1801.

2. Report from the Select Committee appointed to consider the state of the Woollen Manufacture of England 1806. Evidence of Richard Fawcett.

3. G. H. Brown, *On Foot Round Settle* (Settle, 1896), p. 143.

4. An Act for settling disputes that may arise between masters and workmen engaged in the cotton manufacture in that part of Great Britain called England. Statutes 39 and 40, Geo.III, 90, 1800.

5. Minutes of evidence taken before the committee for settling disputes between masters and workmen engaged in cotton manufacture. 1802–3 (114), Vol VIII.

6. Minutes of evidence, op. cit., p. 16.

7. S. Rayner, *The History and Antiquities of Pudsey* (London, 1887).

8. LI 5 March 1792.

9. LI 26 Nov. 1798.

10. LI 21 Feb. 1803.

11. HJ 14 March 1809.

12. HJ 24 March 1810.

13. LM 5 Sept. 1818.

14. HJ 22 March 1806.

15. SunCR59/765430/1805.

16. HJ 12 Dec. 1807, 2 Jan. 1808.

17. LI 18 May 1807.

18. D. A. Farnie, *The English Cotton Industry and the World Market 1815–1896* (Oxford, 1979), p. 281.

19. Report from the Select Committee on Hand Loom Weavers Petitions, 1834, X, p. 5.

20. I. Holmes, *From Hand Industry to the Factory System* (Bradford, 1914), p. 8.

21. G. Giles and I. H. Goodall, *Yorkshire Textile Mills*, 1992, p. 11.

22. LI 12 Nov. 1792.

23. LI 24 June 1805.

24. B&HC 27 Apr. 1826.

25. LM 6 May 1826.

NOTES TO CHAPTER 7

1. S. D. Chapman, 'Sources and Methods for the Study of Capital Formation in Britain 1750–1850', *Economic History Review*, xxiii (1970), p. 259.

2. J. Hodgson, *Textile Manufacture and Other Industries in Keighley* (Keighley, 1879), various references.

3. SunCR9/640651/1795.

4. R. S. Fitton and A. P. Wadsworth, *The Strutts and the Arkwrights, 1758–1830* (1958), footnote, p. 60.

5. J. Hodgson, op. cit., p. 213.

6. Ibid, p. 85.

7. LI 11 May 1801.

8. Badgery MSS 145/398 WYAS Leeds.

9. RE32A/173390/1800.
10. RE32A/156289/1797.
11. H. J. Priestley, 'Mills of the Ryburn Valley', *HAS* (1933).
12. R. E. Greenwood, 'A History of the Greenwoods of Haworth' (unpublished monograph, 1993).
13. Hodgson, op. cit., p. 219.
14. LM 22 Aug. 1818.
15. E. M. Sigsworth, 'William Greenwood and Robert Heaton', *Journal of the Bradford Textile Society* (1951–2), pp. 61–72.
16. Heaton MSS, BCRL.
17. RE32A/156585/1797.
18. Robert Heaton's Diary, Heaton MSS, BCRL.
19. Robert Heaton's Diary, Heaton MSS, BCRL.
20. HJ 23 Apr. 1808.
21. Hodgson, op. cit., p. 232.

NOTE TO CHAPTER 8

1. E. Baines, *History of the Cotton Manufacture in Great Britain* (1835), p. 429.

NOTES TO CHAPTER 9

1. LI 9 Sept. 1805.
2. SunOS392/610683/1793.
3. SunCR77/813033/1808.
4. Beverley, Cross & Billiam's mill burned down in 1796 and Markland, Cookson & Fawcett stopped production in 1797.
5. Report from the Committee on the Bull to regulate the labour of children in the Mills and Factories of the United Kingdom 1831–2 (706) XV, p. 163.
6. LI 9 Sept. 1805.
7. SunCR49/737384.

NOTES TO CHAPTER 10

1. LI 12 Nov. 1792 and also the *Universal Directory*, 1793.
2. Report from the Select Committee appointed to consider the state of the Woollen Manufacture of England 1806 (268) III.
3. HJ 10 Feb. 1810.
4. W. Cudworth, *Round About Bradford* (Bradford, 1876), p. 482.
5. SunCR108/894878/1815.
6. LI 5 Nov. 1810.
7. LI 29 July 1805.
8. H. Speight, *Old Bingley* (London, 1897), p. 222.

NOTES TO CHAPTER 11

1. Factories Enquiry Commission, Supplementary Report 1834, C2. Mill Number 110.
2. SunCR9/646209/1795.
3. Factories Enquiry Commission, Supplementary Report 1834, C2. Mill Number 126.

NOTES TO CHAPTER 12

1. D. A. Heptinstall, 'Cotton and the Upper Ryburn Valley 1790–1841', Unpublished M.A. Thesis, Huddersfield University (1984) p. 1.
2. Ibid, p. 6.
3. A. Newell, *A Hillside View of Industrial History* (Todmorden, 1925), p. 253.
4. Fielden J. *The Curse of the Factory System* (second edn, London, 1969), p vii.
5. Number of Power Looms used in Factories in the manufacture of woollen, cotton, silk and linen respectively in each county of the United Kingdom 1836 (24) XLV.
6. *Universal Directory* 1793.
7. *Holdens Triennial Directory* 1811, Vol 2.

NOTES TO CHAPTER 13

1. Factory Returns 1835.
2. LI 1 Feb. 1792.
3. Factories Enquiry Commission, Supplementary Report 1834, C2. Mill Number 217.

NOTES TO CHAPTER 14

1. Report from the Select Committee appointed to consider the state of the Woollen Manufacture of England 1806 (268). Evidence of John Buckley.
2. A. Wrigley, *Saddleworth: Chronological Notes from 1200 to 1900* (Stalybridge, 1941), p. 27.
3. M. T. Wild, 'Documents and Sources IV. The Saddleworth Parish Registers', *Textile History*, 2 (1971), p. 222.

NOTES TO CHAPTER 15

1. H. Speight, *Upper Wharfedale* (London, 1900), p. 336.
2. LI 16 Jan. 1792.
3. LI 10 Dec. 1792.
4. LI 11 July 1796.

5. A. Raistrick, *Old Yorkshire Dales* (London, 1967), p. 99.

NOTES TO CHAPTER 17

1. Sun CR50/737014/1803.
2. Sun CR13/653203/1796.
3. Sun CR1/622019/1793.
4. RE32A/155968/1797.
5. J. Jackson, *History of Barnsley* (Barnsley, 1858), p. 167.
6. E. Baines, *Directory of the County of York* (1822).

NOTES TO CHAPTER 18

1. E. Baines, *Directory of the County of York* (1822), p. 271.
2. Ibid, p. 505.
3. J. James, *History of the Worsted Manufacture in England* (London, 1857), p. 327.

NOTES TO CHAPTER 19

1. LI 3 Aug. 1812.
2. J. Hodgson, *Textile Manufacture and other Industries in Keighley* (Keighley, 1879), p. 40.

3. Factory Returns, 1835.
4. J. H. Priestley, 'Mills of the Ryburn Valley', Part II *HAS* (1934), p. 54.
5. E. Parsons, *History of the West Riding* (Leeds, 1836), p. 214.
6. Ponden Mill and Crossland Factory. See Crompton's Survey 1811.
7. Reports of the Society for Bettering the Conditions and Increasing the Comforts of the Poor, Vol IV (London, 1805).
8. Factories Enquiry Commission, Supplementary Reports 1834, Part II, C1. Mill Number 16.
9. A. Raistrick, *The Pennine Dales* (London, 1968), p. 121.
10. BC&WRA, 2 Feb. 1826.
11. J Parker, *Illustrated Rambles from Hipperholme to Tong* (Bradford, 1904), p. 88.
12. E. M. Sigsworth, *Black Dyke Mills* (Liverpool, 1958), p. 44.
13. W. Smith, *Morley Ancient and Modern* (London, 1886), p. 305.
14. D. T. Jenkins, 'The Cotton Industry in Yorkshire 1780–1900', *Textile History*, 10 (1979), p. 88.

Sources

A S FEW RECORDS EXIST which relate directly to the mills identified in this study and little has been written so far about the industry or individual mills other sources have had to be used. A primary source has been insurance records and where these are quoted they are usually from the Sun and Royal Exchange Insurance Office registers which are held at the Guildhall Library in London. Considerable information has been gathered from local newspapers, particularly the *Leeds Intelligencer*, *Leeds Mercury* and *Halifax Journal*. Other local newspapers have also been used and a list of abbreviations appears below.

Further records are kept at the North Yorkshire County Record Office (NYCRO) and the various branches of the West Yorkshire Archive Service (WYAS). In addition the various reference libraries of the major towns have documents relating to mills in their area. The Boulton & Watt collection is held in the Central Library, Birmingham and Samuel Crompton's notebooks are kept at Bolton Civic Centre Museum. Both the Dartmouth Estate Office in Slaithwaite and the Bolton Abbey Estate Office have some records relating to early mills built on their respective properties.

Interest in industrial development in at least two areas has led to the collection of data on reference cards. Saddleworth Historical Society and Saddleworth Museum share a collection of records for what was

part of the West Riding of Yorkshire. A similar collection exists for the Upper Calder Valley. Early cotton mills in the Keighley area are well covered in John Hodgson's *Textile Manufacture in Keighley* which was published in 1879 and mills in the Ryburn Valley near Halifax are described in a series of articles which appeared some years ago in the Journal of the *Halifax Antiquarian Society*.

The growth and industrialisation of the textile industries brought conflict and concern which resulted in the appointment of Parliamentary Select Committees and legislation. The resulting Parliamentary Papers are a useful source of information about the local cotton industry although it is necessary to abstract the information relating directly to events and mills in Yorkshire.

For those interested in industrial archaeology the Ordnance Survey six inch to one mile maps, which were surveyed between 1845 and 1854 and published between 1847 and 1857, are invaluable. Many of the mills described in this work can be located with the help of the early OS maps although many had stopped working or become part of a larger complex of buildings by 1850.

NB Recently the Guildhall Library have added the following prefixes to the call numbers of the Sun Old and Country Series and Royal Exchange Series respectively: Sun OS; RE.

Abbreviations

LI	*Leeds Intelligencer*
LM	*Leeds Mercury*
HJ	*Halifax Journal*
BHC&WRA	*Bradford & Huddersfield Courier & West Riding Advertiser*
NYCRO	North Yorkshire County Record Office
WYAS	West Yorkshire Archive Service
HAS	Halifax Antiquarian Society
SHS	Saddleworth Historical Society
B&W	Boulton & Watt Collection
BCRL	Birmingham Central Reference Library
LRL	Leeds Reference Library
BRL	Bradford Reference Library
CRL	Calderdale Reference Library
KRL	Kirklees Reference Library
LUL	Leeds University Library
FEC	Factories Enquiry Commisssioners

Reports from commissioners appointed to collect information in the manufacturing districts, relative to employment of children in factories, and as to the propriety and means of curtailing the hours of their labour; with minutes of evidence and reports of district commissioners: First, Second and Supplementary Reports. 1833 – 1834.

Part 2 References

The references in Part 2 are related to each mill site and for identification purposes the Ordnance Survey Grid Reference is also given as well as the dates when the mill was used for cotton spinning. If any of the dates are uncertain this is indicated by (?). Where (+) is added to the final date the mill continued to be used for cotton spinning after 1835 or the date given.

For some of the mills where extensive insurance records exist only some examples have been given. Full details are available in the Ph.D thesis on which this work is based. Again, for most references, only brief details have been provided as the full title can be found in the Bibliography.

The author is well aware that there will be mistakes relating to the detailed references to mill sites and welcomes additions and corrections, particularly to establish the exact dates when particular mills were built or first used for cotton spinning.

Leeds

Scotland Mills. SE278381. 1792–1808. Sun OS386/598138/1793. LI 19.9.1801, 14.9.1801, 5.12.1808

Waterloo Mill. SE314313. 1787–1816 Sun OS343?530723/1787. Sun OS343/530724/1787. Sun CR36/711544/1800. Sun CR53/744499/1804. Sun CR112/914158/1816. LI 11.1.1802. Ard Walker Account Book DB23 WYAS Leeds.

Bank Low Mill. SE312329. 1790–1833? RE32A/154787/1796. Sun CR38/713824/1801. Sun CR57/758808/1804. LI 1.6.1795, 4.7.1796, 10.1.1803, 5.12.1803, 13.2.1804, 28.5.1804. B & W MSS Portfolio 255. Cotton Mill Returns 1803–1806 WYAS Wakefield. Crompton 1811. Baines 1822. FEC 1834

At Hunslet. ? 1791–1835 RE20/120909/1791. RE32A/183768/1801. Sun CR86/831406/1810 LI 3.8.1779, 25.1.1802, 31.10.1803, 21.1.1805. LM 6.6.1818, 22.5.1830. Crompton 1811. Baines 1822.

At Hunslet. ? 1792–1796 SunOS386/599400/1792. Sun CR7/638891/1795. Sun CR11/649649/1795. B & W MSS Portfolio 96. Hills 1973.

At Far Bank Mills. SE313327 1792–1797 Sun CR8/636966/1795. Sun CR13/653223/1796.

RE32A/164042/1798. B & W MSS Portfolio 91. LI 13.1.1789, 21.4.1792, 1.2.1796. Crump 1931.

At Far Bank Mills. SE313327 1795–1800 RE31/152714/1796. RE32A/157565/1797. RE32A/163965/1798 LI 1.1.1798, 3.3.1800, 11.4.1803. B & W MSS Portfolio 129.

Mabgate or Sheepscar Mill. SE305348 1793–1816 Sun OS392/610103/1793. Sun CR31/701021/1800. Sun CR89/840103/1810. B & W MSS Portfolio 254. LI 2.5.1796, 27.5.1816. Crompton 1811.

Nether Mills. SE304333 1793–1814 Sun OS386/598139/1792. LI 1.5.1787, 8.3.1791, 9.12.1793, 25.5.1795, 30.9.1799, 12.9.1814.

At Steander Mills. SE304333? 1803–1805 Sun CR57/758053/1804. Sun CR63/771341/1805. LI 8.8.1803.

At Simpson Fold. SE304332 1803–1808 Sun CR64/776270/1806. Sun CR69/788561/1807. Sun CR72/791047/1807. LI 14.3.1803, 21.4.1807, 16.5.1808.

Low Fold Mill, Hunslet. ? 1811?–1832? LI 15.7.1811. Baines 1822. FEC 1834.

At Morley. ? 1785–1805 Sun OS319/489479/1795. LI 11.10.1805.

At Wortley. ? 1796–1796 B & W MSS 143

At Sandal. ? 1804–? LI 5.3.1804

At Thorner. SE374395? 1794–1835 +. Sun

CR3/629212/1794SunCR125/947440/1818LI
9.9.1805. Baines 1822. Returns of Mills and Factories 1839 (41) XLII

Bradford

Castlefields Mill, Bingley. SE099403 1791–1835
+. Sun OS318/589673/1792. Sun
CR19/671056/1797SunCR77/810866/1808.
SunCR250/1287901/1838.LI27.9.1791,
13.1.1794, 29.7.1805, 9.8.1806. Cotton Mill Returns 1803–06 WYAS Wakefield. Crompton
1811. FEC 1834.

Providence Mill, Bingley. SE105394 1801–1814
LI 7.8.1809, 20.6.1814. HJ 30.14.1808. Crompton 1811. FEC 1834. Speight p222. Dodd p90.

At Bingley. SE107392 1800?–? Speight 1897.

At Bingley. Site not found. 1800?–? Speight
1897.

Elmtree Mill, Bingley. SE111391? 1800?–?
Speight 1897.

Cottingley Mill. SE127374 1830–? Speight 1897

Hewenden Mill. SE077360 1790?–1800? LI
26.5.1800. Speight 1897.

Eller Carr Mill, Cullingworth. SE065370 1795?–
1835 +. LI 6.6.1808, 6.5.1816. HJ 4.6.1808,
13.8.1808. Parson & White 1830. White 1837.
Cudworth 1876. Giles and Goodall 1992.

Bent Mill, Wilsden. SE077365 1799?–1835 +. LI
7.8.1809. LM 13.6.1818. B & HC 14.9.1826,
15.3.1827. Crompton 1811. Baines 1822. White
1837. Cudworth 1876.

Hallas Bridge Mill, Wilsden. SE078365 1800?–
1829 HJ 11.9.1802, 25.7.1807. LI 30.12.1805.
LM 13.6.1818, 12.12.1829. B & HC 29.6.1826.
Crompton 1811. Cudworth 1876.

Goit Stock Mill, Wilsden. SE079373 1792–1835
+ RE 25/130994/1792. RE 31/150555/1796. Sun
CR211/1185319/1834.Crompton1811.LI
30.4.1804, 4.3.1811. Cudworth 1876.

Wilsden Mills. SE088377 1792–1820 LI
2.1.1792. LM 21.11.1818, 1.7.1820, 3.7.1822.
Cotton Mill Returns 1803–1806 WYAS Wakefield. Crompton 1811. Cudworth 1876.

Old Mill, Wilsden. ?1793?–1812? Crompton
1811. Cudworth 1876. Universal Directory
1793.

Over Mill, Hawksworth. SE155416 1795?–1810?
LI 29.3.1791, 4.4.1803

Low Mill, Hawksworth. SE153413 1795?–1835 +
LI 29.3.1791. White 1837.

Clifton Mill, Baildon. SE170390 1791?–? LI
29.3.1791. Cudworth 1876.

Idle Cotton Mill. Site not found. 1795?–1826
Sun CR68/784468/1806. LI 10.7.1809,
11.6.1810. HJ 1.7.1809, 16.6.1810. Cudworth
1876.

Knight's Mill, Great Horton. SE141314 1806–
1827 B & HC 2.11.1826, 21.6.1827, 13.9.1827.
Bowling Papers, Box 4 BCL. Crompton 1811.
Peel 1893. Parker 1904.

At Bradford. ? 1790?–1792 LI 12.11.1792.
Universal Directory 1792

Holme Mill. SE159329 1800–1807? RE
32A/186633/1801. LI 21.9.1801, 14.9.1807,
9.11.1807, 28.12.1807. HJ 26.9.1807. James
1857. Cudworth 1876.

Rand's Mill. SE160328 1803–1815? Sun
CR48/736104/1802. Sun CR60/761923/1805.
LI 22.2.1813.

At Birstall. SE206248? 1811?–1817? LI 4.8.1817

Clough Mill, Birstall. SE240270 1824–1833?
FEC 1834

Gomersal Hall Mill. SE205262 1788?–1811? RE
29/143438/1795. RE 31/151921/1796. RE
32A/158699/1797. RE 32A/183784/1801. LI
28.7.1789, 18.4.1796, 24.10.1803, 31.8.1807,
18.1.1808. LM 14.3.1786. Crompton 1811.
FEC 1834. Cadman 1930. Cudworth 1876.

Heckmondwyke Cotton Mill. ? 1800?–1805? LI
11.10.1802. Cotton Mill Returns 1803–1806
WYAS Wakefield.

Birkenshaw Mill. SE203288 1805?–1826 RE
23/132771/1793. Sun CR64/778839/1806. Sun
CR89/840117/1810. LI 14.10.1811, 23.1.1819,
2.10.1830. Crompton 1811. FEC 1834.

Huddersfield

The Factory, Marsden. SE053117 1795–1808?
Sun CR7/638227/1795. Sun
CR17/663236/1796. LI 14.1.1808, 21.3.1808.
HJ 2.21808, 5.3.1808. Cotton Mill Returns
1803–1805 YWAS Wakefield.

Upper End Mill, Marsden. SE 053117 1792?–
1808? LI 21.3.1808. HJ 15.2.1806, 5.3.1808,
18.11.1809. LM 26.6.1830. Cotton Mill Returns 1803–1806 WYAS Wakefield.

New Mill, Marsden. SE052117 1796–1830? Sun
CR11/651448/1796. Sun CR17/663236/1796.
HJ 15.2.1806, 5.3.1808. LM 26.6.1830. Cotton
Mill Returns 1803–1806 WYAS Wakefield.

Frank's Mill, Marsden. SE052117 1804?–1808?
HJ 5.3.1808, 18.11.1809. Cotton Mill Returns
WYAS Wakefield.

Crow Hill Mill, Marsden. Site not found. 1795?–

1810? Sun CR7/638227/1795. Sun
CR17/663236/1796. HJ 10.2.1810. Cotton
Mill Returns 1803–1806 WYAS Wakefield.

Smithy Holme Mill, Marsden. SE052120 1795–
1805? Sun CR7/638227/1795. Sun
CR17/663236/1796. LI 6.1.1806.

Lingards Wood Bottom Mill, Marsden. SE054122
1800?–1806? RE 32A/155107/1797. HJ
7.6.1806. Cotton Mill Returns WYAS Wakefield.

At Marsden. (Woodhead) Site not found. 1793?–
1802? Sun OS388/603622/1793. RE
32/153590/1796. RE 32A/188840/1801. LI
16.9.1799, 14.6.1802. HJ 12.6.1802.

Jumble Mill, Marsden. SE045116. 1800?–1805?
Sun CR59.765442/1805. HJ 25.12.1802,
10.1.1807. LI 19.8.1805.

At Marsden. (Gill) Site not found. 1800?–1805?
HJ 16.2.1805. Cotton Mill Returns 1803–1806
WYAS Wakefield

At Marsden. (Davenport) Site not found. 1795?–
1807? LI 27.7.1795, 18.1.1796, 21.1.1805,
7.9.1807, 30.1.1809.

Holme Mill, Marsden. SE063130. 1812?–1833?
FEC 1834

Bridge Mill, Netherthong. SE143088. 1800?–
1808? HJ 30.4.1808

Steps Mill, Honley. SE141126 1800?–1805? LI
28.3.1803, 17.6.1805. HJ 22.6.1805, 3.3.1810.

Crossland Factory, South Crossland. SE117122
1805?–1812 Sun CR64/774801/1805. LI
1.8.1803, 4.3.1811. FEC 1834.

Shaw Carr Wood Mill, Slaithwaite. SE072135
1787–1833 + Dartmouth Terrier 1805. LI
2.9.1793. Cotton Mill Returns 1803–1806 WYAS
Wakefield. FEC 1834. Pigot 1828.

Black Moor Holme Mill, Slaithwaite. SE063130?
1796–1815? Dartmouth Terrier 1805. RE
32A/180517/1800. Crompton 1811. Cotton
Mill Returns 1803–1806 WYAS Wakefield.
FEC 1834.

Phoenix or Waterside Mill, Slaithwaite. Site not
found.1802–1835? Dartmouth Terrier 1805.
Sun CR58/755728/1804. Sun
CR61/768024/1805. LI 9.4.1804, 19.2.1810.
Cotton Mill Returns 1803–1806 WYAS Wake-
field. Crompton 1811. FEC 1834.

Meltham Mills, Meltham. SE109108 1805–1835
+ Pigot 1816. Baines 1822. Parson & White
1830. FEC 1834.

At New Street, Huddersfield. SE144167?. 1800?–

1828 LM 29.4.1815, 10.6.1815. Crompton
1811. Sykes 1898. Mayhall 1860.

At Huddersfield. Site not found. ?–1833? FEC
1834.

Colne Bridge Mill. SE178203 1793?–1835 + Sun
OS388/605087/1793 SunCR9/646209/1795.
SunCR216/1201409/1835. LI 17.3.1800. LM
28.2.1818. Baines 1822. FEC 1834. Brook 1968.

Sheppard's Factory, Dewsbury. Site not found.
1804–1835 + LI 25.6.1804, 5.12.1808,
24.10.1814. HJ 23.6.1804, 18.7.1804. Pigot 1828.
Parson & White 1830.

Halifax

Todmorden

Pudsey Mill. SD908264 1810?–1820? Crompton
1811

Frieldhurst Mill. SD911261 1810?–1835 +
Crompton 1811. White 1837. Travis 1896.

Barewise Mill. SD915258 1813?–? Land Tax Assessment. CRL.

Kitson Wood Mill. SD920257 1810?–1822 +
Crompton 1811. Baines 1822.

Lydgate Factory. SD921256 1804?–1820? Land
Tax Assessment. CRL.

Ewood or Malt Kiln Mill. SD930248 1805?–
1811? Crompton 1811.

Holme Mill. SD929250 1795?–1811? Crompton
1811

Cross Lee Mill. SD930251 1804?–1835 +
Crompton 1811. Baines 1822. White 1837.

Greenhurst Mill. SD940255 1804?–? Land Tax
Assessment. CRL.

Haugh Stone Mill. SD939253 1805?–1811 +
Crompton1811.

Hole Bottom Mill. SD938251 1796?–1835 +
Crompton 1811. LI 23.5.1818, 20.2.1819.

York Field Mill. SD938251? 1801–1811? Crompton 1811

Ridgefoot Mill. SD936243 1790?–1835 + Crompton 1811.

Waterside Mill. SD934238 1795?–1835 + Crompton 1811. Pigot 1816. Baines 1822. White 1837.
Travis 1901.

Salford Mill. SD934239 1810?–1815? Crompton
1811.

Swineshead Mill. Site not found. 1795?–? Misc
165/18/4 CRL.

Todmorden to Hebden Bridge

Lumbutts Mill. SD956234 1794?–1835 + HJ 26.6.1802. Crompton 1811. Travis p135. Binns 1972.

Jumb Mills. SD955234 1801–1835 + Crompton 1811. Baines 1825. White 1837.

Causeway Mill. SD953235 1810?–? Crompton 1811

Causeway Wood Mill. SD951237 1826–1835 + Baines 1822. Parson & White 1830. FEC 1834.

Oldroyd or Folly Mill. SD951239 1794–1835 + RE 32A/187370/1801. LI 20.5.1805. Crompton 1811. Baines 1822. Law HAS 1954.

Woodhouse Mill. SD951244 1832–1835 + Giles and Goodall 1992.

Castle Clough Mill. SD950246 1796?–1810? RE 32A/180992/1801. LI 4.8.1800.

Cinderhills Mill. SD951246 1810?–1835 + LM 3.1.1818. Sutcliffe MSS 133(61) WYAS Leeds. FEC 1834.

Millsteads Mill. SD951245 1811–1835 + Crompton 1811. Parson & White 1830. FEC 1834.

Clough Hole Mill. SD955248 1800?–? Stansfield Valuation Book, CRL.

Lobb Mill. SD954246 1809?–1835 + RE 32A/187370/1801. LI 13.5.1805, 24.2.1817. Holden 1811. Baines 1822. White 1830. Law, HAS 1954.

Stoodley Bridge Mill. SD963250 1808?–1835 + B & HC 13.4.1826. Holden 1811. Baines 1825. FEC 1834.

Guteroyd Mill. SD961241 1800?–1808 HJ 30.7.1803. Misc 165/15/2, 165/19/8, CRL.

Eastwood Mill. SD961258. 1785?–1835 + HJ 9.3.1805. LM 6.11.1819. Crompton 1811. Baines 1822. Parson & White 1830. Newell 1925.

Cockden Mill. SD964255 1791?–1820? HJ 3.8.1805, 6.2.1808, 15.10.1808.

Burnt Acres Mill. SD968259 1810?–? Crompton 1811.

Staups Mill. SD961269 1805?–1835 + HJ 20.7.1805. Crompton 1811. Baines 1822. Pigot 1834.

Cowbridge or Underbank Mill. SD965264 1794–1835 + LM 23.1.1819. Baines 1822. White 1837. Law HAS 1954.

Spa Mill. SD966264 1788?–1835 + Armitage HAS 1967. Baines 1825. White 1837. MP 16/1, CRL.

Jumble Hole Mill. SD968264 1788?–1835 +

Pigot 1816. Baines 1825. White 1837. Armitage HAS 1967.

Winters Mill or Marsh Factory. SD971268 1809–1835 + HJ 18.11.1809. LM 28.2.1818. HG 8.3.1834. Baines 1822.

Near Hebden Bridge

Calderside Mill. SD980270 1824–1835 + Pigot 1828. Walker 1845.

Rodmer Clough or Lister Mill. SD953291 1793?–1835 + RE 32A/186874/1801. LI 28.1.1793. HJ 5.6.1802. Crompton 1811.

Edge Mill. SD953291 1807?–? Township Valuation Book, CRL.

Land Mill. SD954289 1796–1835 + HJ 3.12.1808. Crompton 1811. Baines 1822. Walker 1845.

Hudson Mill. SD965282 1786?–1835 + LI 22.4.1799. Crompton 1811.

Slater Ing or Bob Mill. SD969281 1800?–1835 +? LI 4.3.1805. HJ 16.2.1805, 10.6.1809, 9.9.1809. Baines 1822.

Lumb Mill. SD975281 1801?–1835 + Holden 1811. Crompton 1811. Baines 1822.

Mytholm Mill. SD983273 1788?–1835 + Sun OS391/607866/1792. Sun CR65/774104/1806. LI 21.3.1796, 24.3.1810. Bailey 1781. Crompton 1811. Baines 1822.

Bankfoot Mills. SD986272 1805?–1835 + LI 24.10.1808, 20.12.1813. HJ 2.1.1808, 21.5.1808. Crompton 1811. Baines 1825. Parson & White 1830.

Clough Mill. SD992290 1801–1835 + RE 32A/182200/1801. Crompton 1811. Baines 1825.

Gibson Mill or Lord Holme Mill. SD972298 1803–1835 + HJ 9.2.1805, 2.3.1805. Crompton 1811. Baines 1822. FEC 1834.

Greenwood Lee. SD970295 1802?–1805? LI 18.2.1805.

New Bridge Mill. SD988290 1796?–1835 + RE 32A/187268/1801. Crompton 1811. LI 10.1.1814. FEC 1834.

At Midgehole. SD991288? 1800?–1835 +? HJ 18.6.1803. Crompton 1811. Baines 1825.

At Midgehole. SD992289? 1783?–1835 + FEC 1934.

Lee Mill. SD992284 1815?–35 + LM 25.3.1820. Baines 1825. FEC 1934.

Stubbing or Hebble End Mill. SD988272 1805?–1835 + HJ 1.2.1806, 1.3.1806. Baines 1825. FEC 1834.

Kershaw Mill. SD987273 1834–1835 + Pigot 1834. White 1838. Robson 1840.

Foster Mill. SD992277 1808?–1835 + LI 13.3.1787. HJ 31.12.1808. Pigot 1816. Baines 1822. White 1837.

Nutclough Mill. SD994275 1797–1835 + LI 10.4.1797. BO 24.8.1837. Crompton 1811. Baines 1822.

Mayroyd or Gemland Mill. SD995268 1794–1802 RE 32A/165773/1798. HJ 10.4.1802, 10.3.1804. Law HAS 1954.

At Ibbotroyd. SE002279 1798–1835 + LI 17.12.1798. Walker 1845.

Hawksclough Mill. SE006264. 1825–1835 + Baines 1825. White 1837.

Mytholmroyd and Cragg Vale

Rudclough Mill. SD977231 1801–1835 + LI 10.9.1810. HJ 13.6.1801, 11.8.1804. Baines 1822. White 1837.

Marshaw Bridge Mill. SE000233 1790?–1835 + RE 32A/148516/1795. LI 13.1.1794. FEC 1834.

Castle Mill. SE004235 1820?–1835 + Baines 1822. White 1837. HG 4.2.1865.

Cragg Mill. SE005238 1798?–1835 + LI 27.10.1798, 17.12.1798. HJ 18.12.1802, 28.5.1808. Baines 1822, 1825. White 1837.

Luddenden Valley

Spring Mill. SE046297 1800–1825 Crompton 1811. LM 12.6.1819. Garnett 1951.

Hoyle Bottom or Square Mill. SE048292 1815?–1825 Harwood HAS 1953. Garnett 1951.

Lumb or Stones Mill. SE046288 1803?–1828 LM 10.10.1818. B&HC 10.8.1826, 7.9.1826. Garnett 1951. Harwood, HAS 1952.

Wainstalls Upper Mill. SE044286 1804?–1830? LI 18.7.1814, 18.8.1817. LM 7.10.1815, 26.8.1820, 27.1.1821. Crompton 1811. Garnett 1951. FEC 1834. HXT798, CRL.

Jowler or Holme House Mill. SE039280 1810?–1818 LM 7.3.1818. Crompton 1811. Box 103, CRL.

Dean Mills. SE043273 1792–1835 + HJ 5.5.1804. LI 28.5.1804. Baines 1822. FEC 1834.

Luddenden Mills. SE041259 1804?–1815? HJ 21.9.1805, 10.5.1806, 5.8.1809. Crompton 1811. Pigot 1816.

Luddenden Foot Mill. SE036248 1795–1807 RE 32A/170437/1799. HJ 18.12.1802, 28.5.1808. LI 30.5.1808.

Brearley Mill. SE026259 1800?–1807 LI 13.2.1797, 8.8.1803, 10.2.1812. HJ 6.8.1803, 20.6.1807.

Cooper House Mill. SE039243 1794–1832 RE 29/146023/1795. RE 32A/157244/1797

Longbottom Mill. SE042240 1792–1797 RE 23/127189/1792. RE 32A/155221/1797. DB23, WYAS Leeds.

Higgin Chamber or Ing Head Mill. SE034241 1788–1835 + LI 19.3.1804, 30.3.1805. LM 1.6.1822. BCWRA 1.9.1825. HG 30.11.1833. Crompton 1811. Baines 1822.

Boulder Clough or Swamp Mills. SE037240 and 036239 1790–1820 RE 32A/180078/1800. RE 32A/187371/1801. HJ 1.12.1804, 30.7.1808. LM 28.1.1815, 26.6.1819.

Sowerby Bridge to Halifax

Jumples Mill. SE065278 1816?–1835 + RE 31/152238/1796. LI 11.10.1813. LM 9.9.1815. FEC 1834. Baines 1822. Hanson 1920. Trigg HAS 1933.

Hebble Mill. SE069268 1804?–1835 + LI 18.9.1809. LM 9.1.1830. Crompton 1811. Trigg, HAS 1933.

Holme Field Mill. Site not found. 1797–1812? RE 32A/179281/1800. LI 17.8.1812, 24.8.1812. Trigg, HAS 1933. Crompton 1811.

Shaw Lane Mill. SE083277 1793–1835 + FEC 1834. Baines 1822. White 1837. Trigg HAS, 1933.

Grove Mill. SE081268 1818?–1835 + Sun CR210/1180942/1834. Baines 1822. Trigg HAS, 1933.

Old Lane Mills. SE086263 1793?–1827? RE 32A/157463/1797 LI 11.10.1791, 28.1.1793, 2.12.1793, 5.6.1797, 27.5.1799, 27.7.1812. HJ 6.3.1802. Crompton 1811.

New House Mill. Site not found. 1810?–1820 Crompton 1811. Baines 1822. Trigg, HAS 1933.

North Bridge Mill. SE094297 1798–1816 +? RE 32A/164953/1798. RE 32A/183172/1801. LI 11.3.1799. HJ 27.4.1805, 22.10.1808, 10.6.1809. Crompton 1811. Pigot 1816.

New Bank Mill. SE095257 1832–1835 + Sun CR200/1159190/1832Sun CR228/1223615/1836. White 1838.

Lee Bridge Mill. SE085259? 1800?–1811? HJ 2.4.1803, 14.4.1804, 25.11.1809.

Shibden Mill. SE102273 1803–1815 LI 9.3.1801, 17.1.1803. HJ 8.1.1803, 26.4.1806. Crompton 1811.

Bowling Dyke Mill. SE094257 1801–1815 RE 32A/187358/1801. RE 32A/187723/1801. HJ 13.3.1801. Crompton 1811.

Hoyle House Mill. SE053242 1792–1810 RE 24/128488/1792. RE 32A/185799/1801. HJ 17.11.1804, 5.5.1810.

Willow Hall Mills. SE065243 (Upper) 1797–1835 + SE064241 (Lower) 1783–1835 + RE 32A/157243/1797.RE32A/157907/1797.RE 32A/160941/1798.RE32A/168241/1799.Sun CR4/629366/1794.LI31.12.1782, 8.2.1813,11.10.1813.Crompton1811.CottonMill Returns 1803–1806 WYAS Wakefield. B&W 16/1802. B&W 311/1802. FEC 1834.

Wharfe Mill. SE064237 1800?–1835 + HG 21.6.1845. Crompton 1811. Baines 1822.

Stern Mills. SE077234 1800?–1832? HJ23.10.1802. LM 22.8.1818. Holden 1811. Crompton 1811.

Copley Mill. SE082225 1810?–1815? Crompton 1811. LM 14.2.1835.

Brow Mill. SE064238 1800?–1835 +? HJ 31.12.1803,22.2.1806.LM24.11.1832.Crompton 1811.

Marshall Hall Mill. 1805?–1822? HJ 5.9.1807. Holden 1811. Crompton 1811. Baines 1822.

Thornhill Briggs Mill. SE146236 1797–1835 + LI 19.1.1826. B&HC 7.9.1826. LM 22.1.1830. Crompton 1811. Baines 1822. Walker 1845.

At Hartshead. Site not found. 1800?–1810? Cotton Mill Returns 1803–1806 WYAS Wakefield.

Ryburn Valley

Temple Mill. SE030165 1799–1835 + RE 32A/170591/1799.SunCR194/113536/1832.Sun CR247/1292068/1838.LI10.3.1800,21.4.1800, 7.6.1802. HJ 11.2.1804, 2.11.1805. Crompton 1811. Heptinstall 1984.

Booth Bridge Mill. SE033165 1794–1835 + RE 29/146947/1795.RE32A/158613/1797.HJ 17.7.1802, 11.6.1803. LM 18.4.1818. Baines 1822. Priestley, HAS 1933, 1941.

Booth Wood Mill. SE035169 1800?–1805? HJ 25.5.1805, 1.10.1808. Priestley, HAS 1933.

Spring Mill. SE035069 1800–1835 + Baines 1822. White 1837. Priestley, HAS 1933. Heptinstall 1984.

Slithero Mills. SE035187 1783?–1835 + RE 32A/161553/1798.RE32A/187043/1801.Crompton 1811. Priestley, HAS 1933, 1941. Heptinstall 1984.

Thrum Hall Mill. SE018191 1805?–1835 +

Crompton 1811. Baines 1822. Priestley, HAS 1934. Heptinstall 1984.

Hanging Lee Mill. SE019190 1802–1835 + LI 21.6.1813. LM 26.8.1820. Heptinstall 1984.

Hazel Grove Mill. SE020183 1792–1835 + RE 32A/167265/1799.LI24.3.1794,10.3.1800, 5.6.1802. Priestley, HAS 1941. Heptinstall 1984.

Swift Place Mill. SE027188 1803–1835 + HJ 6.1.1810. Priestley, HAS 1934. Heptinstall 1984.

Lower Swift Place Mill. SE026186 1803–1810 LI 14.11.1803. Priestley HAS 1934.

Stones Mill. SE033187 1800–1835 + LM 18.9.1819, 12.8.1820. Holden 1811. Crompton 1811. FEC 1834.

Dyson Lane Mill. SE033192 1803–1835 + Priestley, HAS 1934, 1941. Heptinstall 1984.

Lower Dyson Lane Mill. SE036192 1820?–1830? Baines 1822. Priestley, HAS 1933, 1934.

Hollings Mill. SE036192 1788–1835 + Priestley, HAS 1933, 1934, 1941.

Smallees Mill. SE037196 1801–1835 + LI 20.8.1782, 30.5.1803. HJ9.6.1804, 12.1.1805, 5.10.1805, 5.4.1806. Holden 1811. Priestley, HAS 1957.

Ripponden Mill. SE042205? 1793?–1835 + LI 20.8.1782, 11.5.1801, 23.5.1808, 8.1.1810. Priestley, HAS 1925, 1934. FEC 1834.

Ripponden Wood Mill. SE043206? 1792–1800? RE 25/130993/1792. RE 32A/160109/1797. HJ 16.8.1806. Priestley, HAS 1934, 1941.

Greenhead Mill. SE017206 1822?–1830 Baines 1822. Heptinstall 1984. HJ 24.9.1803.

Clough Mill. SE026204 1792?–1835 + Baines 1822. Priestley, HAS 1934, 1941. FEC 1834.

Severhills Mill. SE035212 1799–1835 + HJ 23.6.1804. LM 21.3.1818. Priestley, HAS 1934, 1941. Heptinstall 1984.

Soyland Mills. SE033211 1799–1822 RE 32A/168699/1799. LI 20.2.1787, 22.9.1789, 23.4.1804, 13.5.1805. HJ 6.11.1802. Crompton 1811. MSS 156/1, CRL.

Lumb Mills. SE030214 1805–1835 + LI 12.8.1799, 17.2.1812. Priestley, HAS 1934. FEC 1834.

Damside Mill. 1800?–1805? HJ 14.12.1805, 16.8.1806. Priestley, HAS 1934.

Kebroyd Mills. SE042213 1790?–1835 + RE 23/132770/1793.SunCR172/108632/1828.LI 20.8.1782, 4.3.1811. HJ 17.9.1803, 22.3.1806. Priestley HAS, 1934, 1941. FEC 1834. Baines 1822.

Watson Mill. SE055229 1804–1816 LI
21.10.1799. HJ 12.1.1805, 22.3.1806. Pigot
1816.

Tom Hole Mill. 1803–1810?

Keighley

Royd House Mill. SE039358 1791–1810 HJ
23.4.1808. Heaton MSS BRL. FEC 1834.

Bridgehouse Mill. SE035369 1784?–1810? RE
26/132532/1793. LM 19.7.1785. LI 5.5.1789.
Cotton Mill Returns 1803–1806 WYAS Wakefield.

Mytholm Mill. SE034379 1791–1831 HJ
2.7.1808, 7.4.1810. LI 29.3.1813. Cotton Mill
Returns 1803–1806 WYAS Wakefield. Crompton 1811. Universal Directory 1792.

Ponden Mill. SD998373 1791–1835 + RE
29/143460/1795.RE32A/156585/1797.Cotton
Mill Returns 1803–1806 WYAS Wakefield.
Crompton 1811. White 1837.

Griffe Mill. SE007374 1792?–1820 LI 14.2.1803.
LM 22.4.1820. Cotton Mill Returns 1803–
1806. Crompton 1811. Giles and Goodall 1992.

Lumb Foot Mill. SE014374 1797?–1805 RE
32A/155302/1797. B&HC 22.3.1827. Hodgson
1879.

Spring Head Mill. SE029377 1786?–1812? RE
32A/187365/1801. Cotton Mill Returns 1803–
1806 WYAS Wakefield. HJ 14.1.1804,
27.2.1808, 2.4.1808. Crompton 1811.

Higher Providence Mill. SE033383 1801?–1814?
Hodgson 1879. Keighley Parish Valuation
Book, KRL?.

Vale Mill. SE037383 1792–1835 + Cotton Mill
Returns 1803–1806 WYAS Wakefield. Crompton
1811. FEC 1834. Hodgson 1879.

Damens Mill. SE051388 1789–1824 LM
1.1.1820. FEC 1834. Hodgson 1879.

Wire Mill. SE053396 1801–1810? Keighley Parish Valuation Book, KRL?

Grove Mill. SE054396 1795?–1820 RE
29/144418/1795. HJ 7.5.1810. LM 14.11.1818.
Cotton Mill Returns 1803–1806 WYAS Wakefield. FEC 1834.

Ingrow Corn Mill. SE053398 1801–1815? HJ
24.12.1808. LM 15.5.1819. Crompton 1811.
Hodgson 1879.

Hope Mill. SE059404 1800–1812 RE
32A/179777/1800. RE 32A/183170/1801.
Hodgson 1879.

West Greengate Mill. SE062404 1784–1818 RE
10/96228/1786. RE 32A/179776/1800. Cromp-

ton 1811. Cotton Mill Returns 1803–1806
WYAS Wakefield. Ingle 1974.

East Greengate Mill. SE062406? 1795?–1810
Hodgson 1879.

Walk Mill. SE063406 1783–1812 RE
29/148093/1795. RE 32A/154847/1796. Cotton Mill Returns 1803–1806 WYAS Wakefield. Hodgson 1879. Crompton 1811.

Low Bridge Mill. SE063409 1800?–1821 LI
22.1.1810. Crompton 1811. Hodgson 1879.

Low Mill. SE065413 1780–1835 + RE
15/109974/1788.RE30/144716/1795.RE
32A/157445/1797. Cotton Mill Returns 1803–
1806 WYAS Wakefield. FEC 1834.

Dalton Mill. SE070414 1791?–1835 + Sun
CR125/947473/1818LI9.11.1790,17.6.1793,
26.8.1793, 4.2.1805. Crompton 1811. Hodgson
1879.

Aireworth Mill. SE075421 1787–1818. RE
23/131587/1792. RE 32A/155222/1797. RE
32A/170161/1799. Sun CR127/955629/1819.
LI 21.7.1789, 18.6.1810. Crompton 1811.

Higher Newsholme Mill. SE018395 1793?–1810?
Hodgson 1879. Universal Directory 1792.

Lower Newsholme Mill. SE021395 1793?–1820?
Hodgson 1879. Universal Directory 1792.

Brow End Mill. SE027405 1791–1805. RE
29/148093/1795. RE 32A/154847/1796.
Hodgson 1879. Universal Directory 1792.
Brigg Collection (75) KRL.

Goose Eye Mill. SE027405 1797–1805? LI
20.8.1798. BC 26.1.1826. Hodgson 1879.

Wood Mill. SE033407 1795?–1815? Hodgson
1879.

Holme House Mill. SE046408 1792?–1804. HJ
29.9.1804. Universal Diorectory 1792. Hodgson 1879.

Castle Mill. SE052409 1783–1815? RE
32A/187756/1801. HJ 21.9.1805, 4.4.1807.
LI 6.5.1807. Crompton 1811. Hodgson 1879.

Beckstones Mill. SE057409 1810–1820? Sun
CR111/905609/1816. Sun
CR114/920560/1817. HJ 22.9.1810. Hodgson
1879.

Damside Mill. SE057407? 1795?–1820? Sun
CR77/808670/1808. Sun
CR104/878254/1813. Sun
CR124/939459/1818. LI 19.1.1807, 16.1.1809.
Crompton 1811.

Northbrook Mill. 1782?–1805? Sun
OS335/514679/1786. Hodgson 1879.

Cabbage Mill. SE062407 1793–1835 + FEC 1834. Hodgson 1879.

Sandywood Mill. SE061415 1800–1819. Hodgson 1879.

Woodhead Mill, Riddlesden. SE068434 1805–1809. HJ 17.12.1808. LI 1.7.1811.

Upper Mill, Morton. SE102426 1789–1835 +. SunCR49/739167/1803.LI30.7.1798,27.1.1800, 1.5.1809. FEC 1834. B & HC 21.6.1827. Baines 1822.

Morton Mill, Morton. SE102423 1792–1815? SunOS384/593321/1792. RE32A/186725/1801. LI 2.2.1801, 1.3.1802, 6.6.1808. HJ 24.12.1808. Crompton 1811.

Dimples Mill, Morton. SE102418 1793–1821? RE32A/154781/1796. LM 3.10.1818, 9.6.1821. Crompton 1811.

At Steeton. SE033442? 1788–1810? RE30/144717/1795. Re32A/188763/1801. LI 19.12.1796, 6.5.1811. Chapman, IAR 1982–3. Universal Directory 1793.

Sutton Mill. SE006441? 1807?–1820? SunCR87/838544/1810. LI 1.2.1792, 8.9.1800, 13.3.1809.

Cowling Cotton Mill. SESD967429 1830?–1835 + Baines 1822. White 1837.

Ickornshaw Mill. SD968429 1791–1812? RE32A/170160/1799. Badgery MSS 141/487, WYAS Leeds.

Saddleworth

Calf Hey Mill. SD974103 1790–1816. Barnes, SHS Bulletin Vol 9 No 4 Winter 1979

Horest Mill. SD068095 1821–1835 +. Baines 1822. Parson & White 1830. FEC 1834. Gurr and Hunt 1989.

Old Tame Mill. SD971094 1814–1820? Commercial Directory 1814

Pingle Mill. SD979081 1805–1815. Gurr and Hunt 1989.

Woodhouse Mill. SD983079 1802–1818. LI 9.9.1805.

Hull Mill. SD988085 1787–1830. Barnes, SHS Bulletin Vol 9 No 4 Winter 1979

Shore Mill. SD986079 1788–1830. Barnes, SHS Bulletin Vol 9 No 4 Winter 1979

Lumb Mill. SD988076 1804–1810? Sun CR57/756625/1804. LI 13.2.1804, 1.4.1805. Pigot 1816.

Gatehead Mill. SD986072 1781–1789. Barnes,

SHS Bulletin Vol 9 No 4 1979. Gurr and Hunt 1989.

Knarr Mill. SD981069 1824–1835 +. Barnes, SHS Bulletin Vol 9 No 4 1979. Baines 1824.

Yew Tree Mill. SD004085 1818?–? Barnes, SHS Bulletin Vol 9 No 4 Winter 1979

Old Mill. SD997057 1817–1825? SHS Records Ref H/BB/7/12

Andrew Mill. SD999041 1813–1835 +. Pigot 1816. Baines 1822. FEC 1834.

Bentfield Mill. SD997041 1800?–1807. LI 5.1.1807. Barnes, SHS Bulletin Vol 9 No 4 1979

Charlotte Mill. SD983043 1828–1835 +. LI 8.2.1796. Pigot 1828. FEC 1834.

Wright Mill. SD979037 1810–1835 +. LI 30.7.1798. FEC 1834.

Woodend Mill. SD957043 1792–1835 +. Baines 1822. Barnes, SHS Bulletin Vol 10 No 1 Spring 1980

Carr Hill Mill. SD976026 1805–1835 +. LI 27.12.1813. Pigot 1816. Baines 1824. White 1837.

Hopkin Mill. Site not found. 1816–? Pigot 1816.

Strines Mill. SD955062 1802–1821? Sun CR46/729463/1802. Pigot 1816. Baines 1822.

Lowbrook Mill. SD961062 1786–1821? RE 95848/1786. Crompton 1811. Barnes, SHS Bulletin Vol 10 No 1 1980.

Waterhead or Mill Bottom Mill. SD955060? Pigot 1816. Baines 1822.

Austerlands Mill. SD958055 1819–1835 +. Baines 1822. Pigot 1828. FEC 1834.

Scouthead Mill. SD967056 1811?–1835 +. Crompton 1811. Pigot 1816. FEC 1834.

Newhouses Mill. SD972057 1820–1835? Gurr and Hunt 1989.

Pastures Mill. SD967056 1808–1835 +. Pigot 1816. Parson & White 1830. FEC 1834.

Woodbrook Mill. SD967053 1811–1835 +. Pigot 1816. FEC 1834.

Woodbrook Old Mill. SD967052 1814–1835 +. Barnes, SHS Bulletin Vol 10 No 1 Spring 1980

Radcliffe Mill. SD962043 1810?–1815? Crompton 1811.

Shelderslow Mill. SD961051 1810?–1835 +. LI 13.5.1805. Crompton 1811. Baines 1822. FEC 1834.

Spring Mill. SD997056? 1811?–1835 +. Crompton 1811. Baines 1822.

County End Mills. SD956045 1789–1835 +.
Barnes, SHS Bulletin Vol 10 No 1 Spring 1980

Lydgate Mill. SD973042 1795?–1835 +. Barnes,
SHS Bulletin Vol 10 No 1 Spring 1980

Quick Edge Mill. SD967039 1810?–? Barnes,
SHS Bulletin Vol 10 No 1 Spring 1980

Quick Mill. SD974037 1824–1835 +. Baines
1824. FEC 1834.

Valley Mill. SD968027 1823–1835 +. Barnes,
SHS Bulletin Vol 10 No 1 Spring 1980

Brookbottom Mills. SD972026 1811–1835 +.
Crompton 1811. Pigot 1828. FEC 1834.

West Craven

Lothersdale Mill. SD959459 1792–1835.
RE25/135977/1793. RE30/148268/1795.
LI 3.12.1798. Wilson 1972.

Mitchell's Mill. SD875465 1800–1835 +.
RE32A/178856/1808unCR100/868468/1812.

Lower Parrock House Mill. SD875466? 1808–
1835 +.SunCR77/810867/1808.LI20.12.1813.
SunCR191/1132627/1831

Gillins Mill. SD873456 1790–1835 +. Sun
OS361/560710/1798unCR191/1132632/1831.
HJ27.8.1808.

At Barnoldswick. Site not found. 1821? LM
8.12.1821.

At Elslack. SD935490? 1789–1813?
RE32A/186632/1801. Drapers Record
18.9.1897.

At Marton in Craven. Site not found. 1801–
1810? RE32A/187054/1801. HJ19.8.1809. LI
12.11.1810.

Booth Bridge Mill. SD914477 1798–1835 +.
RE32A/162298/1798. Cotton Mill Returns 1803–
6 WYAS Wakefield. LI 23.5.1814.

At Earby. SD916465? 1805?–1810? HJ
22.3.1806. LI 2.4.1810.

Kelbrook Cotton Mill. SD906446 1815?–1835 +.
LM 2.5.1818. White 1837.

Ickornshaw Mill. Site not found. 1791–1810.
Wood 1980.

Howgill Cotton Mill. SD824462 1790?–1835 +.
SunCR12/648787/1795.CottonMillReturns
1803–6 WYAS Wakefield. Baines 1822.

Feazor Mill. SD727451 1792–1835 +.
SunCR168/1071608/1828.Baines1822.Rothwell
1990.

Holden Cotton Mill. SD774495 1796–1805? LI
4.3.1799, 1.7.1799. Rothwell 1990.

Low Moor Mill, Clitheroe. SD729419 1782–1788–
1832? LM 21.4.1832. Rothwell 1990.

Grindleton Mill. SD760451 1792–1796–1831?
Sun187/1120127. Rothwell 1990.

Bottom Factory. SD760503 1795? Rothwell 1990.

Sawley Corn Mill. SD776463 1795?–1811. Rothwell 1990.

East Yorkshire

Great Ayton. Site not found. 1795–1800?
RE32A/159666/1797.

Wansford Mill. Site not found. 1790–1823?
RE23/127968/1792. RE24/132420/1793.
Baines 1822.

At York. Site not found. 1793–? LI 18.2.1793.

At Thirsk. SE430822?. 1797–1809? LI 7.8.1809,
18.12.1809. Land Tax Returns NYCRO.

At Pocklington. Site not found. 1792–? LI
25.6.1792, 11.7.1796, 4.9.1796. York Herald
30.6.1792.

At Easingwold. Site not found. 1800?–1823? LI
19.5.1791. Baines 1822 Vol 2.

South Yorkshire

At Balby. Site not found. 1790?–?
LI 12.11.1792, 7.1.1793.

Sheffield Cotton Mill. SK354882 1788–1815.
SunCR1/622019/1793.
SunCR111/902671/1815. Cotton Mill Returns
1803–6 WYAS Wakefield. LM 6.5.1815.
Baines 1822.

At Sheffield. Site not found. 1811? Crompton
1811.

At Conisborough. Site not found. 1792?–1802?
Manchester Mercury 21.2.1792. LI 6.7.1795,
19.4.1802

At Burton Smithies. Site not found. 1810?–? LI
25.5.1812.

At Tickhill. Site not found. 1803? Cotton Mill
Returns 1803–6 WYAS Wakefield.

Ecclesfield Cotton Mill. SK356945? 1796–1815?
Sun CR13/653203/1796. Sun
CR112/914177/1816.

Stocksbridge Cotton Mill. SK273986? 1800?–
1807? HJ 1.9.1804, 8.8.1807. LI 20.6.1807.

Hartcliffe Mill. Site not found. 1805?–1810? HJ
22.10.1808. LI 11.6.1810.

The Yorkshire Dales

Lonsdale

Millthorpe Mill. SD661913 1790–1835 +, SunOS391/606852/1792. SunCR211/1185320/1834.LI28.12.1790.

Birks Mill. SD658914 1809?–1835 +, SunCR89/840140/1810. SunCR212/1185572/1834.

Burton-in-Lonsdale Cotton Mill. SD649718 1797–1835 +.RE32A/155788/1797.LI29.9.1800, 15.4.1805,1.11.1808.

Westhouse Mill. SD671737 1800?–1835 +. LI 2.3.1812,9.5.1814, 17.2.1817. Baines 1822.

Ingleton Mill. SD694733 1788?–1807. RE32A/158377/1797. LI9.2.1807, 18.4.1808. Baines 1822.

Clapham Mill. SD727688 1786–1807? SunOS341/525208/1786. LI 9.2.1807.

Clapham Wood Mill. SD718668 1795?–? LI 17.9.1798

Austwick Cotton Mill. SD779694 1792?–1835 +. LI 29.6.1795, 17.2.1817.JA/25 WYAS Leeds. Phoenix Fire Office September 1797.

Bentham Cotton Mill. SD649693 1795?–1810? Cotton Mill Returns 1803–6. Birkbeck MS WYAS Leeds.

Swaledale

Richmond Cotton Mill. NZ1700? 1792–1796. Hatcher 1978.

Wensleydale

Gayle Mill. SD870892 1784–1813? LI 20.12.1813. Hartley and Ingilby 1953.

Askrigg Mill. SD942911? 1784?–1820? LI 10.3.1789. Hartley and Ingilby 1953.

Yore Mill. SE004888 1784–1815? RE13/102200/1787. RE20/118521/1790. LI21.10.1811, 17.10.1814.

At Masham. SE430823? 1800?–? Tuke 1800.

Bishopton Mill, Ripon. SE303707 1793–1810? RE32A/155331/1797. HJ 13.10.1810.

High Mill, Ripon. SE310711 1795? Universal Directory 1793. HJ 13.10.1810

Duck Hill Mill, Ripon. Site not found. 1800? Information supplied by Mr M H Taylor of Ripon.

Nidderdale

Wath Mill. SE146678 1800? Lucas 1882.

Hollin House Mill. 1795–1818. RE28/144005/1795. RE32A/155695/1797. Jennings 1967.

Scotton Mill. SE316586 1795?–1798. LI 18.8.1794, 1.6.1795. Jennings 1967.

Wreaks Mill. SE245597 1793–1835 +. SunOS391/605590/1793. SunCR227/1235592/1836.LI21.1.1799,1.7.1799, 28.5.1804. Cotton Mill Returns 1803–6 WYAS Wakefield. FEC 1834.

Castle Mill, Knaresborough. SE347568 1791–1815? LI 22.4.1793. Cotton Mill Returns 1803–6 WYAS Wakefield. Jennings 1967.

Littondale

Halton Gill Mill. Site not found 1801? Churley, Yorkshire Bulletin Vols 5–6.

Arncliffe Cotton Mill. SD930718 1792–1835 +. RE25/131083/1792SunI27/965003/1820LM 18.10.1785,4.3.1815.LI13.12.1791,20.7.1801. FEC 1834.

Wharfedale

Kettlewell Cotton Mill. SE972722 1805?–1835 +. FEC 1834. Raistrick 1967.

Scaw Gill Mill, Grassington. SE000644 1792?–1812? HJ 29.4.1809, 28.6.1813.

Linton Low Mill. SE007633 1800?–1813? HJ 16.6.1804, 8.3.1806, 27.12.1813.

Hebden Mill. SE027624 1790?–1835 +. RE28/143711/179$unCR244/1258327/1837. LI 3.3.1800, 13.2.1809, 2.3.1812. HJ 25.4.1807, 18.12.1809.

Hartlington Mill. SE042609 1787?–1835. RE32A/173537/1800. LI 12.5.1789, 27.1.1812. LM 8.6.1822. Cotton Mill Returns 1803–6 WYAS Wakefield.

Skyreholme Mill. SE066603 1800?–1835 +. HJ 1.12.1810. FEC 1834. White 1837.

High Mill, Addingham. SE081503 1787–1835 + RE13/104521/1787. SunOS358/552473/1789. LI 14.8.1797, 16.10.1797, 30.4.1798.

Townhead Mill, Addingham. SE073499 1799–1835 +. SunCR27/693628/1799. B & W PF 736. Crompton 1811. LM 2.9.1820

Fentiman's Mill. SE078498 1802–1835 +. RE25/135978/1793.Mason1989.

Beamsley Mill. SE077526 1799?–1818? LI 14.1.1799. LM 3.1.1818.

Low Mill, Addingham. SE092494 1824–1835 +.
SunCR161/1060009/1825SunCR/1303326/1839.
FEC 1834. LM 11.5.1822.

Ilkley Cotton Mill. SE117473 1788?–1835 +.
SunOS351/540956/1788.LI29.1.1788,6.5.1793,
13.11.1797.

Spinning Mill, Burley Woodhead. SE148445
1795–1810? RE28/143710/1795. LI28.4.1800

Hazel Grove Mill, Burley Woodhead. SE156443
1795–1815? RE28/143710/1795.
SunCR80/813901/1809. LI3.2.1817.

Greenholme Mill, Burley-in-Wharfedale. SE168468
1792–1835 +.RE24/131235/1792.
SunCR100/868473/1812.HJ9.3.1805.Crompton
1811. FEC 1834. Baines 1822. New Times
7.5.1819.

At Eller Gill, Otley. SE177443 1788–1800?
RE18/121890/1790. RE32A/180512/1800.
LI12.4.1802. Chapman, IAR 1981–2.

Otley Mills. SE194453 1783–1810.
SunOS391/606853/1792.
SunCR17/663871/1797. LI22.4.1793,
10.7.1809, 27.1.1817. Universal Directory 1793.
FEC 1834.

Silver Mill, Otley. SE208449 1784–1820?
RE13/102201/1787. Chapman, IAR 1981–2.
LI28.10.1799, 3.2.1800.

Tadcaster Cotton Mill. Site not found. 1792–
1805? LI 23.1.1792. LM20.5.1815, 19.10.1822.

Washburn Valley

Low Mill, West End. SE152578 1791–1812?
RE32A/155200/1797. RE32A/186172/1801.
Cotton Mill Returns 1803–6 WYAS Wakefield.

High Mill, West End. SE134582 1800?–1812.
Cotton Mill Returns 1803–6 WYAS Wakefield. Jennings 1967.

Little Mill, West End. SE143582 1805?–1812?
LI 2.3.1807, 19.9.1807. Crompton 1811.

Ribblesdale

Stainforth Mill. SD817673? 1793–1800? Universal Directory 1793. Brayshaw and Robinson
1932.

Langcliffe Mill. SD816650 1784–1835 +.
RE15/109974/1788SunCR124/941912/1818.
Cotton Mill Returns 1803–6 WYAS Wakefield.
Baines 1822. FEC 1834. Chapman, IAR 1981–2.

Settle Bridge Mill. SD817642 1785–1835? LM
22.11.1785. LI 14.7.1800. Universal Directory
1793.

Higher Mill, Settle. SD814637 1793–1835 +.
SunCR12/648969/1795. Baines 1822. FEC 1834.

Runley Bridge Mill, Settle. SD811623 1788–1835
+SunOS351/524977/1788.
SunCR12/648969/1795. LM 2.3.1784. LI
25.3.1793. Cotton Mill Returns 1803–6 WYAS
Wakefield.

At Giggleswick. SD808644? 1800?–1820? Brayshaw and Robinson 1932.

Giggleswick Mill. SD812640? 1793?–1816. LM
2.3.1784. LI 25.3.1793. HJ 8.6.1805. Universal
Directory 1793. Brayshaw and Robinson 1932.

At Rathmell, near Settle. SD804592? 1790?–
1800? Speight 1892.

At Rathmell, near Settle. SD796599 1790?–1800?
Speight 1892.

Lower Mill, Long Preston. SD832575 1790?–
1814? Universal Directory 1793. LI 3.8.1812,
24.1.1814.

Fleet's Mill, Long Preston. SD832575 1790?–
1814? Universal Directory 1793. LI 3.8.1812,
24.1.1814.

Airedale

Malham Mill. SD898633 1785–1835 +. LI
28.4.1800, 14.9.1807. Chapman, IAR 1981–2.
Badgery MSS 145/398, WYAS Leeds. Baines
1822. Raistrick 1967.

Scalegill Mill. SD898617 1794–1835 +.
SunCR57/756288/1804.
SunCR149/1025721/1825.LI1.1.1788.

Airton Mill. SD904593 1786?–1835 +.
RE32A/158649/1795SunCR118/926535/1817.
LI16.3.1795.

Bell Busk Mill. SD905563 1794–1835 +.
SunCR7/636029/1795SunCR169/1083376/1828.
FEC 1834.

Eshton Bridge Cotton Mill. Site not found.
1797?–1800? LI 24.12.1798

At Otterburn. Site not found. 1790?–? Universal
Directory 1793

High Mill, Gargrave. SD925538 1793?–1835 +.
SunCR124/939461/1818SunCR126/955165/1819.
Universal Directory 1793. Pigot 1818. Baines
1822. White 1837.

Low or Goffa Mill, Gargrave. SD935539? 1800?–
1811? RE32A/179691/1800.
SunCR88/842448/1810. LI16.11.1807,
16.12.1811. HJ 8.3.1806

Airebank Mill, Gargrave. SD940540 1791–1835
+.SunCR143/1007586/1823.B&HC27.4.1826.
Pigot 1818. Baines 1822.

Threaplands Mill. SD985605 1800–1818?
RE32A/176879/1800. SunCR99/865307/1812.
LM 21.2.1818.

Rilston Mill. SD959579 1794–1806? Chambers
MSS, WYAS Leeds. HJ 31.12.1803. Bolton Estates MS.

Hetton Cotton Mill. Site not found. 1795?–
1800? LI 9.3.1795, 2.4.1798.

High Mill, Skipton. SD992522 1782–1835 +.
SunOS366/566921/1791.
SunCR215/1190281/1834.LI1.12.1789,
25.5.1790, 22.3.1791. Chapman, IAR 1981–2.
FEC 1834.

Belle Vue Mills. SD986516 1831–1835 +.
SunCR131/969416/1820.
SunCR212/1185573/1834.DrapersRecord
18.9.1897.

Whitfield Syke Mill, Embsay. SD998546 1795?–
1835 +RE32A/180514/1800.
SunCR134/984570/1821HJ23.5.1807,
25.11.1809. FEC 1834. Pigot 1828.

Good Intent Mill, Embsay. SE002544 1810?–
1835 +SunCR127/965001/1820.
SunCR198/1162098/1833.Pigot1816.

Crown Spindle Works, Embsay. SD002542 1826–
1835 +. Wharton, Embsay with Eastby–Booklet
in Skipton Library.

Sandbeds Mill, Embsay. SE006534 1794–1835 +.
RE28/143709/1795RE32A/170023/1799.
SunCR187/1120124/1831LI19.9.1794,
28.10.1805.LM24.6.1815.

Millholme Mill, Embsay. SE006535 1793–1835
+.RE26/132536/1793.RE32A/171516/1799.LI
11.11.17934.12.1797.

Eastby Cotton Mill. SE017545 1794–1835 +.
RE32A/158824/1795SunCR135/158824/1821.
HJ 5.8.1809. Badgery BD246 WYAS Leeds.
White 1837.

Draughton Cotton Mill. Site not traced. 1788?–
1822? LI 14.8.1797, 30.4.1798, 8.7.1799. Baines
1822.

Cononley Cotton Mill. SD993468? 1790?–1810?
RE31/152763/1796. LI 29.1.1810.

Bibliography

J. Aikin, *A Description of the Country from Thirty to Forty Miles around Manchester* (1795)

C. B. Andrews (ed.), *The Torrington Diaries* (1936)

H. Armitage, The Rawdon Family, *HAS* (1967)

Anon., *The Century's Progress* (1893)

T. S. Ashton, *The Industrial Revolution* (1948)

C. Aspin, *James Hargreaves and the Spinning Jenny* (1964)

——, *The Cotton Industry* (1981)

Bailey, *Northern Directory* (1781)

E. Baines, *History, Directory and Gazetteer of the County of York* (1822), vols 1 and 2

——, *History, Directory and Gazetteer of the County of Lancaster* (1825)

——, *History of the Cotton Manufacture in Great Britain* (1835)

P. Barfoot and J. Wilkes, *The Universal British Directory* (1793)

B. Barnes, 'The Early Cotton industry in Saddleworth', *Saddleworth Historical Society Bulletin*, vol. 9, nos 3 and 4, vol. 10, no. 1

J. Bentley, *Old Ingleton* (1976)

J. Bigland, *Yorkshire* (1815)

G. R. Binns, 'Water wheels in the upper Calder valley', *Trans. Halifax Antiquarian Society* (1972)

T. Brayshaw and R. M. Robinson, *The History of the Ancient Parish of Giggleswick* (1932)

R. Brook, *The Story of Huddersfield* (1968)

G. H. Brown, *On Foot Round Settle* (1896)

D. Bythell, *The Handloom Weavers. A study in the English Cotton Industry during the Industrial Revolution* (1969)

H. A. Cadman, Gomersal, *Past and Present* (1930)

H. Catling, *The Spinning Mule* (1970)

S. D. Chapman, *The Early Factory Masters* (1967)

——, 'Fixed capital formation in the early British cotton industry', *Economic History Review*, 2nd series, xxiii (1970) (2)

——, 'The cost of power in the Industrial Revolution in Britain: the case of the textile industry', *Midland History*, 1 (1971)

——, 'The Arkwright mills - Colquhoun's Census of 1788 and archaeological evidence', *Industrial Archaeology Review*, 6 (1981–2)

S. Chapman, *The Lancashire Cotton Industry* (1904)

J. Clough, *History of Steeton* (1886)

W. B. Crump, *The Leeds Woollen Industry* (1931)

—— and G. Ghorbal, *History of the Huddersfield Woollen Industry* (1935)

F. Crouzet (ed.), *Capital Formation in the Industrial Revolution* (1972)

W. Cudworth, *Round about Bradford* (1876)

——, *Rambles round Horton* (1886)

G. W. Daniels, *The Early English Cotton Industry* (1920)

——, 'Samuel Crompton's census of the cotton industry in 1811', *Economic History* (1930)

W. S. Dawson, *History of Skipton* (1882)

I. Dewhirst, *A History of Keighley* (1974)

——, *Yorkshire through the Years* (1975)

E. E. Dodd, *Bingley* (1958)

S. Dumbell, 'Early Liverpool Cotton Imports and the Organisation of the Cotton Market in the Eighteenth Century', *Economic Journal*, xxxiii (1923)

M. M. Edwards, *The Growth of the British Cotton Trade 1780–1815* (1967)

D. A. Farnie, *The English Cotton Industry and the World Market 1815–1896* (1979)

J. Fielden, *The Curse of the Factory System*. Second Edition (1969)

G. Firth, 'The Genesis of the Industrial Revolution in Bradford, 1760–1830', Ph.D Thesis, University of Bradford (1974)

R. I. Fitton and A. P. Wadsworth, *The Strutts and the Arkwrights 1758–1830* (1958)

M. Fox, 'Saddleworth Textile Mills', *Saddleworth Historical Society Bulletin*, Vol. 19, No. 2

W. O. Garnett, *Wainstalls Mills: The History of I & I Calvert Ltd* (1951)

C. Giles and I. H. Goodall, *Yorkshire Textile Mills* (1992)

R. Guest, *A Compendious History of the Cotton Manufacture* (1823; reprinted 1968)

D. Gurr and J. Hunt, *The Cotton Mills of Oldham* (1989)

T. W. Hanson, *The Story of Old Halifax* (1920)

M. Hartley and J. Ingleby, *Yorkshire Village* (1953)

H. W. Harwood, 'Wainstalls Mills', *HAS* (1953)

——, 'Wainstalls – Some Industrial History', *HAS* (1953)

——, 'Lower Stock, Wainstalls', *HAS* (1953)

J. Hatcher, *The Industrial Architecture of Yorkshire* (1985)

D. A. Heptinstall, 'Cotton and the Upper Ryburn Valley, 1790–1841' MA, Huddersfield (1984)

D. Hey, *The Village of Ecclesfield* (1968)

R. L. Hills, *Power in the Industrial Revolution* (1970)

——, *Richard Arkwright and Cotton Spinning* (1973)

——, 'Hargreaves, Arkwright and Crompton. Why Three Inventors?', *Textile History* 10 (1979)

J. Hodgson, *Textile Manufacture and other Industries in Keighley* (1879)

Holden, *Triennial Directory* (1811)

I. Holmes, *From Hand Industry to the Factory System* (1914)

G. Ingle, 'A History of R. V. Marriner Ltd, Worsted Spinners, Keighley', M.Phil Thesis, Leeds (1974)

——, 'The West Riding Cotton Industry, 1780–1835', Ph.D Thesis, Bradford (1980)

G. Jackson, *Hull in the Eighteenth Century* (1972)

J. James, *History of the Worsted Manufacture in England from the Earliest Times* (1857)

D. T. Jenkins, 'The Validity of the Factory Returns 1833–50', *Textile History*, vol. 4 (1973)

——, *The West Riding Wool Textile Industry 1770–1835* (1975)

——, 'The Cotton Industry in Yorkshire 1780–1900', *Textile History*, vol. 10 (1979)

B. Jennings (ed.), *A History of Nidderdale* (1967)

—— (ed.), *Pennine Valley* (1992)

H. P. Kendall, 'The Two Willow Halls', *HAS* (1908)

B. R. Law, 'Calder Mill Owners 1792–94', *HAS* (1954)

J. Lawson, *Letters to the Young on Progress in Pudsey During the last Sixty Years* (1887)

R. Lawton, 'The Economic Geography of Craven in the Early Nineteenth Century', *Transactions of the Institute of British Geographers* 28 (1954)

R. E. Leader, *Old Sheffield* (1876)

J. Mayhall, *Annals of Leeds and the Surrounding District* (1860)

W. Marshall, *The Review and Abstract of the County Reports to the Board of Agriculture* (1808; reprinted 1968)

K. M. Mason, *Woolcombers, Worsteds and Watermills, Addingham's Industrial Revolution* (1989)

G. A. Miller, *Blackburn. The Evolution of a Cotton Town* (1951)

A. Newell, *A Hillside View of Industrial Society* (1925)

J. Parker, *Illustrated Rambles from Hipperholme to Tong* (1904)

E. Parsons, *History of the West Riding*, vol. II (1836)

Parson and White, *Directory of the Borough of Leeds and the Clothing District of Yorkshire* (1830)

F. Peel, *Spen Valley, Past and Present* (1893)

E. Phillips, *The Story of Doncaster* (1921)

J. H. Priestley, 'Mills of the Ryburn Valley', *HAS* (1933, 1934, 1941)

——, 'The Growth of a Township – Soyland', *HAS* (1955)

——, 'The Town of Rishworth', *HAS* (1956)

——, 'Smallees in Soyland', *HAS* (1957)

——, 'Mills of the Blackburn Valley', *HAS* (1963)

A. Raistrick, *Old Yorkshire Dales* (1967)

——, *The Pennine Dales* (1968)

——, *West Riding of Yorkshire* (1970)

——, *Industrial Archaeology* (1973)

Robson, *Commercial Directory* (1840)

D. Roberts, 'The Development of the Textile Industry in the West Craven and the Skipton District of Yorkshire', M.Sc Dissertation, LSE (1956)

G. Shutt, 'Wharfedale Water Mills. Unpublished M.Phil thesis, Leeds (1979)

E. M. Sigsworth, 'William Greenwood and Robert Heaton. Two Eighteenth-Century Worsted Manufacturers', *J. Bradford Textile Soc.* (1951–2)

H. Speight, *The Craven and North-West Yorkshire Highlands* (1892)

——, *Old Bingley* (1897)

——, *Upper Wharfedale* (1900)

J. Sutcliffe, *A Treatise on Mills and Reservoirs* (1816)

D. F. E. Sykes, *The History of Huddersfield* (1898)

J. Tann, *The Development of the Factory* (1970)

——, 'The Textile Millwright in the Early Industrial Revolution', *Textile History*, vol. 5 (1974)

J. Travis, *Notes of Todmorden and District* (1896)

——, *Chapters of Todmorden History* (1901)

W. B. Trigg, 'The Industrial Water Supply of Ovenden', *HAS* (1933)

J. Tuke, *A General View of the Agriculture of the North Riding of Yorkshire* (1800)

J. H. Turner, *Haworth, Past and Present* (1879)

G. Unwin, *Samuel Oldknow and the Arkwrights* (1924)

A. P. Wadsworth and J. de L. Mann, *The Cotton Trade and Industrial Lancashire, 1600–1780* (1965)

W. White, *History, Gazetteer and Directory of the West Riding* (1837)

M. T. Wild, 'Documents and Sources IV. The Saddleworth Parish Registers', *Textile History*, vol. 2 (1971)

K. Wilson, *The History of Lothersdale* (1972)

R. G. Wilson, *Gentlemen Merchants, The Merchant Community in Leeds 1700–1830* (1971)

A. Wrigley, *Saddleworth: Chronolgical Notes from 1200 to 1900* (1941)

Index

References in **bold** are to illustrations

Addingham, 93, 103
Airebank Mill, Gargrave, 50, 238
Aireworth Mill, Keighley, 13, 96, 170–1, **171**
Airton Mill, 97, **234**, 234–5
Allen, John & Richard, 93
Andrew Mill, Saddleworth, 182
Arkwright, Richard, 5, 15, 18, 66, 68, 168, 245
Arncliffe Mill, 210–11, **210**, **211**
Askrigg Cotton Mill, 174, 206, **206, 207**
Austerlands Mill, Saddleworth, 184
Austwick Cotton Mill, 27, 204

Baildon, Clifton Mill, 118
Bailey, William & Co., 126
Bainton, Boyes & Co., 92
Balby Mill, Doncaster, 52, 93, 195–6
Bankfoot Mills, Heptonstall, 140
Banks, 50, 58
Banks, John, 25–6
Bank Mill, Leeds, **106**
Barker, A, 132, 136
Barker, J & H, 78, 116, 176
Barewise Mill, Stansfield, 132
Barnsley, mill, 197
Barnoldswick, mill, 189
Bates, Joshua, 60
Bateson M & J, 111
Beamsley Corn Mill, 49, 218, **218**
Beaumont G & W, 92, 126–7
Beckstones Mill, Keighley, 174–5
Bell Busk Mill, 235, **236**
Belle Vue Mills, Skipton, 239–40, 245
Bent, James, 136
Bent Mill, Bingley, 90, 116–17
Bentfield Mill, Saddleworth, 182–3
Bentham Mill, 205
Beresford & Whitmore, 128
Beverley, Cross & Billiam, 5, 30, **31**, 74, 108
Bingley, 85, 103, 114, 115–16
Binns, Thomas, 96, 170, 179
Birks Mill, Sedbergh, 202, **202**
Birkbeck, William, 58, 68, 206
Birkenshaw Mill, 20, 42, 92, 121–2
Birstall, mill, 120

Bishopton Mill, Ripon, 207–8
Black Moor Holme Mill, Slaithwaite, 127
Blackburn, 6, 77, 79, 81–3, 86, 98
Blagborough & Holroyd, 41, 52, 109–10
Blakey, John, 55, 67
Blakey, Joseph, 55, 67
Blubberhouses, 7
Booth & Co., 30
Booth Bridge Mill, Earby, **189**, 190
Booth Bridge Mill, Rishworth, 154
Booth Wood Mill, Rishworth, 154
Bottom Factory, Holden, 191
Boulder Clough or Swamp Mill, Sowerby,
    147
Boulton & Watt
    gas apparatus, **45**, 47, 152
    steam engines, 30–4, 107, 108, 109, 110,
        111, 149, 151, 152
Bowling Dyke Mill, Northowram, 150
Bowling Iron Works, 31, 41, 99
Bracken, J, 145
Bradford, 9, 85, 89–90, 93, 103, 113–15
    Mill, 119
Bradley, Robert & Co., 8, 51
Bramley, Walter, 214
Brayshaw, Richard, 96
Brearley Mill, Midgeley, 49, 146
Brennand & Hinchliffe, 110
Bridgehouse Mill, Keighley, 53, 162
Briggs, J, 99
Brocklehurst & Winterberry, 80
Brook Brothers, Meltham, 50, 123, 128
Brookbottom Mills, Saddleworth, 185
Brookes, Hannah, 50
Brow Mill, Skircoat, 152–3
Brow End Mill, Keighley, 173, 244
Buckley, John, 113, 133
Bucktrout, Mr, 35
Burnley, William, 82
Burnsall, 7
Burnt Acres Mill, Erringden, 136
Burton-in-Lonsdale Cotton Mill, 50, 202–3
Busfield, W & J A, 115

Butterworth & Co., 247

Cabbage Mill, Keighley, 97, 175–6
Calderside Mill, Stansfield, 138
Calf Hey Mill, Saddleworth, 181
Calvert, Lodge, 89, 165
Cardwell, Birley & Hornby, 83
Carr, Thomas, 120–1
Carr, William, 68
Carr Hill Mill, Saddleworth, 183
Cartwright, Edmund, 93
Castle Mill, Keighley, 54, 174
Castle Mill, Knaresborough, 50, 209–10
Castle Mill, Sowerby, 144
Castle Clough Mill, Stansfield, 135
Castlefields Mill, Bingley, 27, 74, 97, 115
Causey or Midgehole Mill, Langfield, 134
Chamberlain, William, 36, 92
Charlotte or High Grove Mill, Saddleworth, 183
Chippendale, Netherwood & Carr, 50, 58
Cinderhill Mill, Langfield, 135
Clapham Mill, 204
Clapham Wood Mill, 204
Claytons & Walshman, 5, 24, 66, 69, 70, 95, 168–70
Clough Mill, Birstall, 120
Clough Mill, Heptonstall, 140
Clough Mill, Soyland, 158
Clough Hole Mill, Stansfield, 135
coal and iron, 11
Cockden Mill, Stansfield, 136
Cockshott & Lister, 82
Collier, Samuel, 61
Colne, 96, 98, 113–14
Colne Bridge Mill, 46, 47, 128–9
Colquhoun, Patrick, 13–16
Conisborough, mill, 197
Cononley Cotton Mill, 242
construction, 65–6
Cooper House Mill, Warley, 146
Copley Mill, 152
Corlas, Thomas, 76, 166
corn mills, 10, 49, 65
County End Mills, Saddleworth, 184–5
Coupland, Thomas & Co
Cowbridge or Underbank Mill, Stansfield, 136–7
Cowling Mill, 179
Crabtree & Green, 110
Cragg Mill, Erringden, 144
Craven, 2, 88, 98, 104, 186
Craven, Brigg & Shackleton, 54, 67, 167
Craven, Edward, 97, 116
Cromack, John, 69

Crompton, Samuel, 6, 19, 20
Cross Lee Mill, Stansfield, 132
Crossland Mills, 27, 126–7
Crow Hill Mill, Marsden, 124–5
Crown Spindle Works, Embsay, 241
Cullingworth, Bradford, 114
Curtis, Driffield & Co., 209

Dale End Mill, Lothersdale, 27–8, 187
Dalton Mill, Keighley, 58, 170
Damens Mill, Keighley, 56, 97, 165
Damside Mill, Keighley, 53, 97, 175
Damside Mill, Soyland, 159
Dartmouth, Lord, 56
Davenport, William, 70
Dean Mills, Midgley, 145
Devonshire, Duke of, 8, 51, 56
Dewhurst & Co., 84, 91
Dewhirst, W & L, 148
Dimples Mill, Morton, 27, 58, 178
Doncaster, 93
Draughton Cotton Mill, 242
Driver, Joseph & Co., 13, 78, 206
Duck Hill Mill, Ripon, 208
Dyson Lane Mill, 96, 156–7

Earby, mill near, 190
Easingwold, 194
Eastby Cotton Mill, 41, 241–2
Eastwood, James, 132
Eastwood, Thomas, 136
Eastwood Mill, Stansfield, 136
Ecclesfield Cotton Mill, 197
Edmondson, Allen, 111
Ellar Carr Mill, Bingley, 27, 97, 116
Ellis, Lister, 97
Ellis, William, 73
Elslack Mill, 189
Embsay, 41
Emmett & Co., 30, 147
Eshton Bridge Mill, 235–6
Ewood Mill or Malt Kiln Mill, Stansfield, 132

Factories Inquiry Commission 1834, 24
Factory Acts, 16
Factory, The, Marsden, 123–4
Far Bank Mills, Leeds, 109
Fawcett, Richard, 88
Feilden Brothers, 130, 133
Fielden, Joshua & Son, 82
Feazor Mill, Waddington, 190
Fentiman, Anthony, 217–18
Fentiman's Mill, Addingham, 217–18
Fenton, Murray & Wood, 41, 45, 46

Ferrand, Benjamin, 117
fire, damage to mills, 47
   insurance, 61, 98
Firth House Mill, Scammonden, 38
Firth & Haworth, 134, 135
Fleet's Mill, Long Preston, 232
Folly Mill, Langfield, 134
Foster & Co., 117
Foster & Sugden, 141
Foster Mill, Wadsworth, 142
Franks Mill, Marsden, 27, 124
Frieldhurst Mill, Stansfield, 131–2
fulling mills, 49

Garforth, Peter, 115
Garforth & Sidgwick, 51, 70, 71, 73, 91
Gargrave, 93
Garnett, James, 113
Gartside & Parkin, 125
Gatcliffe, William, 110–11
Gatehead Mill, Saddleworth, 182
Gaukroger, T & J, 141
Gayle Mill, Hawes, 205–6, **205**
Gibson, Abraham, 141
Gibson Mill, 51, 140–1, 244
Giggleswick, mill, 231
Giggleswick Mill, 10, 231
Gill & Ashton, 126
Gillians Mill, Barnoldswick, 187–8, **188**
Goit Stock Mill, Bingley, 117–18
Gomersal Hall Mill, 20, 53, 120–1
Good Intent Mill, Embsay, 41, 240
Goose Eye Mill, Keighley, 97, 173–4
Gowland & Co., 109
Graham Brothers, 84
Great Ayton, 192
Green, Mawson & Dobson, 80
Greengate Mill, Keighley, 55, 67, 166, **166**, **244**, 245
Greenholme Mill, Burley-in-Wharfedale, 97, 221–3
Greenhurst Mill, Stansfield, 132
Greenwood & Ellis, 24, 115, 175
Greenwood, James, 55, 67
Greenwood, John, 54, 67, 73, 96–7, 116, 165, 173, 175–6
Greenwood, Robert, 132–3
Greenwood, T & J, 148
Greenwood & Whitaker, 47, 80, 84
Greenwood Lee, Heptonstall, 28, 141
Griffe Mill, Haworth, 164
Grindleton Mill, 191
Grove Mill, Keighley, 165
Grove Mill, Ovenden, 148

Guteroyd Mill, Langfield, 136

Hadwen, J & T, 96
Haggas, John and William, 12, 88, 116, 164
Haigh Brothers, 70, 84, 96, 124, 125
Hallas Bridge Mill, Bingley, 90, 117
Halifax, 2, 40, 77, 85, 88, 94, 103, 130–1
Halifax Piece Hall, 85, 98
Halstead, Coupland & Wilkinson, 107–8
Halton Gill Corn Mill, 210
Hammond & Tattersall, 41
Hanging Lee Mill, Soyland, 155
Hargreaves, James, 4, 18
Hargreaves, Robert & Co., 66
Harrison, Joseph, 116
Hartcliffe Mill, Denby, 197–8
Hartley, W, C & T, 115
Hartlington Mill, Burnsall, **214**, 215–16
Hartshead Mill, 153
Hattersley, Richard, 68
Hawksclough Mill, Wadsworth, 143
Hawksworth
   Over Mill, 118
   Low Mill, 118
Hawksworth, Thomas, 82
Hazel Grove Mill, Burley Woodhead, 220–1, **221**
Hazel Grove Mill, Rishworth, 96, 155–6
Heathfield & Co., 50, 84
Heaton, Dinah, 51
Heaton, John, 51, 85, 164
Heaton, Robert & William, 59, 74, 80, 83, 97, 99, 162
Hebble Mill, Ovenden, 148
Hebden Mill, 50, **213**, 213–15
Heckmondwyke Mill, 121
Hegginbottom, J, 147
Hegginbottom, William & Brothers, 93
Helliwell, T & J, 132
Hetton Cotton Mill, 38, 239
Hewenden Mill, Wisden, 116
High Mill, Addingham, 13, **37**, 38, **216**, 216–17
High Mill, Gargrave, 236, **237**
High Mill, Ripon, 208
High Mill, Skipton, 239
High Mill, West End, 226, **226**
Higher Providence Mill, Keighley, 12, 164
Higgin Chamber or Ing Head Mill, Sowerby, 146–7
Hillhouse Bank, Leeds, 107
Holden Cotton Mill, 191
Holdforth, John, 107
Hole Bottom Mill, Stansfield, 132
Hollin House Mill, Nidderdale, 208, 213
Hollings Mill, Soyland, 96, 157

Hollings & Ross, 164
Hollinrake, James, 134
Holme Mill, Bradford, 59, 90, 119–20
Holme Mill, Marsden, 126
Holme Mill, Stansfield, 132
Holme Field Mill, Ovenden, 148
Holme House Mill, Keighley, 54, 174
Holroyd, John, 50
Holroyd, J & T, 118
Holt, Ralph, 96
Hope Mill, Keighley, 18, 76, 166
Hopkin Mill, Saddleworth, 183
horse mills, 12, 35–8, 41
Horsefall, Jeremiah, 93
Horest Mill, Saddleworth, 181
Horsfall, J & T, 117
hosiers, 86
Howard, John, 111
Howgill Cotton Mill, 190
Hoyle, Elkanah, 60, 157
Hoyle, Wheelright John, 96
Hoyle Bottom or Square Mill, Warley, 144
Hoyle House Mill, Warley, 150–1
Huddersfield, 103, 123
    mill at New Street, 128
    mill, 128
Hudson, Betty, 55, 78, 175
Hudson, Joseph, 111
Hudson Mill, Stansfield, 138–9
Hull, 77
Hull Mill, Saddleworth, 181
Hunslet, Leeds, mill, 107–8
Hunslet Moor, mill, 108–9

Ibbotroyd Mill, Wadsworth, 143
Ibbotson, Sir James, 120
Ickornshaw Mill, 179
Idle Cotton Mill, Bradford, 69, 118–19
Illingworth, Ann, 55, 165
Ilkley Cotton Mill, 219–20
Ingham, Richard, 135
Ingham & Fox, 114
Ingleton Cotton Mill, 204
Ingrow Mill, Keighley, 165–6
insurance policies, 3, 61, 62

Jowler or Holme House Mill, Warley, 145
Jubb, John, 41, 68, 111
Jumb Mill, Langfield, 133–4
Jumble Mill, Marsden, 125–6
Jumble Hole or Low Underbank Mill,
    Stansfield, 137
Jumples Mill, Ovenden, 30, 147–8

Kebroyd Mills, Soyland, 91, 96, 159–60

Keighley, 40, 54, 55, 95, 99, 103–4, 161–2
Kellbrook Cotton Mill, 190
Kershaw, J & J
Kershaw Mill, Heptonstall, 142
Kettlewell, 92
Kettlewell Mill, 212
King's Mill, Settle, 10, **229**, 229–30
Kitson Wood or Naylor Mill, Stansfield, 132
Knarr Mill, Saddleworth, 182
Knight, John & Benjamin, 90, 119, 177
Knights Mill, Bradford, 69, 119
Knowles, John & Co., 90, 116, 117

labour
    skills, 64–5, 66, 74–5, 247
    child,, 8, 16, 19, 68–71, 72–3
Lancashire
    connections with, 1, 2, 10, 77, 88, 95–6
Lancaster, 77
Land Mill, Stansfield, 138
Langcliffe Mill, Settle, 10, 24, 68, **227**, 227–8
Leach, Rachael, 55, 170
Leach, Thomas, 178
Lee, Thomas, 92
Lee Mill, Heptonstall, 142
Lee Bridge Mill, Halifax, 150
Leeds, 103, 105–6
    merchants, 48, 77, 95
Lees, James, 61
Lees, R, J & J, 130
Lingards Wood Mill, Marsden, 125
Linton Mill, Grassington, 7
Linton Low Mill, Grassington, **212**, 212–13
Little Mill, West End, 226
Liverpool, 77, 99, 103
Lobb Mill, Langfield, 135
Lobley, Richard, 110
Lockwood, Joshua & Co., 13, 128
Lodge, Edmund, 96, 151
Lodge, T & H, 146, 151
Lomas, Thornton & Co., 50
London, 77, 82, 85
Longbottom Mill, Warley, 40, 146
looms, see weaving
Low Bridge Mill, Keighley, 24, 40, 167
Low Mill, Addingham, 93, 218–19, **219**
Low Mill, Gargrave, 236–8, **237**
Low Mill, Keighley, 6, 30, 68, 167–70, **168**
Low Mill, West End, **225**, 225–6
Low Moor Iron Company, 41
Low Moor Mill, Clitheroe, 191
Lowbrook Mill, Saddleworth, 183–4
Lower Mill, Long Preston, 232
Lowe, G & J, 82
Lower Parrock House Mill, Barnoldswick, 187

Lowfold Mill, Leeds, 111
Luddenden Mills, Warley, 145–6
Luddenden Foot Mill, Warley, 146
Lumb Mill, Heptonstall, 139
Lumb Mill, Saddleworth, 181–2
Lumb Mills, Sowerby, 159
Lumb Mill, Warley, **143**, 145
Lumb Foot Mill, Stanbury, 164
Lumb, Soloman, 59–60, 78, 153–4, 155–6
Lumbutts Mill, Langfield, 28, **29**, 133
Lydgate Mill, Saddleworth, 185
Lydgate Factory or Low Mill, Stansfield, 132

Mabgate Mill, Leeds, 109–10
Malham Mill, 96, 232–3
Mallalieu & Platt, 137
Manchester, 77–8, 79, 81, 82, 84, 85, 86, 98, 103
Markland, Cookson & Fawcett, 27, 52, 109
Marriner, William, 165, 174
Marriner, William & Benjamin, 89, 166
Marsden, 123
   mills, 125, 126
Marsh Factory or Winters Mill, Stansfield, 72, 137–8
Marshall Hall Mill, Elland, 153
Marshall & Lister, 38, 49
Marshall, Fenton & Co., 41
Marshall & Mounsey, 86
Masham, 207
Marshaw Bridge Mill, Erringden, 143–4
Marsland, Edward, 90
Marton-in-Craven, mill, 189–90
Mason, Joseph, 50
Mayroyd or Gemland Mill, Wadsworth, 142–3
Meltham Mills, Meltham, 50, 128
merchants, 85–6
   investing in mills, 56–8, 103
   cotton, 77–80
Merryweather, John, 86
Merryweather & Whitaker, 61, 70
Midgehole Mill, Hebden Bridge, 53, 141, 142
Millholme Mill, Embsay, 41, 241
Millsteads Mill, Stansfield, 135
Millthorpe Mill, Sedbergh, 201–2
Mitchell's Mill, Barnoldswick, 187
Morley, 41, mill, 111
Morton or Oldside Mill, Morton, 27, 40, 177–8
Morton Valley, 27
Morvill, Mary, 116
mules, spinning, 19, 24, 38–41, 52
Murgatroyd, John, 98, 162
Murgatroyd, Nathanial, 113, 119
Musgrave, William & Co., 86, 110
Myers, Fielding & Co., 78

Myers, William, 82
Mytholm Mill, Keighley, 162–3
Mytholm Mill, Stansfield, 139–40, 177
Nether Mills, Leeds, 7, 110
New Bank Mill, Halifax, 149–50
New Bridge Mill, Heptonstall, 141
New Mill, Marsden, 124
New House Mill, Ovenden, 149
Newhouses Mill, Saddleworth, 184
New Street Mill, Huddersfield, 128
Newsholme, W & Sons, 163
Newsholme Mills, Keighley, 56, 97, **172**, 173
North Bridge Mill, Halifax, 149
Northbrook Mill, Keighley, 67, 97, 175
Nutclough Mill, Wadsworth, 142

Old Mill, Saddleworth, 182
Old Lane Mill, Northowram, 30
Old Lane Mills, Ovenden, 148–9
Old Tame Mill, Saddleworth, 181
Oldroyd Mill, Langfield, 134–5
Otley, 35, mill at Ellar Gill, **222**, 223
Otley Mills, 223–4
Otterburn, 236
Ovenden, Halifax, 7

Paley, Richard, 31, 50, 96, 107
Parker, Thomas, 72, 78, 175, 210
Parker & Smaley, 82
Pastures Mill, Saddleworth, 184
Pearson, Robert, 78
Peel, Yates & Co., 83, 85
piece halls, 49, 85, 120
Pilkington, James, 78
Pingle Mill, Saddleworth, 181
Pocklington, 193–4
Ponden Mill, Keighley, 56, 98, **163**, 163–4
Priestley, B, 162
Providence Mill, Bingley, 18, 42, **114**, 115
Pudsey, 91
Pudsey Mill, Todmorden, 131

Quick Mill, Saddleworth, 185
Quick Edge Mill, Saddleworth, 185

Radcliffe Mill, Saddleworth, 184
Ramsbotham, Henry, 119
Rand, John, 90, 120
Rand's Mill, Bradford, 120
Rangely & Tetley, 84, 92, 122
Rathmell, Settle, 231–2
Rawdon, C & J, 136
Rawstorne & Co., 128–9
Redley & Crompton, 79
Richmond Cotton Mill, 205, 208

Ridgefoot Mill, Todmorden, 133
Rilston Cotton Mill, 51, 238–9
Ripponden Mill, 96, 157
Rishworth Mill, 96
Robinson, Thomas, 67
Robson, Edmondson & Co., 69, 119
Rodmer Clough or Lister Mill, Stansfield, 138
room and power, 54, 180
Royd House Mill, Keighley, 56, 98, 162
Rudclough Mill, Erringden, 143
Runley Bridge Mill, Settle, **230**, 230–1
Ryburn House, Soyland, 28, 157–8
Ryburn Mill, 96

Saddleworth, 49, 54, 61, 93, 103–4, 180–1
Sagar, Richard, 98
Salt Pie or Edge Mill, Heptonstall, 138
Sandal Cotton Mill, Wakefield, 111
Sandbeds Mill, Embsay, 41, 241
Sandywood House, Keighley, 176
Sawley Corn Mill, 191
Scalegill Mill, Kirkby Malham, **233**, 233–4
Scaw Gill Mill, Grassington, **211**, 212
Scholes, Varley & Co., 126, 128
Schorfield, S, 144
Scotland Mills, Leeds, 106
Scotton Mill, 208
Scouthead Mill, Saddleworth, 184
Sedbergh, 117
Settle, 9, 90, 93
Settle Bridge Mill, **26**, 107, **228**, 228–9
Severhills Mill, 96, 158
Shaw Carr Wood Mill, Slaithwaite, 127
Shaw Lane Mill, Ovenden, 148
Sheffield, 196
Sheffield Cotton Mill, 86, 196, 245
Shelderslow Mill, Saddleworth, 184
Sheppard's Factory, Dewsbury, 129
Shibden Mill, Northowram, 150
Sidgwick, William, 58, 84, 90
Silkstone, Barnsley, 7
Silver Mill, Otley, 52, **224**, 224–5
Shaw Carr Wood Mill, 127
Shore Mill, Saddleworth, 181, **182**
Simpson Fold, mill, Leeds, 110–11
Skipton, 9, 35, 103
Skyreholme Mill, **28**, **215**, 216
Slaithwaite, 123
Slater Ing or Bob Mill, Heptonstall, 139
Sleddon, Thomas & Co., 117
Slitheroe Mills, 49, 154–5
Smallees Mill, 60, 157
Smith, Abraham, 55
Smith, Joseph, 13
Smith, Tetley & Co., 118

Smith, Watson, Blakeys & Greenwood, 55, 67, 78, 79, 80, 81, 83
Smithy Holme Mill, Marsden, 125
Soyland Mills, Soyland, 158–9
Spa Mill, Stansfield, 137
Spinning Mill, Burley Woodhead, 220, **220**
Spring Mill, Rishworth, 154
Spring Mill, Saddleworth, 184
Spring Mill, Warley, 144
Spring Head Mill, Haworth, 164
Stainforth Cotton Mill, 226–7
Staups Mill, Stansfield, 136, **137**
Steam Mill, Todmorden, 133
steam power
    number of engines, 34, 42–4
    type of engines, 30, 117
    use in mills, 11, 30–4, 63, 176
    and water power, 33–4
    see also Boulton & Watt
Steander Mills, Leeds, 110
Stedal & Lonsdale, 82
Steeton Mill, 178
Steps Mill, Honley, 126
Stern Mills, Skircoat, 152
Stocksbridge Cotton Mill, 197
Stones Mill, 96, 156
Stoodley Bridge Mill, Langfield, 135–6
Strines Mill, Saddleworth, 183
Stubbing or Hebble End Mill, Heptonstall, 142
Stubbin House Mill, Keighley see Aireworth
    Mill
Suddall, Henry, 82–3
Sutcliffe, G, 139
Sutcliffe John, 26
Sutcliffe, Samuel & John, 13, 128
Sutcliffe, Thomas, 135
Sutton Mill, 178–9
Swaine Brothers & Co., 58–9, 60, 85, 119, 121
Swarcliffe Hall, 97
Swift Place Mills, 96, 156
Swineshead Mill, Langfield, 133

Tadcaster Cotton Mill, 225
Tarboton & Carr, 111–12
Tempest, Joseph, 67, 98–9
Temple Mill, Rishworth, 40–1, 96, 153–4
Thirsk, 193
Thompson, George, 112
Thompson & Naylor, 36–7
Thorner, Leeds, mill, 111–12
Thornhill Briggs Mill, Hipperholme, 153
Thornton, Bradford, 7
Thorpe, Thomas, 91
throstles, 19, 38–9, 52
Throstle Nest, Marsden, **65**

Thrum Hall Mill, Soyland, 155
Tickhill, 197
Tillotson, M & J, 145
Todd & Bosworth, 67
Todmorden, 103
Tom Hole Mill, Soyland, 27, 158
Townhead Mill, Addingham, **32**, 42, 217
Turner & Son, 82
Turner, Bent & Co., 91, 138, 139
turnpike roads, 8, 9, 130
Twyford, Robert, 78

Upper Mill, Morton, 27, 40, 176–7
Upton, James & Co., 117

Vale Mill, Keighley, 97, 164–5
Valley Mill, Saddleworth, 185
Varley, John, 127

Wainhouse, R & W, 152
Wainstalls Mill, Warley, 145
Walk Mill, Keighley, 54, 67, 167
Walker, Ard, 40, 45–7, 68, 106
Wansford, 192–3, **193**
Washburn valley, 7
water power
   mill sites, 7, 27, 55, 130
   dams, 27
   value, 29
   wheels, 25–9, 34, 41
Waterhead Mill, Saddleworth, 184
Waterloo Mills, Leeds, 106–7
Waterside Mill, Langfield, 133
Waterside Mill, Slaithwaite, 127–8
Watson, Rowland, 55, 170
Watson Mill, Norland, 160
weaving
   hand-loom, 8, 81, 89, 90–1, 99, 114
   hand-loom weaving shops, 12, **89**, 92
   power-loom, 3, 81, 93
West Riding, 1, 2, 18, 88
   magistrates, 71
   mill owners, 71
Westhouse Mill, Ingleton, 203–4, **203**

Wetherill & Co., 111
Wharfe Mill, Skircoat, 152
Whitaker, Jonas, 55, 107
Whitaker, J & J & Co., 107
Whitaker & Merryweather, 61, 72–3
White, Thomas, 52, 113, 119
Whiteley, James & Sons, 106
Whiteley, John, 96, 156
Whitfield Syke Mill, Embsay, 41, 240
Wilkinson, Thomas, 107
Willett, M W, 84, 115, 209
Willow Hall Mills, Skircoat, 96, 151–2
Wilsden, Bradford, 114
   Old Mill, 118
Wilsden Mill, 118
wind mills, 13
Winters Mill see Marsh Factory
Wire Mill, Keighley, 165
Wood Mill, Keighley, 54, 174
Wood, Winstanley & Co., 68
Woodbrook Mill, Saddleworth, 184
Woodend Mill, Saddleworth, 183
Woodhead, George, 125
Woodhead Mill, Keighley, 176
Woodhouse Mill, Langfield, **134**, 135
Woodhouse Mill, Saddleworth, 181
Wortley, Leeds, mill, 111
Wreaks Mill, Birstwith, 10, 97, 208–9
Wright, John, 91, 142
Wright Mill, Saddleworth, 183
Wrigley, W, 153

Yew Tree Mill, Saddleworth, 182
Yore Mill, Aysgarth, 68, 75, 78, 206–7
York, mill, 193
York Field or Oak Hill Clough Mill,
   Stansfield, 132–3
Yorkshire Dales, 1 2, 9, 13, 16, 88, 103–4,
   199–201
Yorkshire
   East, 192
   South, 195
Young, William, 111